41st Scientific Annual Meeting of Indonesian Association of Geophysicists 2016

Reflection, Challenge, and Innovation

IOP Conference Series: Earth and Environmental Science
Volume 132

Lampung, Indonesia
26 - 29 September 2016

Editors:

Hendra Grandis
Andri D. Nugraha
Zulfakriza
Rexha V. Ry
David P. Sahara
Endra Gunawan
Riky Hendrawan

ISBN: 978-1-5108-6057-5
ISSN: 1755-1307

Printed from e-media with permission by:

Curran Associates, Inc.
57 Morehouse Lane
Red Hook, NY 12571

Some format issues inherent in the e-media version may also appear in this print version.

Printed by Curran Associates, Inc. (2018)

For permission requests, please contact the Institute of Physics at the address below.

Institute of Physics
Dirac House, Temple Back
Bristol BS1 6BE UK

Phone: 44 1 17 929 7481
Fax: 44 1 17 920 0979

techtracking@iop.org

Additional copies of this publication are available from:

Curran Associates, Inc.
57 Morehouse Lane
Red Hook, NY 12571 USA
Phone: 845-758-0400
Fax: 845-758-2633
Email: curran@proceedings.com
Web: www.proceedings.com

Table of contents

Volume 132

41st HAGI Annual Convention and Exhibition 2016

26–29 September 2016, Lampung, Indonesia

Accepted papers received: 5 March 2018
Published online: 29 March 2018

Preface

Papers

IOP Conf. Series: Earth and Environmental Science **132** (2018) 011001 doi:10.1088/1755-1315/132/1/011001

Preface

Hendra Grandis[1], Andri D. Nugraha[1], Zulfakriza[1*], Rexha V. Ry[1], David P. Sahara[1], Endra Gunawan[1], Riky Hendrawan[2]

[1]Department of Geophysical Engineering, Faculty of Mining and Petroleum Engineering, Institut Teknologi Bandung, Indonesia
[2]Pertamina, Indonesia

[*]email: zulfakriza@itb.ac.id

This issue of Journal *IOP Conference Series*: *Earth and Environmental Science* contains the outstanding papers presented at The 41th Scientific Annual Meeting of Indonesian Association of Geophysicists. This is a scientific meeting which was organized by Indonesian Association of Geophysicists. It was held on August 26 - 29 September, 2016 in Lampung – Indonesia.

This conference provided a venue for collaboration, sharing and developing new ideas and technologies. The 41th Scientific Annual Meeting of Indonesian Association of Geophysicists created an opportunity for academics, professionals, and students to promote and discuss the scientific results related to recent development in geophysical methods and related science.

Theme of 2016 meeting is “Reflection, Challenge, and Innovation”. This convention is one of the best ways for employees, executives, lecturers, and students, to share their knowledge, ideas, experiences, and success stories even the challenging efforts, face to face. Included in the event are live talks and poster presentations. Oil, Mining, and Service companies are welcome. This Convention will also serve as a great opportunity for you to expose your vision and core technology as the Convention Committee has accepted more than 370 papers.

The main objective of this conference is to lead to develop new ideas and technologies in geophysical methods and related science.

IOP Conf. Series: Earth and Environmental Science **132** (2018) 011001 doi:10.1088/1755-1315/132/1/011001

There were two topics of panel discussion. The 1st Panel Discussion was The Challenge to Synergize a Sustainable Energy in Indonesia. The speakers were (1) **Rida Mulyana (DirJen EBTKE)** *"Opportunity and Challengein Geothermal as Indonesia Energy Backbone", (2)* **IGN Wiratmaja Puja (Dirjen Migas)***"Policy and Strategy Changes Facing Current O&G Industry" (3)* **Andang Bachtiar (Ketua Komite Eksplorasi Nasional)** *"Indonesian Energy Overlapping Synergy Process, O&G Industry and Geothermal (Blueprint Energy Indonesia)" (4)* **Amien Sunaryadi (Kepala SKK Migas)** *"Future Positioning Managing Sustainable Indonesia E&P" (5)* **Syamsu Alam (Direktur Hulu Pertamina Persero)** *"Upstream Pertamina Challenges and Value Creation toward to Indonesian Energy Company". The* 2nd Panel Discussion was Indonesia Geophysical Strength: 40 years of Journey "Volcanic and Tectonic: Earthquake Engineering. The speakers of were: (1) **Andi Eka Sakya (Kepala BMKG)** *"Role and Innovation of Goverment in Earthquake Mitigation " (2)* **Sri Widiyantoro (Global Seismologi – ITB)** *"Updating Hazard Map : Industry and Goverment Implication" (3)* **Kirbani Sri Brotopuspito (Pakar Gunungapi – UGM)** *"Updating and Future Volcano Hazard in Indonesia" (4)* **Heddy Pritasa (Director of Maipark)** *"Risk and Insurance in Earthquake Engineering"*

The organizing committee would like to thank the sponsors of this conference: SKK MIGAS, Pertamina, BMKG, Exxon Mobil, Elnusa, Medco Energy, Paradigm and other sponsors. We would like to express our sincere thanks to all authors and presenters for their valuable research. We hope to see you again during the next event of Scientific Annual Meeting of Indonesian Association of Geophysicists 2018 which will be held in Semarang, Indonesia.

Peer review statement

All papers published in this volume of *IOP Conference Series: Earth and Environmental Science* have been peer reviewed through processes administered by the proceedings Editors. Reviews were conducted by expert referees to the professional and scientific standards expected of a proceedings journal published by IOP Publishing.

IOP Conf. Series: Earth and Environmental Science **132** (2018) 012001 doi:10.1088/1755-1315/132/1/012001

Broadband seismic : case study modeling and data processing

M B Cahyaningtyas[1] and A Bahar[2]

[1] Geophysical Engineering Department, Faculty of Mining and Petroleum Engineering, Institut Teknologi Bandung, Indonesia
[2] Exploration and Engineering Seismology Research Group, Faculty of Mining and Petroleum Engineering, Institut Teknologi Bandung, Indonesia

mariabinarc@outlook.com

Abstract. Seismic data with wide range of frequency is needed due to its close relation to resolution and the depth of the target. Low frequency provides deeper penetration for the imaging of deep target. In addition, the wider the frequency bandwidth, the sharper the wavelet. Sharp wavelet is responsible for high-resolution imaging and is very helpful to resolve thin bed. As a result, the demand for broadband seismic data is rising and it spurs the technology development of broadband seismic in oil and gas industry. An obstacle that is frequently found on marine seismic data is the existence of ghost that affects the frequency bandwidth contained on the seismic data. Ghost alters bandwidth to bandlimited. To reduce ghost effect and to acquire broadband seismic data, lots of attempts are used, both on the acquisition and on the processing of seismic data. One of the acquisition technique applied is the multi-level streamer, where some streamers are towed on some levels of depth. Multi-level streamer will yield data with varied ghost notch shown on frequency domain. If the ghost notches are not overlapping, the summation of multi-level streamer data will reduce the ghost effect. The result of the multi-level streamer data processing shows that reduction of ghost notch on frequency domain indeed takes place.

1. Introduction

There are some important things that effect survey design for seismic acquisition, such as depth of target, resolution, and objective of the seismic survey. Resolution is related to frequency bandwidth of seismic signal. Broader frequency bandwidth will give better resolution. Acquisition technique to get wide frequency bandwidth is known as broadband seismic. One obstacle on marine seismic to get broadband seismic signal is ghost. Ghost is a phenomenon due to depth position of source and receiver below the sea surface. Ghost on seismic records can cause missing certain value of frequency bandwidth (ghost notch) therefore seismic data will become band-limited. To reduce this effect, an acquisition technique for marine seismic is to use receiver or streamer for more than one level of depth (multi-level streamer). The usage of multi-level streamer can reduce the effect of ghost to get broadband seismic data. This method was introduced by Parrack in 1976 and during mid-1980s this method is used to reduce weather noise [1]. In 2010, a study about variable-depth streamer was conducted [2]. The result shows variable-depth streamer acquisition data increase in bandwidth. There was also a study that compare some acquisition technique (conventional streamer, deep streamer, slanted streamer, and dual streamer) to get broadband seismic data [3]. After processing data, the result shows that dual streamer technique provides the broadest bandwidth. In this study, seismic modeling is used to understand ghost phenomenon and technique to reduce ghost to get broadband seismic with multi-level streamer acquisition technique.

IOP Conf. Series: Earth and Environmental Science **132** (2018) 012001 doi:10.1088/1755-1315/132/1/012001

2. Wavelet modeling

Input of wavelet (figure 1), used in this modeling, is a source signature without ghost. In this modeling, source depth is 5m, and the receiver depth varies from 7.5m, 10m, to 15m. The first step of wavelet modeling is calculation of ghost operator for ghost source and ghost receiver, followed by convolution between wavelet and ghost operator to produce wavelet with source ghost and receiver ghost. Then Fourier Transform is done to see the ghost notch in frequency domain.

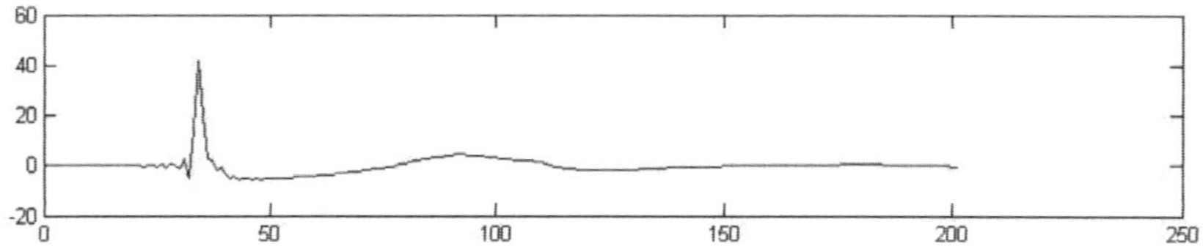

Figure 1. Wavelet without ghost.

3. Synthetic data and data processing

Forward modeling process in this study is done to get synthetic data. A simple geology model consisting of three layers is used in this forward modeling process. The synthetic data contains three data with variation of streamer depth: 7.5m, 10m, and 15m, and each streamer has 120 channels with receiver interval 12.5m. 110 source is towed at depth of 5m with source interval 25m. Figure 2 is data processing workflow to get broadband seismic data.

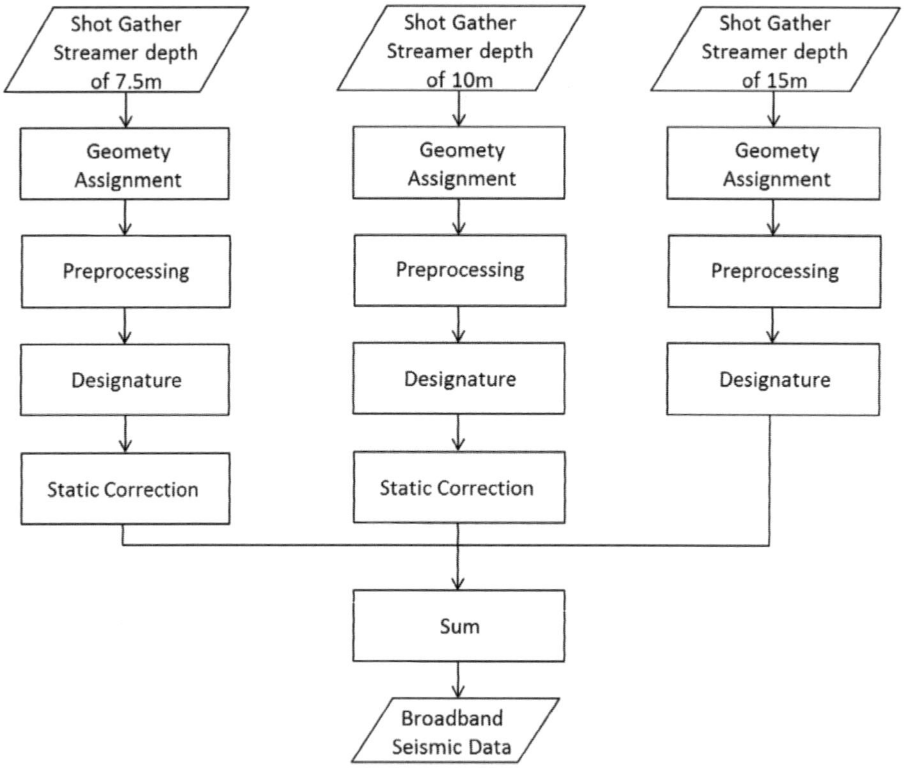

Figure 2. Data processing workflow.

Geometry assignment is a process to combine seismic data and its survey navigation information. Preprocessing is used to reduce noise in data such as direct wave. Designature is done to remove source effect in seismic data. Variation of streamer depth give a consequence as delay of arrival time for deeper streamer, therefore static correction is done as an effort of time shifting to a time reference. After each data has same reference time, summation mode is used to get a broadband data. The idea of this summation mode is the same as stacking, that is averaging the same trace with the same position for every streamer depth.

4. Result and analysis

4.1. Wavelet modeling

Each wavelet that contains source ghost and receiver ghost with variation of streamer depth results in different form of wavelet both in time domain and frequency domain. The result of wavelet modeling in time domain, as shown in figure 3, shows that the deeper the streamer, the longer composite wavelet will be produced, this shows the delay time is greater as the deeper depth.

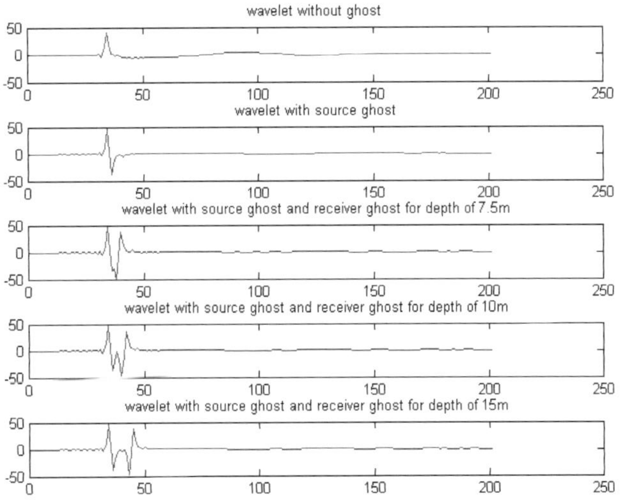

Figure 3. Wavelet modeling result in time domain.

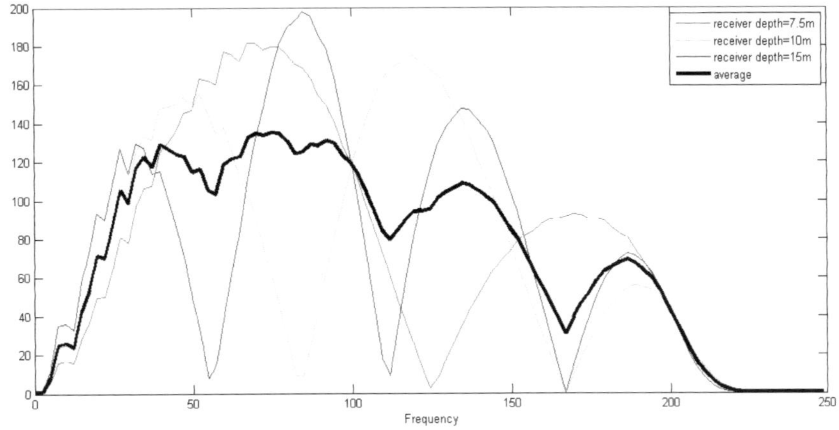

Figure 4. Wavelet modeling result in frequency domain.

In addition, in the frequency domain, as shown in figure 4, shows that variation of streamer depth will result on variation of ghost notch that is not overlapping and there will be more ghost notch for deeper receiver. The average of each frequency spectrum give ghost notch recovery. In addition, averaging spectrum provide increment of frequency content.

4.2. Synthetic data processing
The shot gather show the same result as previous wavelet modeling, data with deeper streamer has a wider composite wavelet recorded on seismic data. After data processing, frequency spectrum shows ghost notch is reduced. From shot gather, broadband seismic data have sharper wavelet. In addition, stack section give better imaging. The result of data processing is shown in figure 5.

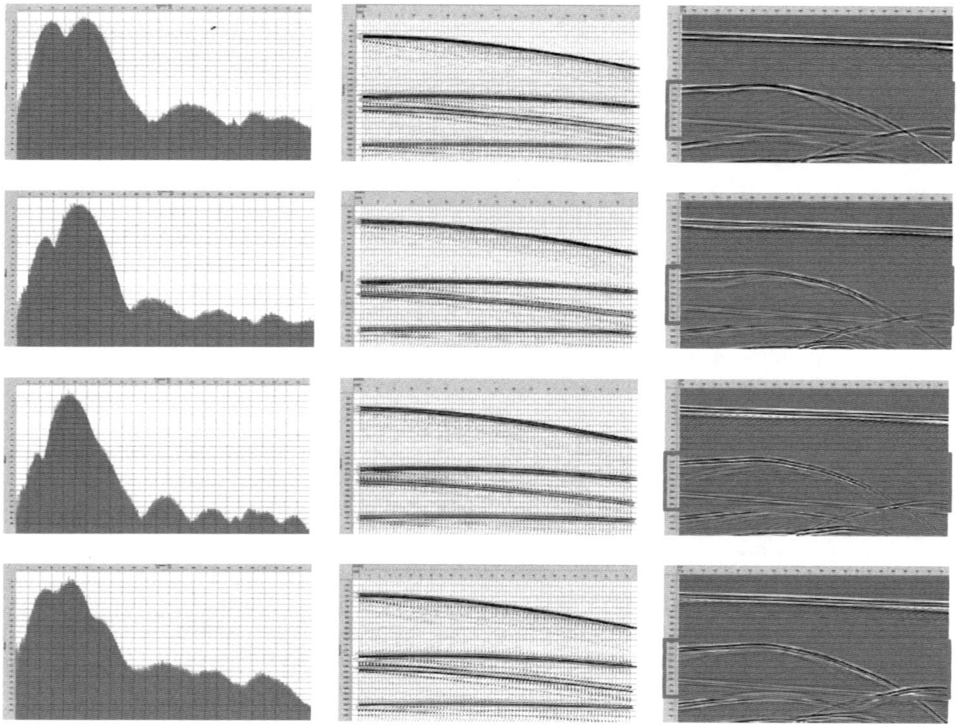

(a) (b) (c)

Figure 5. Comparison of (a) frequency spectrum, (b) shot gather, and (c) stack section for synthetic data (sequentially from top to bottom) with steamer depth of 7.5 m, 10.m, 15m, and the result of summation mode.

5. Conclusion
Ghost appears due to source and receiver position in some range of depth below the sea surface. Variation of its depth will give different seismic response, both in time and frequency domain. Multi-level streamer technique will give us seismic data with variation of ghost notch, therefore by using summation mode, ghost notch is reduced and frequency content is increased. Multi-level streamer technique could be a solution to get broadband seismic data.

References

[1] Moldoveanu N, Combee L, Egan M, Hampson G, Sydora L and Abriel W 2007 *The Leading Edge* **26** 41–58

[2] Soubaras R and Dowle R 2010 *First Break* **28** 89–96

[3] Bayly M P, Chandola S K, Loo C F, Muhamad A A and Ng S L 2013 Broadband marine seismic, does acquisition make a difference? *Petroleum Geoscience Conf. & Exhibition 2013 (Kuala Lumpur, Malaysia)*

IOP Conf. Series: Earth and Environmental Science **132** (2018) 012002 doi:10.1088/1755-1315/132/1/012002

Digital Rock Physics Aplications: Visualisation Complex Pore and Porosity-Permeability Estimations of the Porous Sandstone Reservoir

Handoyo[1], Fatkhan[2], and Fourier DEL[3]

[1]Geophysics Engineering, Faculty Engineering Institut Teknologi Sumatera, Indonesia
[2]Geophysics Engineering, Faculty of Minning and Petroleum Engineering Institut Teknologi Bandung, Indonesia
[3]Earth Science and Complex System, Faculty of Mathematic and Natural Science Institut Teknologi Bandung, Indonesia

handoyo.geoph@tg.itera.ac.id and handoyo_1906@yahoo.com

Abstract. Reservoir rock containing oil and gas generally has high porosity and permeability. High porosity is expected to accommodate hydrocarbon fluid in large quantities and high permeability is associated with the rock's ability to let hydrocarbon fluid flow optimally. Porosity and permeability measurement of a rock sample is usually performed in the laboratory. We estimate the porosity and permeability of sandstones digitally by using digital images from μCT-Scan. Advantages of the method are non-destructive and can be applied for small rock pieces also easily to construct the model. The porosity values are calculated by comparing the digital image of the pore volume to the total volume of the sandstones; while the permeability values are calculated using the Lattice Boltzmann calculations utilizing the nature of the law of conservation of mass and conservation of momentum of a particle. To determine variations of the porosity and permeability, the main sandstone samples with a dimension of 300 × 300 × 300 pixels are made into eight sub-cubes with a size of 150 × 150 × 150 pixels. Results of digital image modeling fluid flow velocity are visualized as normal velocity (streamline). Variations in value sandstone porosity vary between 0.30 to 0.38 and permeability variations in the range of 4000 mD to 6200 mD. The results of calculations show that the sandstone sample in this research is highly porous and permeable. The method combined with rock physics can be powerful tools for determining rock properties from small rock fragments.

Keyword : μCT-scan, sandstone, digital image, permeability, porosity

1. Introduction

Calculations of rock physical parameters can be determined using computer simulation (digital). The digital simulation is performed to model or visualize difference between the pore and the rock matrix and is used to predict the value of porosity, permeability, electrical conductivity, density, and elasticity of the rock. Digital image (3D image) rock can be obtained from scanning using the CT-Scan and combined with digital simulation software. In addition, the digital image has been applied to visualize of rocks and predict physical parameters (e.g., Arns and Knackstedt, 2012; Dvorkin et al, 2009; Handoyo et al, 2014). This paper discusses the digital simulation to visualize complex pore, pore fluid flow and predict value of porosity and permeability of reservoir sandstones that have high porosity values.

2. Theory and Method

A rock (sandstone) generally consists of two main parts: pores and rock solid matrix (Figure 1a). The pore is the space between the rock solid matrix can be filled by a fluid (liquid or gas). Other important physical parameter is permeability that is the ability of a rock to drain the fluid either liquid or gas (Figure 1b).

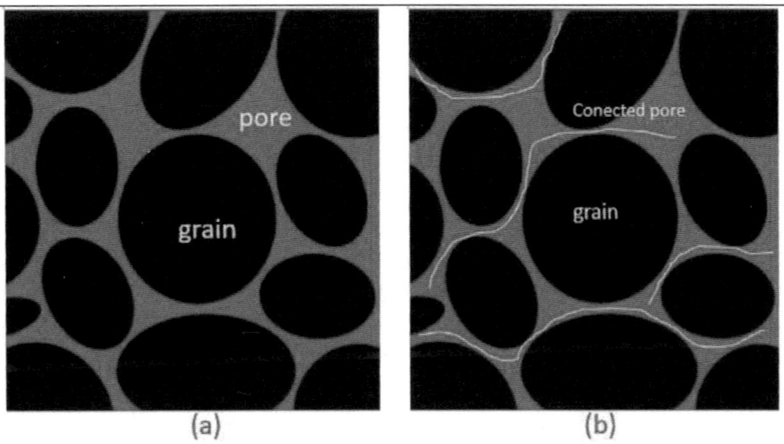

Figure 1. A model clean sandstone. Figure (a) Visualization is shown by the pore spaces between rock solid matrix and Figure (b) Models fluid flow in porous rock interconnected.

Quantity of rock pore is expressed by the value of porosity. Porosity (ϕ) is the ratio between the pore volume (V_{pore}) of the total volume of rock (V_{total}) which is mathematically written as:

$$\phi = \frac{V_{pore}}{V_{total}} = 1 - \frac{V_{matrix}}{V_{total}} \quad (1)$$

Where (V_{matrix}) is the volume of the solid part of the rock. Furthermore, 3-dimensional model of sandstone samples made a cube with size 300 × 300 × 300 pixels. To find variations, we created a subcube 8 units (2^n) with size 150 × 150 × 150 pixels (Figure 2).

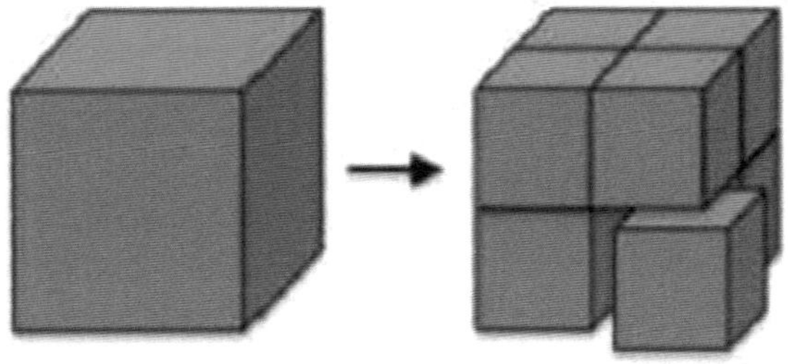

Figure 2. Illustrations of 3-D sandstone model and manufacture of sub-cube variations of the main cube (modified from Dvorkin, 2009).

Absolute permeability appears in Darcy's law as a constant coefficient relating fluid, flow and the porous medium parameters as follows:

$$\frac{Q}{A} = -\frac{k}{\mu}\frac{dP}{dL} \quad (2)$$

Where $\boldsymbol{Q}$ is the fluid debit, $\boldsymbol{A}$ is surface area of pore, $\boldsymbol{dP/dL}$ is the gradient pressure as length function, $\boldsymbol{k}$ is absoulute permeability, and $\boldsymbol{\mu}$ is the fluid viscousity. In order to calculate permeability, we simulate fluid flow using Lattice Boltzmann method that is as follows (Fourier, 2014):

$$\frac{Q}{A} = \frac{\mu\langle v\rangle}{dP/dL} \quad (3)$$

The process of DRP begins with an image of a large rock sample acquired at a relatively coarse resolution to cover a large field of view (cube). At this stage, rock fabrics larger than the image resolution are resolved while smaller ones are unresolved (sub-cube). Information concerning the unresolved rock fabrics is analyzed from additional images acquired at a finer resolution and smaller field of view. The information from resolved and unresolved rock fabrics are average from small-scale image and back into the large-scale image. In general, the stage of the DRP analysis shown in Figure 3.

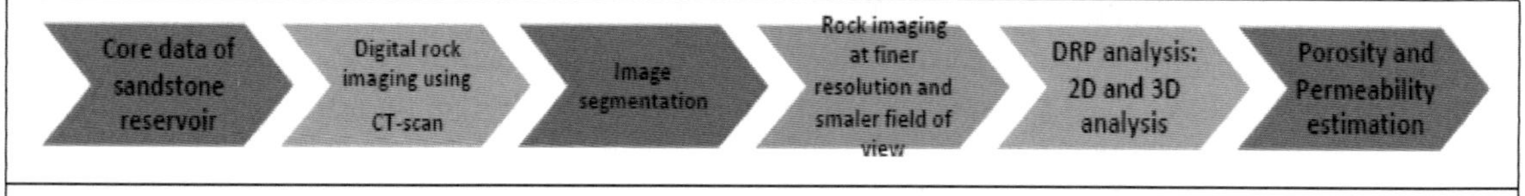

Figure 3. DRP data processing stages on sandstone sample. Stage begins with digital image processing to the calculation porosity and absoulute permeability.

3. Result and Discussions

Results of scanning using a CT scan are shown in Figure 4. Figure 4 (a) is a natural image of the sample, and Figure 4 (b) is the result of the thresholding process with a view to separate the pores and rock solid matrix.

To determine the variation of the porosity and permeability of the sandstone samples, the main cube made into multiple sub-cube. In this study, made eight major sub-cube from main cube (Figure 5). The cubes shows that scale-factor has influence to value of porosity and permeability of digital sample. The variation shown in Figure 6a.

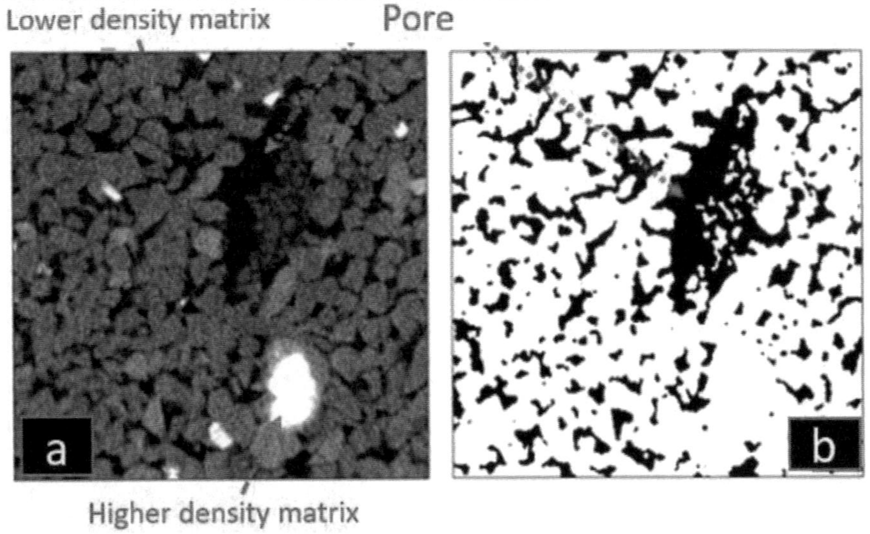

Figure 4. The scans using CT-Scan with dimensions of 300×300 pixels. Figure (a) shows the difference in density of solid rock matrix, where the highest density indicated by the bright white color. Figure (b) shows the color difference between the pore and solid matrix rock. Pore indicated by black, rock solid matrix is indicated by the white part (after treshold process).

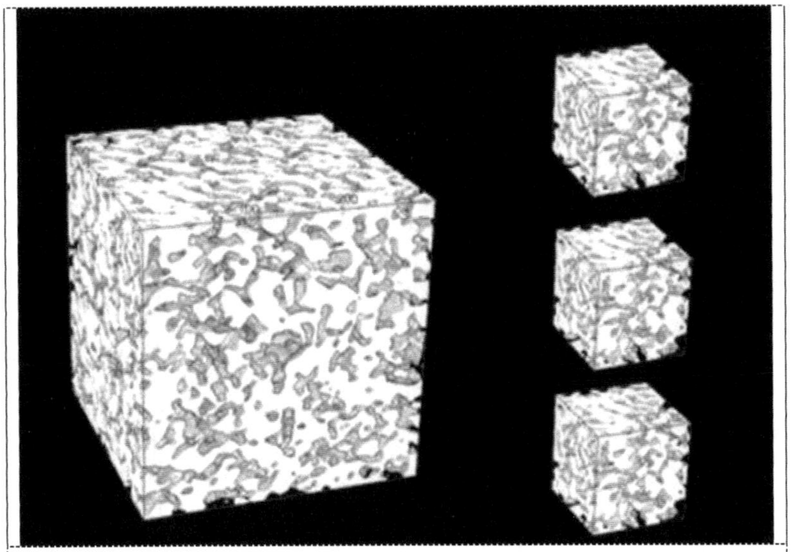

Figure 5. Illustrations of 3-D sandstone model and manufacture of sub-cube variations of the main cube. (a) the sub-cube illustration (Dvorkin, 2009), (b) the main cube with size 300×300×300 pixels, and (c) sub-cubes with size 150×150×150 pixels.

Furthermore, porosity and permeability relationship is shown in Figure 6a. The value of permeability as a function porosity is as an exponential function. The calculations show a high permeability values for higher porosity. Results of digital simulation permeability values are displayed in Figure

6b. Visualization of fluid flow in the pore (streamlined) is expressed in normal velocity (blue) from Latice Boltzman equation.

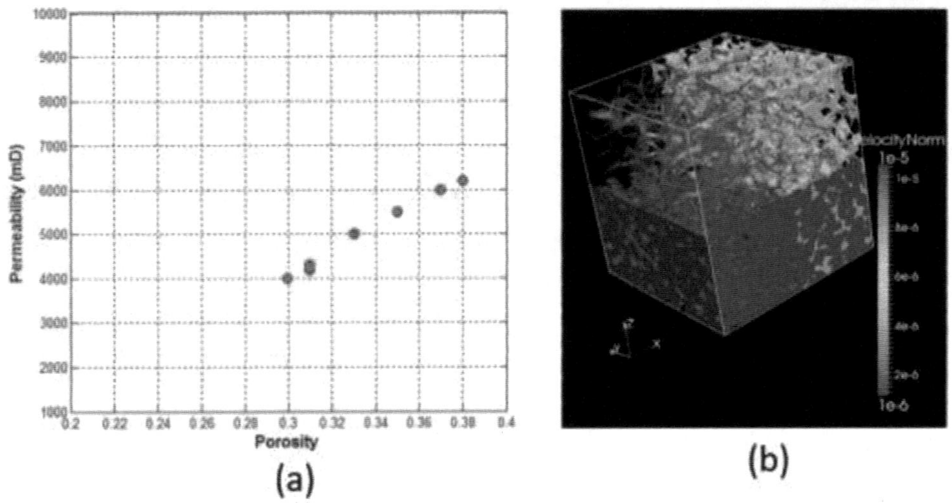

Figure 6. The results of digital simulation calculation of permeability values as a function of porosity (a) and model of fluid flow in porous rock (b).

4. Result and Discussions

- The digital simulations of sandstone samples using CT-Scan are able to separate pores and rock matrix.
- Calculation results show that values of porosity and permeability are relatively high.
- Factor sample size (sub-cube) gives variations of porosity and permeability values are different for each subcube.
- The method combined with rock physics can be powerful tools for determining rock properties from small rock fragments.

References

[1] Andra, H., Combaret N., Dvorkin J., Glatt E., Han J., Krzikalla F., Lee M., Madonna C., Marsh M., Mukerji T., Ricker S., Saenger E.H., Sain R., Saxena N., Wiegmann A. and Zhan X., 2012. Digital Rock Physics Benchmarks-Part I: Imaging and Segmentation. Journal Computers & Geosciences ELSEVIER 50 (2013) 25–32.

[2] Arns, C.H., Knackstedt, M.A., 2002. Computation of Linear Elastic Properties from Microtomographic Images: Methodology and Agreement Between Theory and Experiment Journal of Geophysics, Vol. 67, No. 5, 1396-1405. Australian National University. Canberra.

[3] Dvorkin, J.P., 2009. Relevance of computational rock physics. Geophysics Journal. Geophysics Department, Stanford University. Stanford. USA.

[4] Fourier, D.E.L., 2014. Analysis of Permeability and Tortuosity of Fontainebleau Sandstone and Its Models Using Digital Rock Physics Approach. Physics of Earth and Complex System, Faculty of Mathematics and Natural Sciences, Bandung Institute of Technology, Indonesia.

[5] Handoyo, 2014. Digital Rock Physics Analysis to Pore Structure Parameters Characterization, Fluid Modeling and Parameter Estimation of Elasticity in Saturated Sandstone Samples. Thesis Magister. FTTM. Bandung Institute of Technology. Indonesia.

Indonesian Geomagnetic Maps for Epoch 2015.0 to cover of Indonesian Regions

M Syirojudin[1,a], J Murjaya[2], S Zubaidah[3], Hasanudin[3], S Ahadi[3], N Efendi[3] and T Suroyo[3]

[1] Analyst at Potential Geophysics and Time Keeping, Meteorological, Climatological and Geophysical Agency, Angkasa I street No. 2, Central Jakarta 10720, Indonesia
[2] Centre of engineering seismology, Potential Geophysics and Time Keeping, Meteorological, Climatological and Geophysical Agency, Angkasa I street No. 2, Central Jakarta 10720, Indonesia
[3] Division of Potential Geophysics and Time Keeping, Meteorological, Climatological and Geophysical Agency, Angkasa I street No. 2, Central Jakarta 10720,

[a]syirojudin.bmkg@gmail.com

Abstract. In compliance with the resolutions of IAGA (International Association of Geomagnetism and Aeronomy), Since 1960's, every five years BMKG or Meteorology, Climatology and Geophysics Agency of Indonesia made geomagnetic field maps based on actual measurements in 53 repeat stations. It's the map for more accurate result of Geomagnetic maps Epoch 2015.0, the number of repeat stations has been increased to 68 locations. Analysis data was conducted by spatial analyses using collocated co-kriging and kriging with external drift to map the observation data in five components, such as Declination (D), Inclination (I), Vertical (Z), Horizontal (H), and Total Geomagnetic Field (F). The data reduction used one permanent observatory i.e., Kupang Geophysical Observatory, as a reference standard. The results of this Geomagnetic Maps, that the contour lines of Indonesian geomagnetic declination in range -1 to 4.5 degree, Inclination component are -5 to -37 degree, Vertical component are -4000 to -28000 nT, Horizontal component are 36000 to 42000 nT, and Total Geomagnetic Field are 39000 to 46000 nT. In conclusion, Indonesian Geomagnetic Maps for Epoch 2015.0 can be used to compute geomagnetic data around Indonesian regions until next 5 years.

Keywords: BMKG, Epoch 2015.0, geomagnetic map, Indonesian

1. Introduction

The Indonesia is country archipelago with more than 17,504 islands and rich in mineral resources that spread from Sabang to Merauke. This condition makes the geomagnetic data very important to use, because navigation, mineral exploration, Geology map and defense map are using geomagnetic data in any case. Accurate geomagnetic map is important in many fields ranging from navigation, mineral exploration to defense. The accuracy of such map, in mineral exploration, is about 0.1° in declination and about 50 nT in total intensity [1]. Since the Earth's magnetic field is dynamic, geomagnetic maps need to be revised on regular basis. This process requires direct observation of the geomagnetic field in a number of repeat stations.

In Indonesia, the above activities are carried out by BMKG (Badan Meteorologi Klimatologi dan Geofisika or/Meteorology, Climatology and Geophysics Agency of Indonesia). BMKG conducted limited survey in 1960-1962 and in 1970-1974. In compliance with IAGA (International Association of Geomagnetism and Aeronomy) resolutions number 23/1963 and 9/1995, BMKG conducted more extended survey every five years since 1985 until 2010 in 53 repeat stations at Epoch 2015.0, BMKG add 15 new repeat stations to increase quality and accuracy of geomagnetic map epoch 2015.0 (see Table 1).

.

TABLE 1. History of Indonesia repeat stations expansion

	1960-1962	1970-1974	1985	1990	1995	2000-2001	2005	2010	2015
Sumatra	4	10	15	15	15	13	12	12	17
Java and Bali	7	7	8	8	8	9	8	8	9
Kalimantan	3	4	8	9	9	9	10	10	11
Sulawesi	0	4	9	7	7	7	8	8	11
Nusa Tenggara	1	1	5	6	6	5	5	5	5
Maluku	0	1	4	2	2	3	4	4	5
Irian Jaya	0	5	10	6	6	7	7	6	10

During this time, geomagnetic data from BMKG is rarely used as a reference for analysis especially in the exploration and other research using geomagnetic method by Indonesian researchers. The researchers generally using IGRF (International Geomagnetic Reference Field) which arose from the results of the World Magnetic Survey (WMS) that is produced by IAGA [2]. In principle, IGRF is the Earth's magnetic field model based on spherical harmonics. IGRF, in turn, could be used to create geomagnetic map. The first IGRF version was ratified by IAGA in 1971 and since then it has been revised ten times. The IGRF benefitted from excellent coverage and quality of satellite data provided by Ørsted and CHAMP and is now truncated at degree of 13 [3].

Comparative study between IGRF data and observation data in Trelew, Argentina in 1990-2000 show that there was a big discrepancy that could not covered by IGRF model [4]. Another research conducted in China, which used 29 observation stations in 1960-1990 reported that a sizable discrepancy between observation data and IGRF [5]. In the previous research, it is investigated that there is difference between the geomagnetic maps of BMKG model and IGRF. The difference varies in each epoch from 1985 to 2010 with difference between 300 to 200 nT for the total intensity component and -3.5 to 0.5 degrees for declination component [6].

2. Data and methods

This study uses BMKG geomagnetic data of 2015 and the IGRF-12 model. The BMKG data were derived from about 68 stations all over Indonesia (see Figure 1). Most of the stations are located in airports and other accessible sites that are relatively free from geomagnetic disturbance.

For determination of epoch value in measurement location, first shall do is data correction by using daily geomagnetic variation that recorded in Kupang geomagnetic observatory. After it, the data reduced to the 2015.0 epoch at each measurement location. This correction value calls as term C1 calculated for every datum from direct measurements reduced to Kupang geomagnetic observatory. Mathematically, is described by

$$C_1 = C_e - C_t \tag{1}$$

Where C_1 is data of each geomagnetic component in geomagnetic station for epoch year (C_e) reduced by data value of the same component of measurement location (C_t). The data values of 2015.0 epoch for geomagnetic station were determined from the average data of January until December 2015.

Since the measuring stations were spread unevenly in the archipelago, the data (coordinates and geomagnetic values) could not be used for extrapolation processes. Therefore secondary data derived from IGRF-12, in the form of coordinates and value patterns or trends, were used to complement the BMKG data. To obtain data in locations outside BMKG measuring stations the approaches namely Collocated Cokriging and Kriging with External Drift were used

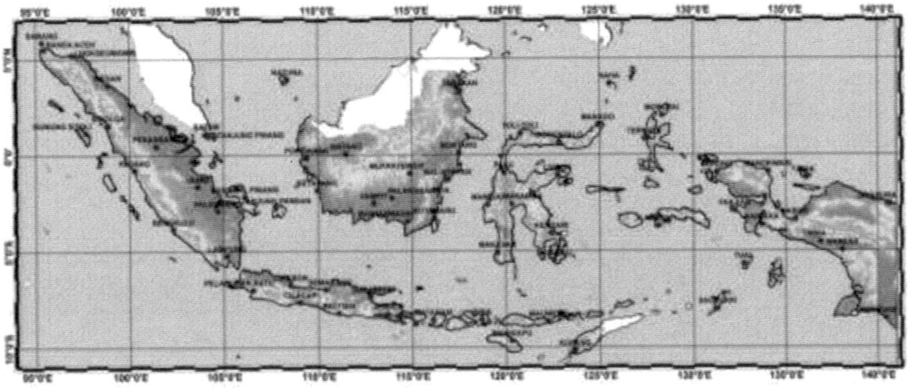

Figure 1. Distribution of 68 BMKG geomagnetic measuring stations in Indonesia

Collocated Cokriging or CC is a simplified cokriging that is used when a target variable (like our BMKG data) is sparsely sampled, compared to a densely sampled auxiliary variable (like IGRF data). In its strict sense, collocated cokriging makes use of the auxiliary variable only at the current point where the target variable is to be estimated [7]. Mathematically, CC is described by

$$Z_{CC}(x_0) = \sum_i \lambda_i Z_1(x_i) + \mu Z_2(x_0) \tag{2}$$

where Z_{CC} (x_0) is the result of CC calculation at a particular position x_0, Z_1 (x_i) is the primary or BMKG data at the ith station, λ_i is the weight of Z_1 (x_i), Z_2 (x_0) is the secondary or IGRF data calculated for position x_0 and μ is the weight for Z_2 (x_0).

Kriging with External Drift or KED is a kriging with a trend model that is used if auxiliary information (like IGRF data) is available at all grid nodes and correlated with target variable (in this case BMKG data) [8, 9]. Mathematically, KED is described by

$$Z_{KED}(x_0) = \sum_i \lambda_i Z_1(x_i) + a \sum_i \lambda_i Z_2(x_i) + aZ_2(x_0) \tag{3}$$

where Z_{KED} (x_0) is the result of KED calculation at a particular position x_0, Z_1 (x_i) is the primary or BMKG data at the ith station, Z_2 (x_i) is the secondary or IGRF data at the ith BMKG station, λ_i is the weight of Z_1 (x_i) and Z_2 (x_i) and a is the slope of secondary data at x_0.

The results from CC and KED were then compared. The one with smaller error variation and the best cross-plot validation were adopted as BMKG geomagnetic map. In each location repeat station shall do measurement absolute geomagnetic component such as component declination (D), component inclination (I) and component geomagnetic field total (F). That measurement shall do in the same time measurement. Thus, value of component geomagnetic horizontal (H) and component geomagnetic vertical can calculate from the formula. Mathematically, is described by:

$$H = F * \cos I \tag{4}$$

$$Z = F * \sin I \tag{5}$$

All of data measurement in location repeat station after reduced with data from station geomagnetic Tondano and reduced again to data in measurement epoch 2010.0 then we set a final data of epoch for each location measurement repeat. Value data of each location for each component geomagnetic plotted in as a map and illustration like a contour for determine a changes data, and that can call as isomagnetic map

3. Results and discussions

This Figure 2a show typical BMKG map for declination component produces in this study. It shows that the contour lines of Indonesian geomagnetic declination component in range -1 to 4.5 degree. For validation accuracy this declination component map, we compare with respective declination component map derived fully from IGRF model as shown at Figure 2b. There is rather similar between the geomagnetic declination component in the two maps for 2015 epoch. This discrepancy has interval about 16 to 20 minutes (see Figure 2c).

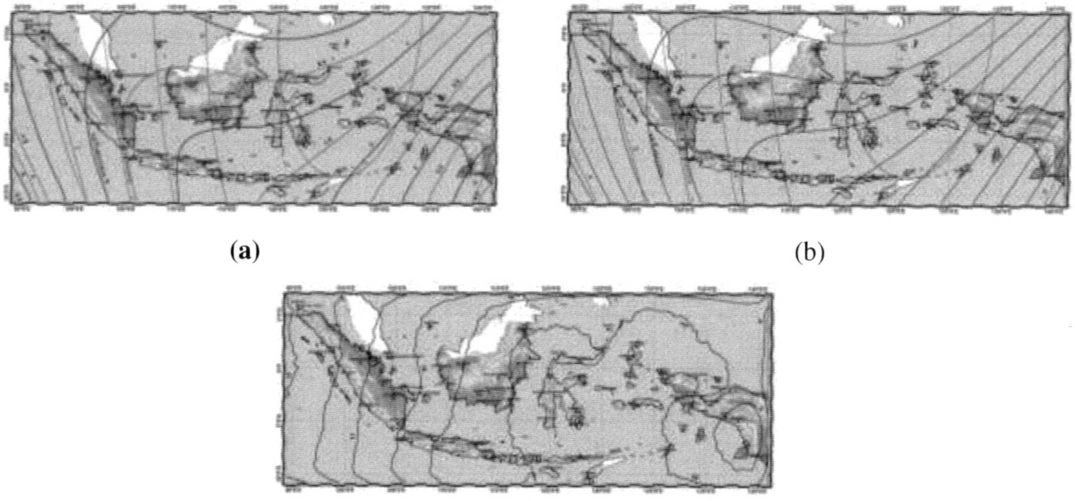

(a) (b)

(c)

Figure 2. Typical geomagnetic maps produced in this study showing the Declination component for epoch 2015.0. The BMKG contour in degree (a) IGRF contour in degree (b) and discrepancy between BMKG and IGRF map in minute (c).

Inclination component have interval contour lines are -5 to -37 degree as shown at Figure 3a. If this component is compare with respective inclination component map derived fully from IGRF model as shown at Figure 3b.There is rather similar discrepancy about 20 to 25 minutes (see Figure 3c).

Figure 4a show typical BMKG map for Total Intensity component have interval contour lines are 39000 to 46000 nT. If this component is compare with respective Total intensity component map derived fully from IGRF model as shown at Figure 4b.There is rather similar discrepancy about 30 to 140 nT (see Figure 4c).

Horizontal component after derived from equation 4, have interval contour lines are 36000 to 42000 nT. Vertical component after derived from equation 5, have interval contour lines are -4000 to -28000 nT. This might be due to the limited number of magnetic observatories in Indonesia used by IGRF, which is only two stations: Tuntungan and Tondano, North Sulawesi [3]. The second factor is tectonic structure in Indonesian Region hasn't been cover by IGRF model. This second conclusionis supported by a research of Pétrélis which shown that the movement mechanism of tectonic crustal has a significant contribution for geomagnetic field [10].

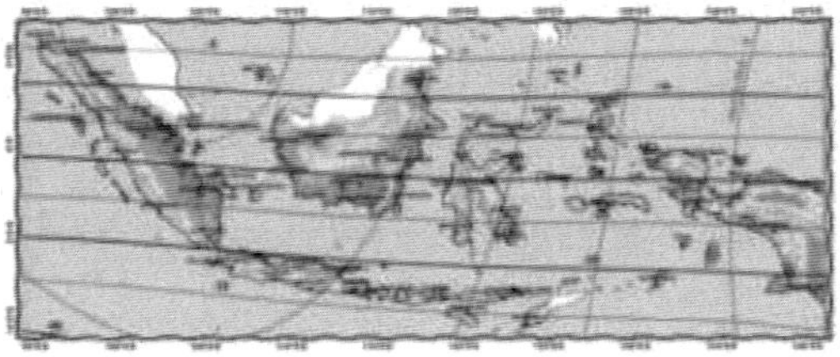

(a)

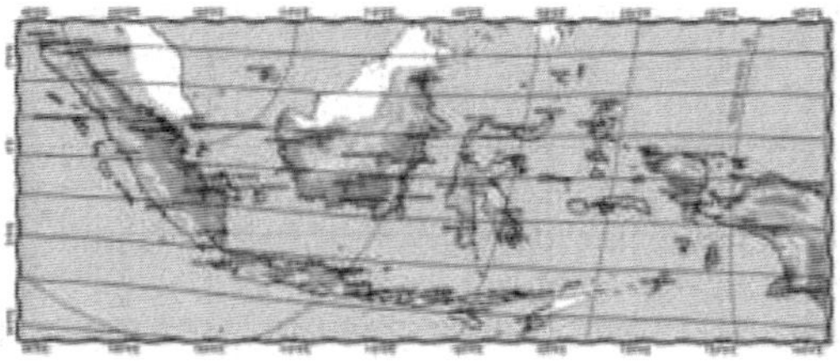

(b)

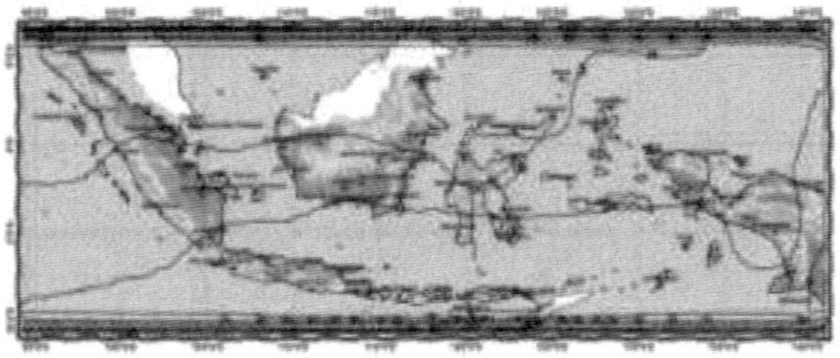

(c)

Figure 3. Typical geomagnetic maps produced in this study showing the Inclination component for epoch 2015.0.The BMKG contour in degree (a) IGRF contour in degree (b) and discrepancy between BMKG and IGRF map in minute (c).

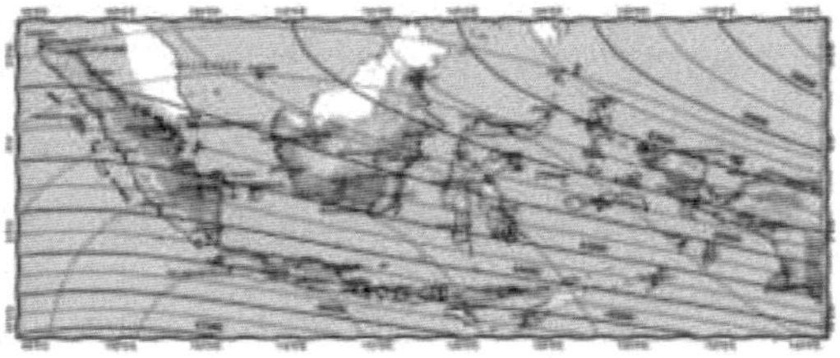

(a)

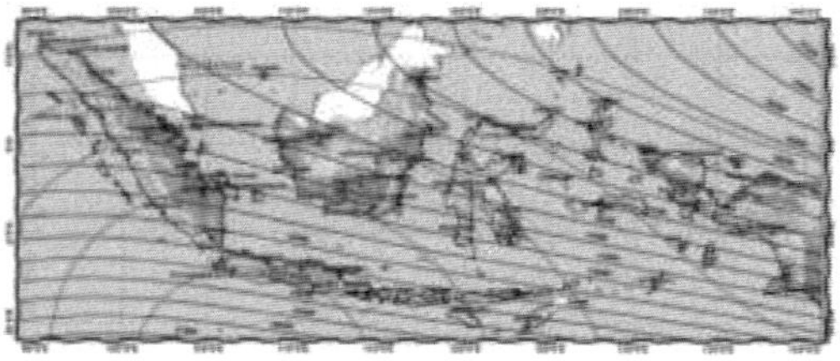

(b)

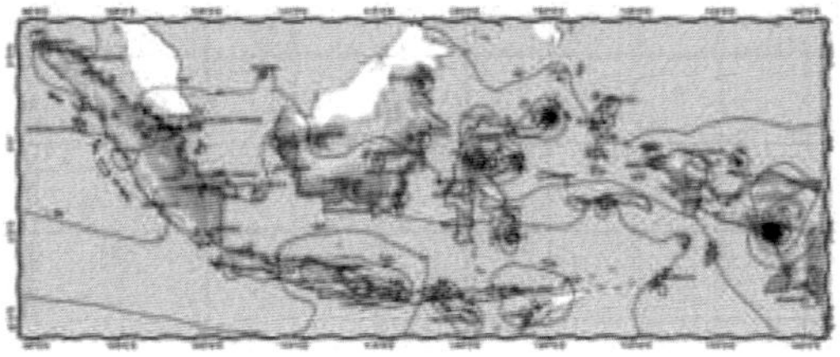

(c)

Figure 4. Typical geomagnetic maps produced in this study showing the Total Intensity component for epoch 2015.0.The BMKG contour in nT (a) IGRF contour in nT (b) and discrepancy between BMKG and IGRF map in nT (c)

4. Concluding remarks

We have shown that BMKG geomagnetic maps of 2015.0have good result. The contour lines of Indonesian geomagnetic declination in range -1 to 4.5 degree, Inclination component are -5 to -37 degree, Vertical component are -4000 to -28000 nT, Horizontal component are 36000 to 42000 nT, and Total Geomagnetic Field are 39000 to 46000 nT. In conclusion, Indonesian Geomagnetic Maps for Epoch 2015.0 can be used to compute geomagnetic data around Indonesian regions until next 5 years.

References

[1] Reay S J, Allen W, Baillie O, Bowe J, Clarke E, Lesur V and Macmillan S 2005 *Ann. Geophys.* **23** 3081
[2] Barraclough D R 1993 *Phys. Earth Planet. Int.* **48** 279
[3] Finlay C C, Maus S, Bondar T, Chambodut A, Golovkov V P, et al. 2010 *Geophys. J Int.* **183** 1216
[4] Gianibelli J C, Köhn J and Ghidella M E 2003 *Geofis. Int.***42** 635
[5] Tanwen W 2003 *Chinese J. Geophys.* **46** 236
[6] Syirojudin M, and Bijaksana S 2013 *AIP Conf. Proc.* **1554** 249
[7] Rivoirard J 2001 *Math Geol.* **33** 2
[8] Hengl T, Heuvelink G B M and Stein A 2004 *Geoderma.* **120** 75
[9] Webster R and Oliver M 2000 *Geostatistic for Environmental Scientists*, New York, John Wiley & sons, 2000, pp. 203-211.
[10] Pétrélis F, Besse J and Valet J P 2011 *Geophys Res Lett.* **38** L19303

Acknowledgments

Authors are thankful to the all of the team members for geomagnetic measurement epoch 2015.0. In additional we also thanks to the Marine Geological Institute for the support of providing the total intensity data for Java sea.

IOP Conf. Series: Earth and Environmental Science **132** (2018) 012004 doi:10.1088/1755-1315/132/1/012004

Rock Physics and Petrographic Parameters Relationship Within Siliciclastic Rocks: Quartz Sandstone Outcrop Study Case

S Syafriyono[1,4] , D Caesario[1] , A Swastika[1] , Q Adlan[1] , I Syafri[2] , A Abdurrokhim[2] , U Mardiana[2] , F Mohamad[2] , M K Alfadli[2] , V M Sari[3]

[1] PT FERG Geosains Indonesia, Cigadung Greenland K55 Bandung 40191, Indonesia
[2] Department of Geological Engineering, Padjadjaran University, Jl. Raya Bandung Sumedang KM 21, Jatinangor 45363, Indonesia
[3] Department of Geophysical Engineering, Institute of Technology Bandung, Jl. Ganesha 10 Bandung 40132, Indonesia
[4] Author to whom any correspondence should be addressed

syafriyono@ferg.co.id or syafriyono_jr@yahoo.com

Abstract. Rock physical parameters value (Vp and Vs) is one of fundamental aspects in reservoir characterization as a tool to detect rock heterogenity. Its response is depend on several reservoir conditions such as lithology, pressure and reservoir fluids. The value of Vp and Vs is controlled by grain contact and contact stiffness, a function of clay mineral content and porosity also affected by mineral composition. The study about Vp and Vs response within sandstone and its relationship with petrographic parameters has become important to define anisotrophy of reservoir characteristics distribution and could give a better understanding about local diagenesis that influence clastic reservoir properties. Petrographic analysis and Vp-Vs calculation was carried out to 12 core sample which is obtained by hand-drilling of the outcrop in Sukabumi area, West Java as a part of Bayah Formation. Data processing and interpretation of sedimentary vertical succession showing that this outcrop comprises of 3 major sandstone layers indicating fluvial depositional environment. As stated before, there are 4 petrographic parameters (sorting, roundness, clay mineral content, and grain contact) which are responsible to the differences of shear wave and compressional wave value in this outcrop. Lithology with poor-sorted and well-roundness has Vp value lower than well-sorted and poor-roundness (sub-angular) grain. For the sample with high clay content, Vp value is ranging from 1681 to 2000 m/s and could be getting high until 2190 to 2714 m/s in low clay content sample even though the presence of clay minerals cannot be defined neither as matrix nor cement. The whole sample have suture grain contact indicating telogenesis regime whereas facies has no relationship with Vp and Vs value because of the different type of facies show similar petrographic parameters after diagenesis.

1. Introduction

Physiographically, the sandstone outcrop in the study area is located in Sukabumi area along the boundary of Bandung Zone and Central Depression Mountain Zone. It is belong to Bayah Formation in Walat Mountain and reliable as a good analogue of sandstone reservoir with good porosity. A method of reservoir characterization had been applied to this outcrop, named rock physics.

Rock physics is a method to identify rock chararteristics based on its physical parameters are P-wave velocity (Vp), S-wave velocity (Vs), bulk density, acoustic impedance, shear modulus and elastic modulus.

2. Geological Overview

West Java had been clearly divided into 5 physiography zone by Van Bemmelen (1949). The study area is located along Bandung Zone and Central Depression Mountain zone based on this zoning. Bandung Zone is formed by intramontaine depressions that consists of Tertiary rocks. Structurally, this zone is a collapsed West Java anticline crest after uplift tectonic and filled by young volcano deposits afterwards while regionally it is located on volcanic-magmatic part of Sunda Arc (Martodjodjo, 1984). Sunda arc lies from Sumatra towards East until Nusa Tenggara as manifestation of interaction between Indo-Australia plate and Eurasian. It is occured through subduction process (Fachri, 2001).

Plate activity has a significant role to configure structural setting and stratigraphy of a certain area. Panigoro (1984) in Martodjodjo (1984) made general structural trend map of West Java based on integrated seismic and gravity data (Figure 1). According to that map, structural trend that developed in West Java is divided into three:

a. Southwest – Northeast Trend (Meratus Trend) with Late Cretaceous to Early Tertiary age that responsible to Tertiary Basin forming along Ciletuh to Meratus.

b. North – South Trend with Eo-Oligocene age. This trend is marked by N-S orientation of normal and transform faults.

c. West – East Trend (Java Trend) with Oligo-Miocene age, marked by W-E orientation of fold and thrust fault which controlled Bogor trough and its turbidite deposits.

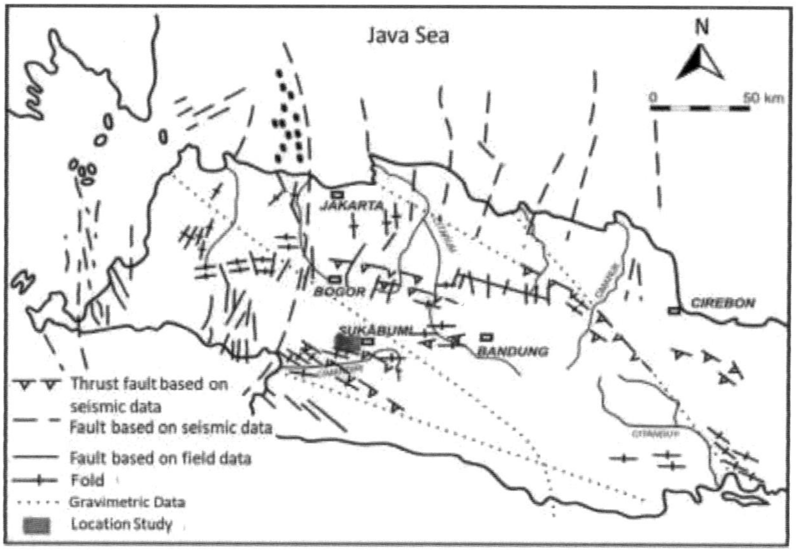

Figure 1. Geological structure trend map of West Java (modified from Panigoro, 1981 in Martodjodjo, 1984)

Bayah Formation consists of quartz sandstone with calcareous shale and coal intercalation that deposited during Middle to Late Eocene (Martodjodjo, 1984) (Figure 2). It lies unconformably by shale of Batuasih Formation which is interfingering with Oligo-Miocene Rajamandala Formation.

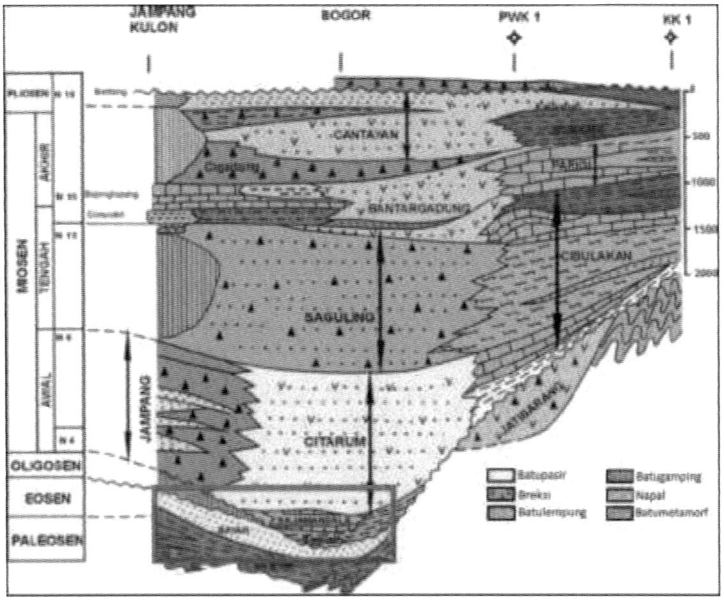

Figure 2. West Java regional stratigraphy (modified form Martodjodjo, 1984)

3. Rock Physics Analysis

Physical rock properties could be defined through elastic parameters e.g. P-wave velocity (Vp), S-wave velocity (Vs), density (ρ), Lambda-Rho (λρ), Mu-Rho (μρ), and Poisson Ratio (σ) that describes subsurface rock conditions and create wave transmission type within the rock as well. These parameters are controlled by matrix, porosity and pore fluids (Biot, 1956).

Every earth material has a certain Vp value. Generally, it will be increasing as well as increasing of medium compacity that passed by. P-wave or compression wave is also known as longitudinal wave that can be vine through solid, liquid or gas medium while it velocity can be calculated as a function of bulk modulus (k), shear modulus (μ), and densitas (ρ) (Gassman, F, 1951) by using this following equation.

$$Vp = \sqrt{\frac{\lambda + 2\mu}{\rho}} = \sqrt{\frac{K + {}^4/_3\,\mu}{\rho}} \quad (1)$$

In other hand, S-wave (shear wave) is defined as transversal wave with particle movement perpendicular from its wave transmission direction. S-wave could not pass liquid and/or gas medium because of zero value of shear modulus. Vs is affected by density (ρ) and shear modulus (μ), without modulus bulk (k), can be expressed as following.

$$Vs = \sqrt{\frac{\mu}{\rho}} \quad (2)$$

IOP Conf. Series: Earth and Environmental Science **132** (2018) 012004 doi:10.1088/1755-1315/132/1/012004

4. Methods

This study began by doing observation of the outcrop and made detail description afterwards (Figure 3). The results used as guidance to choose facies that can be taken for further analysis. Laboratorium activities including rock physics parameters calculation, Routine Core Analysis (RCAL) and petrographic investigation was carried out for each sample from the outcrop (Figure 4). Then, interpretation was done based on integration of these processes to identify sample characteristics and physical properties. Result of integration process is visualized by cross plot to lead a brief conclusion (Figure 5).

Figure 3. The sandstone outcrop in the study area

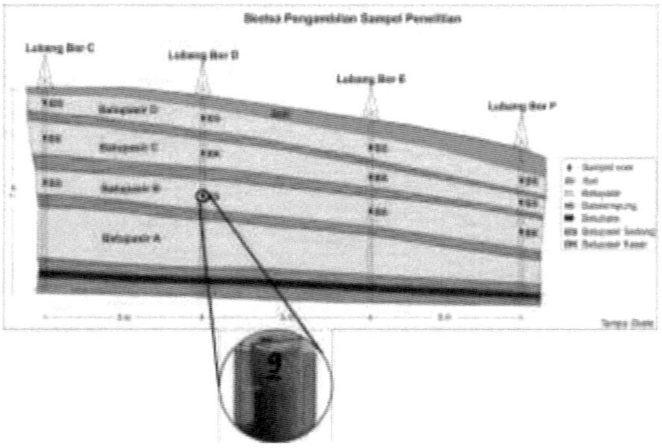

Figure 4. Sampling techniques sketch

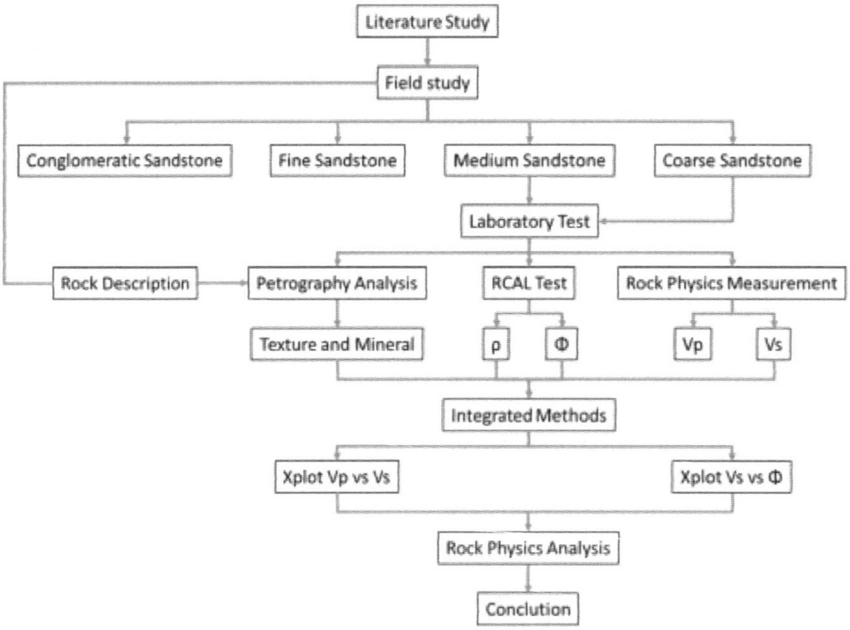

Figure 5. Study workflow

5. Interpretation and Result

The outcrop in the study area is dominated by quartz sandstone interbedded with shale and coal intercalation. It comprises of four lithofacies based on their grain size are conglomeratic sandstone, fine-grained sandstone, medium sandstone, and coarse sandstone. Several sedimentary structures found e.g cross lamination and current ripple. Sedimentary/rock textures (sorting, roundness, fabric, grain contact, and clay content) also be an important part and included to lithology description.

This study is focused on medium to coarse sandstone which is good sorted, close fabric, sub-rounded to sub-angular indicating that grains has been transported far from its source and during a long time. The grain interacts each other and formed some types of grain contact such as long, point, concave-convex and suture contact that could be observed obviously by petrographic observation.

Rock classification in this study refers to Pettijohn's Classification (1975) based on clay (as matrix) percentage of rocks. Petrographic analysis result shows that most of sample is classified into arenite because of low clay content (<15%).

Figure 6 showing the relationship between petrographic parameters and rock physics. Generally, there are 3 major group of sample based on their clay content (0-15%; 16-30%; >30%). Cross plot of Vp and Vs among matrix percentage gives a unique trend (Figure 6). Sample with low matrix percentage has the highest value of Vp and Vs and vice versa. It can be said that the presence of matrix within rocks gives significant effect to Vp and Vs value.

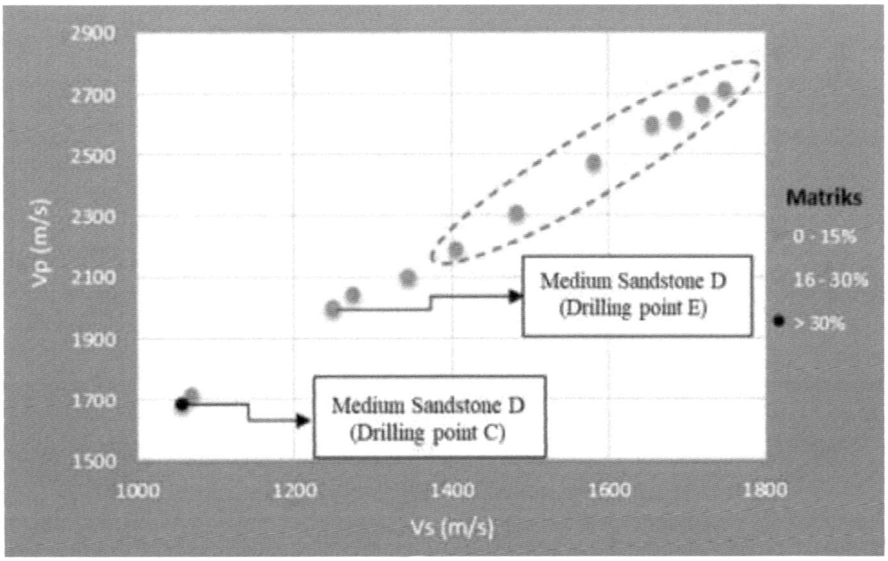

Figure 6. Cross-plot of Vp-Vs and matrix percentage

Fracture is another factor that controlled rock physics parameters (especially for Vs). It can decrease S-wave velocity due to its perpendicular position to wave direction (Figure 7). Sorting and roundness also give contribution to the response of Vp and Vs value. Poorly sorted lithology with subrounded grain shape has value of Vp and Vs lower than well sorted lithology with subangular grain shape whereas grain size is not clearly effected to Vp and Vs value.

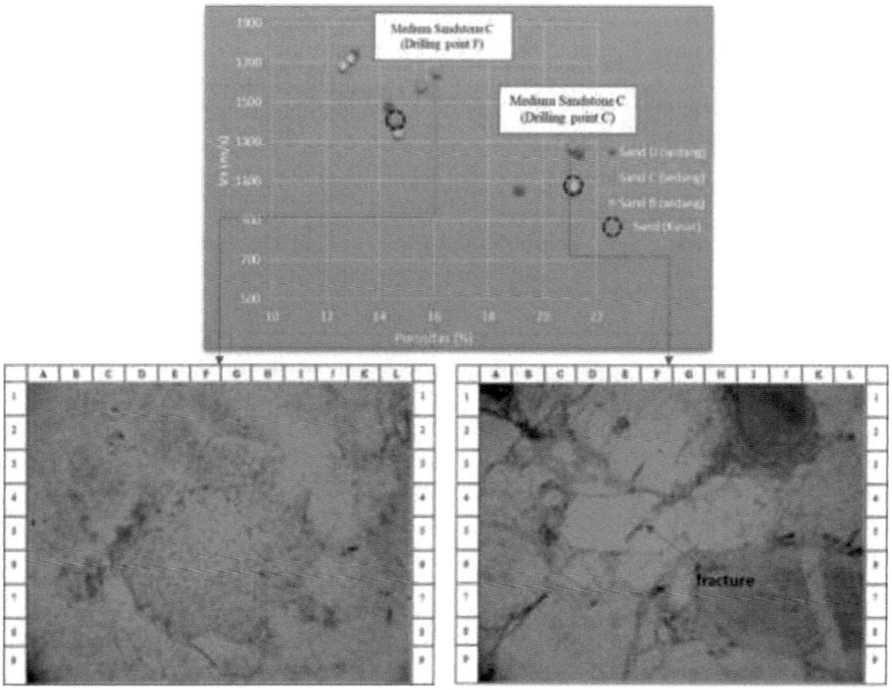

Figure 7. Cross-plot of Vs and fracture porosity

6. Conclusion

Vp and Vs value are strongly affected by matrix content, quartz mineral and fracture also controlled by sorting and roundness. However, grain size has no significant role to Vp and Vs value.

References

[1] Biot, M. A., 1956, *Theory of Propagation of Elastic Wave in a Fluid Saturation porous Solid*, J. Appl. Physics
[2] Gassman, F., 1951, Elastic wave through a packing of spheres, *Geophysics, 16*, p. 673-685
[3] Magnificat, G., Gu´eguen, I., 1998, What does control VP/VSin granular rocks?: *Geophysical Research Letters 25*, p. 381-384.
[4] Martodjojo, S., 1984, *Evolusi Cekungan Bogor, Jawa Barat*, Doctoral Dissertation, Postgraduate Study Program ITB (unpublished)
[5] Pettijohn, F. J., 1975, *Sedimentary Rock 3rd Ed. Harpen & Row*, New York.

IOP Conf. Series: Earth and Environmental Science **132** (2018) 012005 doi:10.1088/1755-1315/132/1/012005

Pliocene Deep Water Channel System of Celebes Molasse as New Exploration Play In Banggai Sula Foreland Basin, Eastern Sulawesi-Indonesia

Anggoro Prima Kurniawan[1*], Galang Purnomo Adi[1], Wahyuni Tri Sundari[1], Maikel Arifin[1], Muharram Jaya Panguriseng[1], Sri Hartanto[2], Neni Herawati[2]

[1] PT. Pertamina EP, Indonesia
[2] JOB Pertamina-Medco E&P Tomori Sulawesi, Indoesia

*Corresponding author's email: anggoro.kurniawan@pertamina.com

Abstract. Banggai-Sula foreland basin in Matindok block is the product of Late Miocene to Early Pliocene collision between Banggai-Sula microcontinent and East Sulawesi Ophiolite (ESO)-magmatic arc of Sundaland. Proven petroleum system plays for Donggi, Senoro, and Matindok gas fields are Miocene build-up carbonate and platform carbonates related to wrench structures, sealed by Plio-Pleistocene Celebes Molasse consists of conglomerate, conglomeratic sandstone, shale and charged during early Pliocene related to collisional tectonic from Miocene marine clastic source rocks. Latest interpretation of 3D & 2D seismic datasets on onshore area of Matindok block shown that there is possibility of NE to SW trending channel systems existed in Plio-Pleistocene Celebes Molasse as post collision sediments. Based on biostratigraphy interpretation, Celebes Molasse interval was deposited in marine environment, generally outer sublittoral to upper bathyal. Channel filled pattern on seismic data expression shown onlap, divergent and mounded-onlap fills on 6 (six) interpreted channel bodies. Fining upward sequences is well recognized based on gamma ray log from several wells that penetrated those interpreted channel. The thicknesses of channel bodies vary from 50 m to 150 m with sand to shale ratio up to 40 % and gas composition reading from C1 to C3. Despite of there is no well testing data on these channels interval; we believed that this Pliocene deep water channel can become the new exploration play although several studies about seal effectiveness of channel bodies as stratigraphic traps, lateral and vertical migration pathways need to be carried out.

1. Introduction

Study area located in the onshore area of Matindok Block, Eastern Sulawesi, Indonesia (figure 1). Pertamina EP started exploration activities in this area from 1999, several gas fields have been discovered such as Matindok, Maleo Raja and Donggi. The main reservoir targets are Miocene build-up and platform carbonate as syn-dfriting deposits.

In 2014, Pertamina EP drilled 2 (two) wells, Matindok-7 and Penyu-1, located 20 km west of Matindok gas field, and discovered gas & condensate in Pliocene Carbonate (M-52 layers) of Celebes Molasse as post-collision deposits. Based on gas geochemistry analysis, gas sample in Matindok-7 is possible thermogenic origin and gas sample in Penyu-1 is possible biogenic origin. These discoveries were significantly important because from previous study Celebes Molasse only act as overburden sediment and regional seal. There is no detailed study about petroleum system of Celebes Molasse.

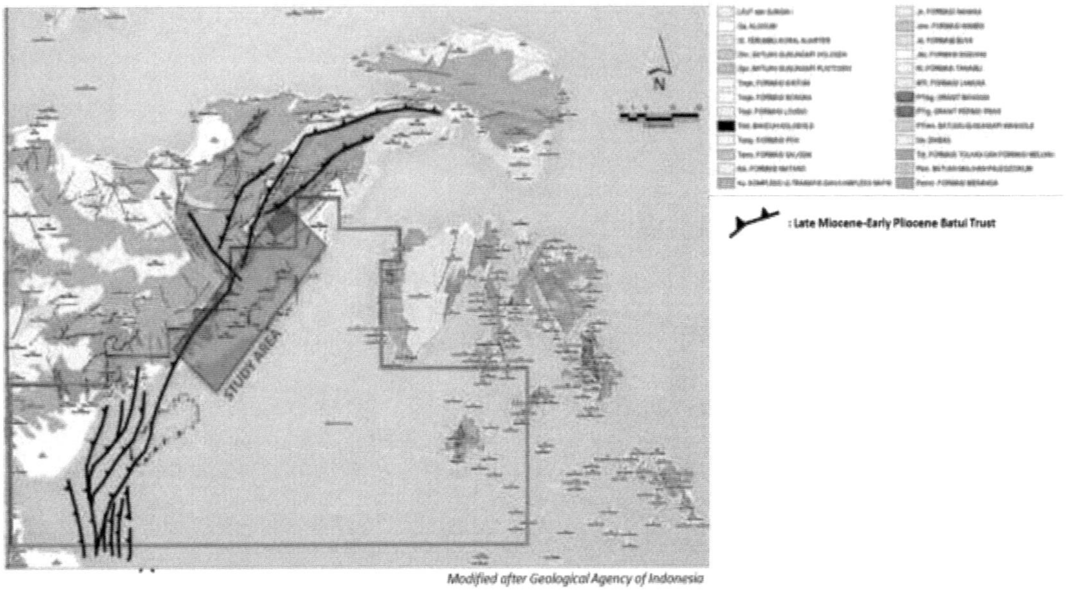

Figure 1. Study area location shown in red boundary area

In this study, using 2D and 3D seismic datasets, we deliver the geological model of NE to SW trending Pliocene channel systems in Celebes Molasse. Hopefully, this model can trigger another possibility of finding hydrocarbon in Pliocene post collision interval.

2. Regional Tectonic Setting

Tectonically, Sulawesi island is the product of several collisions between micro-continents from Sundaland and Australian Area (figure 2a). Garrard (1988), Simandjuntak and Barber (1996), Satyana (2011), Hall (2012) and Sani (2015) describe rifting, drifting and collision stages of Banggai Sula micro-continent toward East Sulawesi (figure 2b). Matindok block is precisely located on the collision border and post collision foreland basin. Tertiary rocks in the Banggai foreland basin along the east coast of Sulawesi comprise Miocene carbonate shelf and reefal limestones as main reservoir. The overlying Pliocene Celebes Molasse sediments contain conglomerate, conglomeratic sandstones and thick shales that provide a good overburden and regional seal (figure 2c). Hasanusi et. al. (2004) explained about Mercury Injection Capillary Pressure analysis of Senoro-1 samples indicated that the Pliocene Kintom shales have excellent sealing capacity with pore throat sizes mostly less than 0.5 micron.

As early as 1985, oil was discovered in Miocene carbonate platform related to imbricate thrust in the offshore Tiaka Field In the Onshore area, subsequent discoveries have been mainly gas in Minahaki and Matindok Fields (figure 2d). Possible source rocks in the Banggai Basin include Lower Miocene carbonaceous shale and argillaceous limestone and Eocene bituminous limestone and shale (Kartaadiputra and Samuel, 1988).

Shale facies of Pliocene Celebes Molasse also have potential to be biogenic gas generation. Subroto et. al (2009) describe biogenoc gas in Matindok Block. Burial history models were useful in identifying wells with shallow intervals having high sedimentation rate and present-day low temperature that are predicted to be favorable for biogenic gas generation, such as the Pliocene celebes molasse. Another result of the geochemical modelling indicates that the biogenic gas found in the Tomori block was deposited 1.7-3.5 million years ago (Plio-Pleistocene).

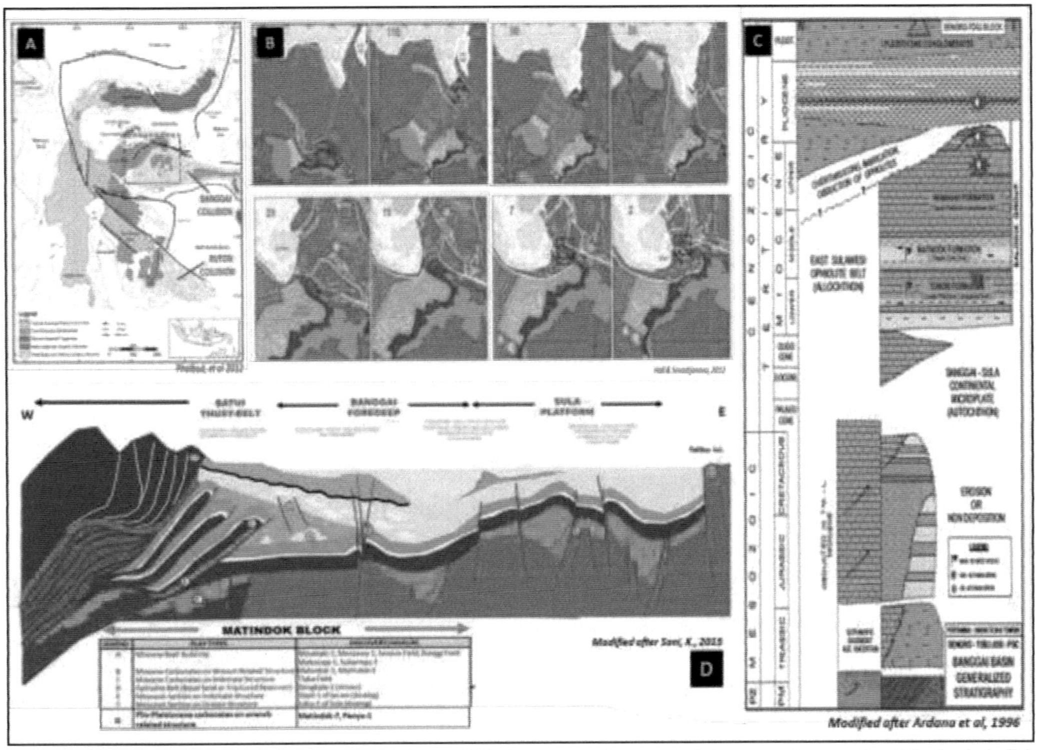

Figure 2. (a) Geological & Structural Map of Sulawesi - the red dot area shows the Matindok Block location (Pholbud et al, 2012), (b) Rift-Drift-Collision stages of Australian Micro-Continent, the black dot box show collision area (Hall and Sevatjanova, 2012), (c) Updated Stratigraphic Column of Matindok Block (modified after Ardana et al, 1996), (d) Petroleum System Play in Matindok Block (modified after Sani, K., 2015)

3. Data and Method

3D Seismic and mostly 2D Seismic lines from several vintages from 1973 to 2005 were used as the main data source in this study. Several well data related to wireline log, mudlog, and biostratigraphy analysis were also used to calibrate the channel system model. Published and unpublished reports about previous study of post collision deposits in other foreland basin were collected and used as model analogue.

4. Result and Discussion

Based on NW-SE 2D regional seismic key line, we interpret 6 (six) NE to SW trending channel system (figure 3). Isochron map of each channel were used to construct channel system model. Using channel fill pattern on seismic data characterization by Mitchum et al (1977), mostly all channels have onlap, divergent and mounded onlap pattern fill that related to low energy sediment in medial to distal area. We interpreted several point of erosion on the NNE area as sediment provenance that filled the deep water channel in the southwest area (figure 4).

Regional well correlation using 4 (four) well data pass through NW-SE 2D Regional seismic line in figure 3 were made, there are several result as follows (figure 5):

- *Biostratigraphy analysis shown that post-collision interval was deposited in marine environment from outer sub littoral to bathyal*
- *Fining upward sequences is detected based on gamma ray log.*
- *Gas reading of channel interval shows C1 to C3 up to 980 ppm. Gas Analysis ratio using gas dryness, wetness, balance, character and pixler plot indicate possible light dry gas.*

- *Possible Low Resistivity – Low Contrast (LRLC) pay due to thin laminated sandstone-shale sequences in channel interval.*
- *Shale Kintom Formation as top and base regional seal*

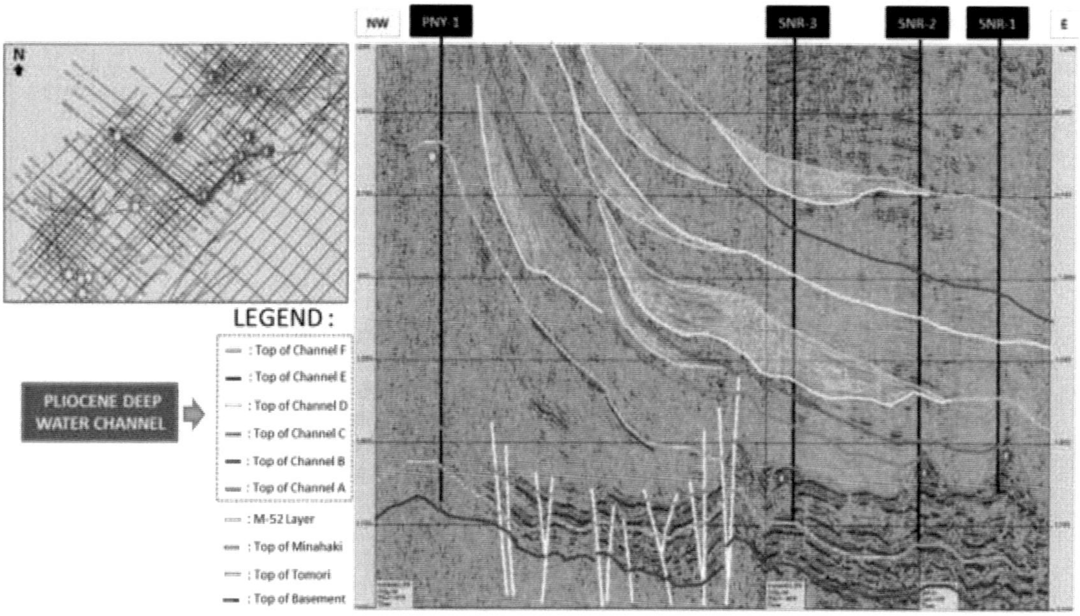

Figure 3. Pliocene Deep Water Channel Top and Base Interpretation in 2D NW-E Composite Seismic line

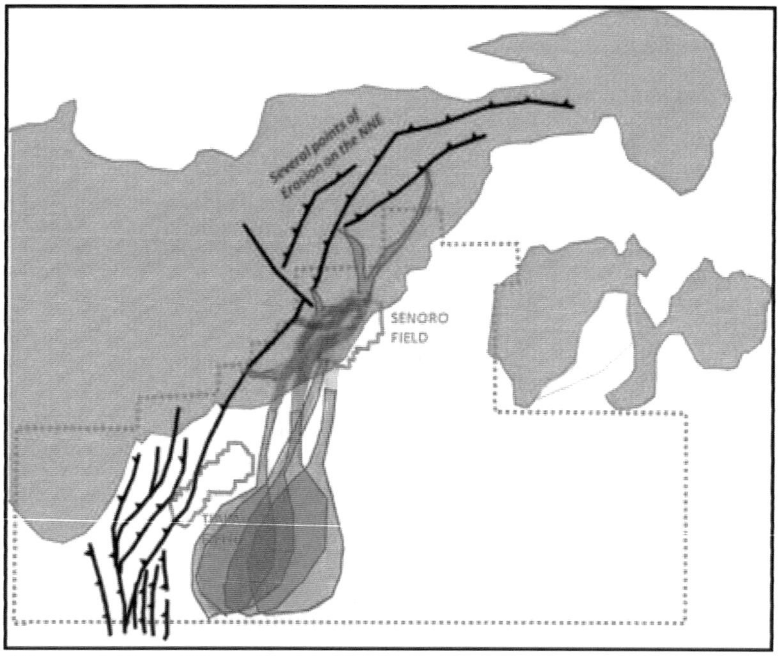

Figure 4. Deep Water Channel Model of Banggai Sula Foreland Basin

In order to calibrate the geometry of deep water channel based on isochron map, we used lithofacies section on each channel. Cibaj (2011) describe correlation of sand and shale ratio and pattern on each channel depositional environment such as overbank/levee, axis, off-axis, and margin. For example; lithofacies section of channel "C" (figure 6) in Minahaki-1 show channel margin to channel off-axis pattern and KPB-1 show channel margin to overbank, this interpretation positively correlate with channel "C" isochron map.

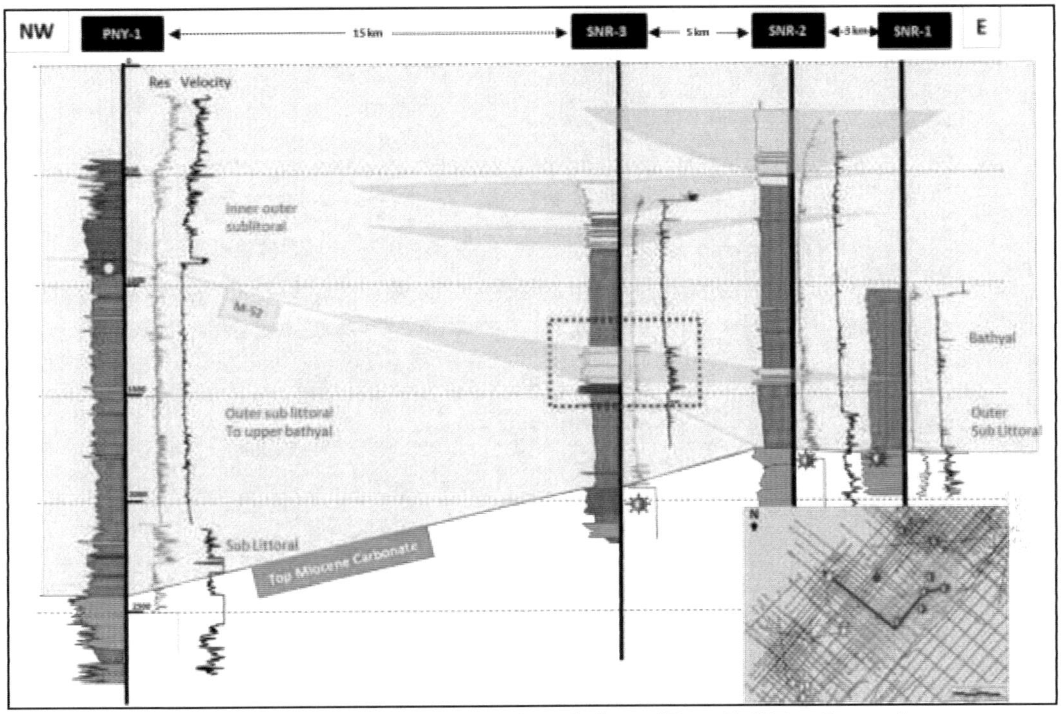

Figure 5. NW-E Regional Well Correlation of Post Collision Interval

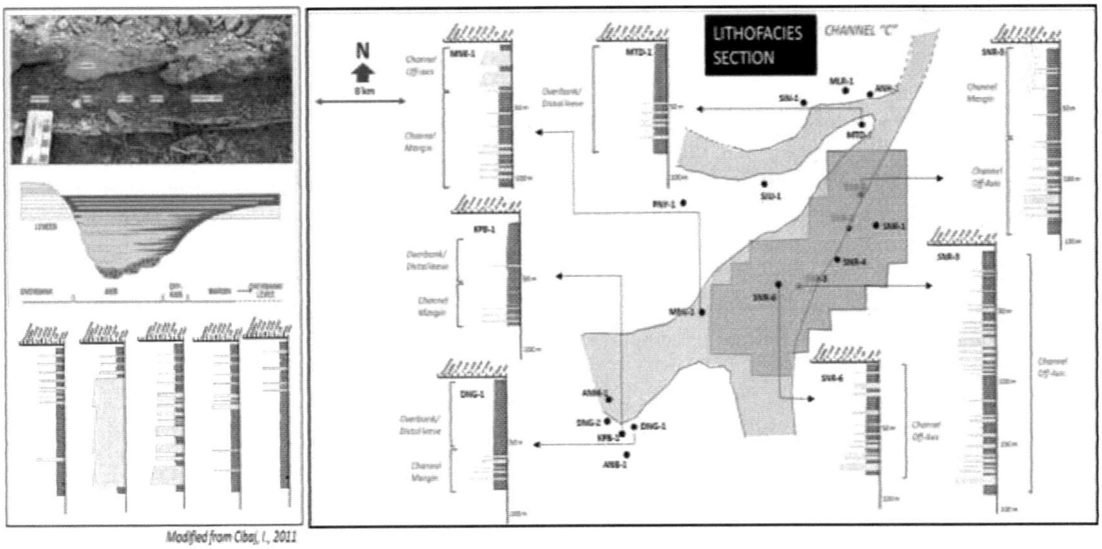

Figure 6. Lithofacies Interpretation of Channel "C"

We also run 3D seismic attributes such as RMS amplitude and spectral decomposition to determine the geometry of the channel. Due to 3D Seismic data set only covered small portion of southern study area, we only able to interpret 2 (two) channel, which are channel C and D (figure 7). Using these seismic attributes, we are able to define much better depositional environment such as oxbow lake, abandoned channel, possible splays or distributary channel.

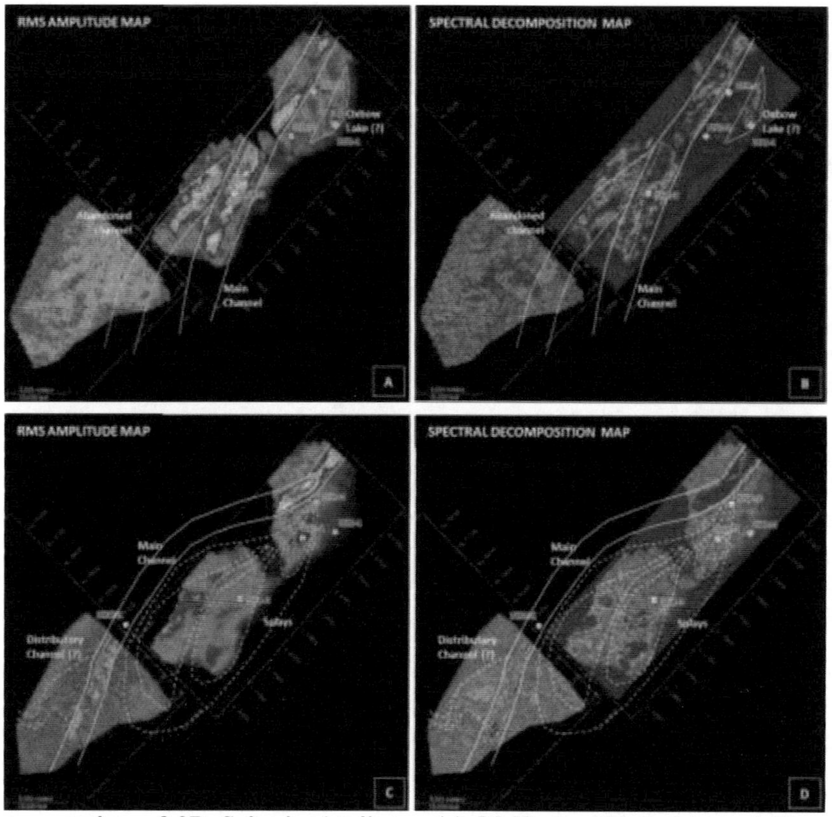

Figure 7. Interpretation of 3D Seismic Attributes (a) RMS Amplitude Map of Channel "C", (b) Spectral Decomposition of Channel "C", (c) RMS Amplitude Map of Channel "D", (d) Spectral Decomposition of Channel "D"

5. Conclusions

Based on our interpretation, NE to SW trending deep water channel existed in Celebes Molasse as Pliocene post collision interval. Existing 2D and 3D seismic data sets are able to delineate each channel geometry and depositional environment. Regional well correlation and lithofacies section positively support deep water channel model.

Despite of there is no well testing data on these channels interval yet; we believed that this Pliocene deep water channel can become the new exploration play.

Further study about biogenic generation and migration, lateral seal effectives of each channel and possibility to acquire new 3D seismic data set in northern area study is needed to give better understanding of deep water channel petroleum system play

6. Acknowlegdements

The Authors appreciate the support from Eastern Indonesia Exploration Team in this study and also would like to thanks Pertamina EP Management for giving permission to publish this data.

References

[1] Alif, S. M., Meilano, I., Gunawan, E. and Efendi, J., 2016, Evidence of Postseismic Deformation Signal of the 2007 M8.5 Bengkulu Earthquake and the 2012 M8.6 Indian Ocean Earthquake in Southern Sumatra, Indonesia, Based on GPS Data, *Journal of Applied Geodesy*, 10(2), 103-108, DOI: 10.1515/jag-2015-0019.

[2] Anugrah, B., Meilano, I., Gunawan, E. and Efendi, J., 2015, Estimation of postseismic deformation parameters from continuous GPS data in northern Sumatra after the 2004 Sumatra-Andaman earthquake, *Earthquake Science*, 28(5-6), 347-352, DOI: 10.1007/s11589-015-0136-x.

[3] Ardika, M., Meilano, I. and Gunawan, E., 2015, Postseismic Deformation Parameters of the 2010 M7.8 Mentawai, Indonesia, Earthquake Inferred from Continuous GPS Observations, *Asian Journal of Earth Sciences*, 8, 127–133, DOI: 10.3923/ajes.2015.127.133.

[4] Cibaj, I., 2011, Channel-Levee Facies and Sea Floor Fan Lobes in the turbidites of Lower Kutei Basin, East Kalimantan : Field Trip, *Sedimentology Newsletter*, 21, p15-21.

[5] Garrard, R., A., et al, 1988, The Geology of Banggai-Sula Microcontinent, Eastern Indonesia. *Proceedings Indonesian Petroleum Association*, 17th Annual Convention, 23-52.

[6] Gunawan, E., Kholil, M. and Meilano, I., 2016, Splay-fault rupture during the 2014 Mw7. 1 Molucca Sea, Indonesia, earthquake determined from GPS measurements, *Physics of the Earth and Planetary Interiors,* 259, 29-33, DOI: 10.1016/j.pepi.2016.08.009.

[7] Gunawan, E., Maulida, P., Meilano, I., Irsyam, M. and Efendi, J., 2016, Analysis of coseismic fault slip models of the 2012 Indian Ocean earthquake: Importance of GPS data for crustal deformation studies, *Acta Geophysica*, 64(6), 2136-2150, DOI: 10.1515/acgeo-2016-0106.

[8] Gunawan, E., Meilano, I., Hanifa, N. R. and Widiyantoro, S., 2017, Effect of coseismic and postseismic deformation on homogeneous and layered half-space and spherical analysis: Model simulation of the 2006 Java, Indonesia, tsunami earthquake, *Journal of Applied Geodesy*, DOI: 10.1515/jag-2017-0009.

[9] Gunawan, E., Ghozalba, F., Syauqi, Widiastomo, Y., Meilano, I., Hanifa, N. R., Daryono and Hidayati, S., 2017, Field Investigation of the November to December 2015 Earthquake Swarm in West Halmahera, Indonesia, *Geotechnical and Geological Engineering*, 35(1), 425-432, DOI: 10.1007/s10706-016-0117-4.

[10] Hall, R. and Sevastjanova, I., 2012, Australian Crust in Indonesia. *Australian Journal of Earth Sciences* vol. 59, p.827-844.

[11] Hasanusi, D., et al, 2004, Prominent Senoro Gas Field Discovery in Central Sulawesi, *Proceedings, Deepwater and Frontier Exploration in Asia & Australasia Symposiu*n, Indonesian Petroleum Association

[12] Ito, T., Gunawan, E., Kimata, F., Tabei, T., Meilano, I., Agustan, Ohta, Y., Ismail, N., Nurdin, I. and Sugiyanto, D., 2016, Co-seismic offsets due to two earthquakes (Mw 6.1) along the Sumatran fault system derived from GNSS measurements, *Earth, Planets and Space*, 68(1), 1, DOI: 10.1186/s40623-016-0427-z.

[13] Kartaadiputra, L., W., and Samuel L., 1988, Oil Exploration in Eastern Indonesia: Facts and Perspectives. *Proceedings Indonesian Petroleum Association*, 17th Annual Convention, 1-22.

[14] Mitchum, R., M., Vail, P., R., and Sangree, J., B., 1977, Seismic stratigraphy and blobal changes in sea level, part 6 : stratigraphic interpretations of seismic reflection patterns in depositional sequences, in Payton, C., E., ed., *Seismic Stratigraphy and Applications to Hydrocarbon Exploration*; AAPG Memoir 26, p.117-133

[15] Pholbud, P., Hall, R., Advokaat, E., Burgess, P., and Rudyawan, A., 2012, A New Interpretation of Gorontalo Bay, Sulawesi, *Proceedings, Indonesian Petroleum Association*, 36th Annual Convention & Exhibition, May 2012.

[16] Pratama, C., Ito, T., Sasajima, R., Tabei, T., Kimata, F., Gunawan, E., Ohta, Y., Yamashina, T., Ismail, N., Nurdin, I., Sugiyanto, D., Muksin, U. and Meilano, I., 2017, Transient rheology of the oceanic asthenosphere following the 2012 Indian Ocean Earthquake inferred from geodetic data, *Journal of Asian Earth Sciences*, 147, 50-59, DOI: 10.1016/j.jseaes.2017.07.049.

[17] Raharja, R., Gunawan, E., Meilano, I., Abidin, H. Z. and Efendi, J., 2016, Long aseismic slip duration of the 2006 Java tsunami earthquake based on GPS data, *Earthquake Science*, 29(5), 291-298, DOI: 10.1007/s11589-016-0167-y.

[18] Sani, K., and Yusuf, B., E., 2015, Eastern Region of Indonesia : Continents Adrif & The Potential New Play Models, in Revisit on Tectonic of SE Asia, Pertamina (unpublished)

[19] Satyana, A.,H., and Purwaningsih M., E, .M., 2011, Collision of Micro-continents with eastern sulawesi: Records from uplifted reef terraces and proven –potential petroleum plays, *Proceedings, Indonesian Petroleum Association*, 35thAnnual Convention & Exhibition.

[20] Simandjuntak, T., O., and Barber, A., J., 1996, Contrasting tectonic styles in the Neogene orogenic belts of Indonesia, in Hall, R. and Blundell, D.J., eds., Tectonic Evolution of SE Asia, *Geological Society of London Special Publication,* 106, 185-201.

[21] Subroto, E., A., Noeradi, D., and Afriatno, B.,Y., 2009, Geochemical Identification of favorable basin for biogenic gas exploration in Indonesia, *Proceedings, Indonesian Petroleum Association*, 33rd Annual Convention & Exhibition.

IOP Conf. Series: Earth and Environmental Science **132** (2018) 012006 doi:10.1088/1755-1315/132/1/012006

New Prespective Paleogeography of East Java Basin; Implicationrespond to Oil and Gas Eksploration at Kujung Formation Carbonate Reservoar

Aprilana C*[1], Premonowati[1], Hanif I S[1], Choirotunnisa[1], A Shirly[2], M K Utama[2], Y R Sinulingga[2], F Syafitra[2]

[1]Universitas Pembangunan Nasional "VETERAN" Yogyakarta
[2]SKK Migas
*aprilana2011@gmail.com

Abstract. Paleogeography is one of critical points that always less considered by explorationist in the world. Almost all of the consideration is focused on trapping mechanism. Paleogeography is guidance in understanding both of physical and chemical of rock characteristic which will correlate with its depositional environment. Integration of various geological and geophysical data such as; tectonic, structural geology, stratigraphy, lithology, and biostratigraphy will lead us to a better understanding of rock characteristics. Six paleogeographic interpretations was made consist of; Early Tertiary (P5-56-55 ma), Middle Eocene (P14-41 ma), Late Oiligocene (P22-25.5 ma), Early Miocene (N7-16.5 ma), Middle Miocene (N9-14.5 ma), and Pleistocene (NN19-1.5 ma). That six paleogeographic interpretations are assumed represent the paleogeographic evolution of East Java Basin time after time. In Middle Eocene time, it would be more than hundred possibilities regarding the location where the formation deposited. This would be controlled by the existence of some local structural paleohighs and horsts which oriented NW-SE followed by their own sedimentary transportation path. With assumption that hydrocarbon generation was occurred in 15 Ma and the depth of maturation window lies on about 2,500 m depth. Therefore, the possibility of source rock maturation is high, due to almost of the clastics sediment of Ngimbang deposited into the series of grabens. The Kujung reef types simplified defines and categorize into; 1) Patch Reef 2) Berrier Reef 3) Pinnacle Reef Over Isolated Reef. Kujung Carbonates were deposited in Early Miocene when regional transgression occurred. The depositional environments were dominated by shallow marine littoral-sublittoral. Generally, the reservoir quality of this Kujung Carbonate shows fair to good quality, in range7-32% porosity, and 1-1400 mD permeability (internal SKK Migas data).

1. Background

Indonesia is the 3rd biggest hydrocarbone reserves (±3.6000 MMSTB) and East Java Basin have proven reserves oil ±550 MMSTB. As well as prospect & lead exploration potential is the 2nd biggest on Indonesia with resources estimation are ±15.000 MMBOE from 80.000 MMBOE on East Java Basin & North East Java Basin (RPS SKK Migas, 2014). Base on those data, we need more explorations to increase national hydrocarbon reserves. As first initiated exploration, paleogeography was used in understanding physical characteristic of rock. Identification of rock distribution, especially fine clastical sediment from Ngimbang formation functioned as hydrocarbon source and Oligo-Miocene Kujung's carbonate rock as key reservoir hydrocarbon explorations on all of East Java Basin.

Fine clastical rock from Ngimbang formation with interbedded claystone and sandstone with interclast coal is present in several sedimentation places with fluvial-deltaic depositional environment. Source rock potential on onshore and offshore East Java areas where mostly oil & gas exploration activities focus on Ngimbang formation was identified by good source rock, none the worse it need more evaluation further to know exploration continuity for further. And neither is carbonate rock dissemination in Oligo-Miocene, it's need paleogeography identification with locally targets which match for reef built-up at that transgressive phase.

Rock characterization comprehension as basic on petroleum system was delivered this study to understand paleogeography as rock genetic come from. Recomendation this areas for explore are our research objective.

2. Method

By definition, paleogeography is visible image of geological condition in the past which focus on physical aspect and environment changing in the past (Lyell, 1919, J.Diana, 1860). This paleogeographic study was made based on several data integration, there are; structure (Tectonic), stratigraphy, lithology, and biostratigraphy. Derivative of biostratigraphy is bathimetry environment, rock sedimentation and fossils age.

Paleogeography also related to sea level change, either on local or global event at specific time. Those are reflected in lithologies which sedimented with fossil association, such as benthonic foraminifera or planktonic foraminifera which is abundant in specific environmnet and bathimetry.

3. Basic of Paleogeography Interpretation

In this study paleogeography built from several data such as focus on bathymetry index fossil and lithology. There many classification made to generate paleogeography. This study focus on bathymetry index fossils as primery data to define paleogeography conditions. This study also made some classification modification for making simple way classification Hedgpeth (1957), Berggren (1978), Ingle (1980), Tipsword et.al.b(1966), Brasier (1980), Bandy (1953), Schroder et al. (1988) research are adapted for this study (Figure 1).

4. Regional Geology

Mudjiono & Satyana (2005) mapped the paleogeography of East Java, in offshore area there are Muriah Trough – Bawean Arch – Tuban Trough – JS-1 Ridge – Central Deep – North Madura Platform and in onshore area there are terdapat West Cepu High – Kening Trough - East Cepu High – Ngimbang Basin – Madura Basin – BD Ridge – South Madura Graben – XX Ridge. The highs and lows are half graben system which fomed by Paleogene tectonic.

As the bigger picture, Java island is active margin Sundaland (West arm Australian microcontinent) Hall et. al. East Java formed by the subduction of Sundaland microcontinent and West arm Australian microcontinent, also affected by the Indian collision. It rotated the West arm and froming RMKS fault.

Rifting in East Java basin started in Paleogene as implication of eastern sundaland detached by Indo-australia and eurasia subduction. Ngimbang clastic dominated almost all the syn-rift depocenters, this formation is mainly recognized as source rock in this area.

After rifting, regional transgresive occured in East Java basin from Oligocene to Miocene. In high areas with shallow marine environment, corals started to grow which will be the source of Kujung formation. The coral clasts deposited in lower areas and become carbonate clastic. Kujung formation become the main reservoir rock in East Java basin.

At the end of transgresive phase, a tectonic event changed the paleogeography of East Java basin significantly. Rembang - Madura - Kangean - Sakala (RMKS) formed by west sulawesi microcontinent, bending of banda arc, australia - timor collision, and indo-australia and eurasia subduction (Figure 2b).

Generaly, hydrocarbon generated in paleogene grabens and migrated to higher areas. Hydrocarbon in offshore area is generated by Bawean Trough, central deep and South Madura. The source rock is rich organic shale from Ngimbang formation (Satyana, 2003)(Figure 2a). This formation mainly deposited in delta, lacustrine and shallow marine environment. Maturity started in middle miocene due to subsidence in low areas which causing enough temperature to generate hydrocarbon. Hydrocarbon migrated to reservoir (Kujung Formation) in high areas through carrier beds lateraly and faults verticaly. The hydrocarbon exploration and exploitation are very intensive in high areas where the hydrocarbon accumulated.

5. Result and Discussion

5.1. Early Tertiary Paleogeografi (P6, 56-55 Ma)

Based on data and references extant now, Early Tertiary or around Paleosen (P6, 56-55 Ma) is great age to know basement configuration before whole East Java Basin rifting event (Figure 3a). Generally basement in this time was terrestrial environment. Mostly basement have terrestrial environment such as goundsmass (paleosoil), breccia conglomerate, and palynomorps (Mayeripollis nayarkotensis). In this time transitional environment have indications such as interbedded carbonaceous shale, mangrove pollen fossils, coal, amber, and glouconitic.

5.2. Middle Eocene Paleogeography (P14, 41 Ma)

Eocene time is petroleum system crucial age in East Java basin. At this age is sedimentation time of Ngimbang Clastic which is syn-rift phase. Base on analysis hydrocarbon there are mostly I & II kerogene type which is terrestrial – transitional depositional environment indication condition (Figure 3b). Fluvial – deltaic – intertidal environment have indications such as interbedded sandstone, carbonaceous shale, amber, coal, borrow palynomorf (*Palmaepollenites lutchensis*), and fossils (*Durio* sp., *Lanagiopolis microrugulatus*, and *Marginipolis concintus*). Inner – outer littoral environment have indications such as skeletal grains (shell & coral pieces), and larger foraminiferas. Inner – outer sublittoral environment have indications such as foraminifera fossils (*Nummulites, Discocyclina, Pellatispira, Spiroclypeus, Operculina, Quinqueloculina, Heterolepa, Eponides,* and *Discorbis*).

5.3. Late Oligocene Paleogeography (P22, 25.5 Ma)

Late Oligocene was the end of carbonate factory in this age. Late Oligocene base on tectonostratigraphy implied in post-rift phase which was Ngimbang Carbonate give-up. There was a base layer for Kujung formation build-up and sedimentation place (Figure 3c). Inner – outer littoral environment was dominantly as packstone, wackstone, grainstone, sandstone, and clystone. Inner – outer sublittoral environment have indications such as foraminifera fossils (*Sphenolithus ciperoensis, Zygrhablithus bijugatus,* H. *recta*, D. *bisecta*., *Lepidocyclina* (N) *parva*, *Lepidocyclina isolepidinoides*, *Miogypsinoides praeubaghsi, Heterostegina bornensis, Heterostegina bantamensis, Nummulites intermedius/fichteli., Globigerina angustiumbilica* (middle neritic), and *Globigerina* spp. (middle neritic, Fadel (2013)) Outer – upper bathyal environment have indications such as foraminifera fossils (*Spiroclypeus, Eulepidina, Miogypsina., Austrillina., Nummulites fichteli, Spiroclypeus* spp., *Lepidocyclina isolepidinoides, Miogypsinoides praeubaghsi, Heterostegina praecursor,* and H. *bantamensis.*), shale, and mudstone.

5.4. Early Miocene Paleogeography (N7, 16.5 Ma)

Early Miocene time. Early Miocene is important to understand Carbonate of Kujung formation. Sedimentation of Kujung formation is post-rift phase (Figure 3d). Intertidal environment was identity foraminifera fossils such as *Haplophragmoides /Trochammina., Ammonia beccarii.,* and *Textularia* spp. Inner – outer littoral environment was foraminifera fossils dominantly such as *Quinqueloculina., Eponides., Lenticulina., Haplophragmoides /Trochammina., Ammonia beccarii., Globigerinoides primolus., Rhodophytes.,* and *Textularia* spp. Inner – outer sublittoral environment have indications foraminifera fossils such as *Bolivina., Nodosaria., Operculina., Uvigerina aculeta., Lenticulina., Globorotalia acostaensis acostaensis, Globigerinoides ruber., Cibicides., Bulimina* spp., *Gyroidina* spp., *Amphistegina, Lepidocyclina*, and *Ammonia*. Outer – upper bathyal environment have indications foraminifera fossils such as *Pyrgo* spp., *Globorotalia acostaensis acostaensis*, and *Globigerinoides ruber*.

5.5. Middle Miocene Paleogeography (N9, 14.5 Ma)

RMKS fault influence appeared Middle Miocene time begun from onshore East Java in the northern part to offshore East Java. Bathimetry East Java offshore middle eocene time was shallow than southern part East Java onshore. Middle Miocene time was a maximum force sea level drop. There happened in the several East Java basin as dominantly fluvial deposite. This era was mostly Ngrayong formation sedimentation time (Figure 3e). Intertidal environment was indicated by foraminifera fossils such as *Angulogerina., Aveolinella.,* and *Haplophragmoides/Trochammina*. Inner – outer littoral have indications such as foraminifera fossils (*Lenticulina., Cibicides., Sphenolithus heteromorphus., Angulogerina.,* and *Ammonia indopacifica*.), chalky limestone, reef banks, larger foraminiferas, Mollusca debris and coral fragments (patch reefs possibly). Inner – outer sublittoral was indicated by foraminifera fossils such as *Bolivina., Uvigerina., Lenticulina., Globigerinoides sicanus., Gaudyrina* spp., and larger foraminiferas.

5.6. Pleistocene Paleogeography (NN19, 1.5 Ma)

Activation RMKS fault effect was clearly present in this era, specially there appeared on northen to east part East Java onshore and Pagerungan. Pleistocene time explained Paciran formation and those equivalents sedimentation as shellow marine environment condition. North - East West trend subduction compresional force in Southern Java made deformation and erotion predominantly (Figure 3f). Intertidal environment have indications such as foraminifera fossils *(Ammonia beccarii., Quinqueloculina.,* A. *trispinosa., Textularia* spp., and *Florilus*.), coaly, claystone with limestone interclast, siltstone, and lignite. Inner – outer littoral environment have indications such as foraminifera fossils (*Textularia* spp., *Elphidium* spp., *Lenticulina., Ammonia beccarii.,* and *Quinqueloculina.*), sandstone, interbedded carbonacous shale, coal, chalky limestone, and corals. Inner – outer sublittoral environment was indicated by foraminifera fossils (*Bolivina., Bulimina., Gyroidina., Lenticulina., Pullenia bulloides., Pulleniatina obliquiloculata., Uvigerina.,* and *Elphidium* spp.), sandstone, claystone, packstone, wackstone, and grainstone. Upper bathyal environment was indicated by foraminifera fossils (*Globigerinoides quadrilobatus immaturus., Pyrgo* spp., and *Oridorsalis umbonatus.*), plankton/bentos ration (50:50 – 70:30), mudstone, wackstone, sandstone, and shale.

Paleogeography and Source and Reservoir Rock Distribution.

Discussion about concept of paleogeography will not complete without the implementation/application in the field. In this study paleogeography will be related to petroleum system, specifically for source and reservoir rock elements. There are many references said that sediment clastics of Ngimbang formation is main source rock in East Java basin. The intercalation between shalestone with sandstone, sometimes with coal seam, makes this formation is rich with organic materials which later can form hydrocarbon such as oil and gas. Clastic rock of Ngimbang Fm. always be analogized as fluvio – deltaic environment or in tectonic setting always close with half-graben faults system. As we look in elevation there is a massive difference and sedimentation space between the highlands and lowlands until from wide to narrow sedimentation space.

One of the objectives from this study is trying to identify distribution of Clastic Rock of Ngimbang formation in East Java Basin. this study was conducted with same concepts of predecessor researchers, but only with larger research area, this study was supposed to give comprehensive understanding about clastic rock of Ngimbang Fm. regionally. Based on paleogeography map which is result of this study, probability of clastic rocks in Middle Eocence was quite widespread. Clastic rock of Ngimbang fm. good distribution from fluvio - deltaic environment, and shallow marine environment is illustrated in Figure 4. Some of interpolation method was conducted based on conture bathimetry which showed 'valley', for helping interpretation on western research area, because of short of data.

Beside of lateral distribution, this study also identifies, according to vertical position of locations where shale deposited, it has potential as source rock since the beginning of hydrocarbon formed (approx. 15 million y.a.) and has been in graben all this time. From some of references, it is known that "hydrocarbon window" in East Java Basin at elevation approx. 2500 m. From this data, the probability of dicovering mature source rock is quite high in all graben.

In reservoir's side, this study focussed at carbonate reservoir of Kujung formation. In Satyana, (2003), one of his publications explained quite detail about Oligo – Miocene's Carbonate rock in Java Island. He classified two models of carbonate rocks forming, which are (1) Deposition on a Land-attached Platform and (2) Deposition on Offshore Isolated Platforms. In same publication, 5 facies was identified in this area, they are (1) Fringing Reef at Rim of Basement, (2) Basinal Mud Mound, (3) Patch Reef over Platform, (4) Shelf-Edge Barrier Reef, and (5) Pinnacle Reef over Offshore Isolated Platform. Meanwhile this study classify in simpler classification, they are (1) Patch Reef, (2) Berrier Reef, and (3) Pinnacle Reef Over Isolated Reef.

In reservoir's terminology, porosity and permeability parameters also show good quality, for example in JS-1 Ridge area, from surrounding well Camar-1, average porosity is 23% and permeability is 473.3 mD. While on North Madura Platform shows porosity 22% and permeability 253.3 mD. Similar with the Offshore area, on East Java's Onshore area also shows good quality of reservoir with average porosity 16% and permeability 335.16mD. Different with three other areas that have been mentioned before, on Madura Strait area, reservoir rock data is limited. But based on geological and bathymetri models which is deepening on the East and limited local highground, then in that result in reservoir quality is less good than other areas.

6. Conclusion

There 6 (six) paleogeography maps have been made in this study comprise of : Early Tersier (P6 or 56 -55 mya), Middle Eocene (P14 or 41 mya), Late Oligocene (P22 or 25.5 mya), Early Miocene (N7 or 16.5 mya), Middle Miocene (N9 or 14.5 mya), and Plistocene (NN19 or 1.5 mya). All six paleogeography is considered showing evolution of East Java Basin paleogeography (Figure 3).

Paleogene is period that has become the key in East Java Basin, in this case become an initial point for basement setting which formed highland and lowland in halfgraben system like we see right now. Manifestation of halfgraben system has appeared since Middle Eocene's paleogeography.

Clastic rock of Ngimbang formation, is dominated by lithology of shale-sand interbeded and some of coal seam, and believed as main source rock in East Java Basin. These rocks was deposited widespread and filled Paleogen's grabens wih sediment material source from vicinity highland. Laterally many places have probability for this roks deposited, because of on Middle Eocene there are some highlands which have Northeast – Southwest bearings, with probability of sediment paths on Western and Eastern of this highlands.

Kujung fm. in East Java Basin could be classifies by several reef types, thy are : Patch Reef, Berrier Reef, and Pinnacle Reef Over Isolated Reef (Figure 3d). This carbonate rocks of Kujung fm. was formed at Early Miocene and regional transgression phase where almost of East Java Basin area, was dominated by lithoral – sublithoral environment. Generally the quality of Kujung fm. carbonate rocks reservoirs shows Fair to Very Good quality.

7. Awknoledgement

This study was initiated by SKK Migas incorporated with University of Pembangunan Nasional 'Veteran' Yogyakarta and also several PSC (K3S) which operated in East Java Basin. This study has a purpose as motivation which is suppossed to all stakeholders in oil and gas industry, from government, PSC, oil and gas contractors, consultant and forth, in facing recent impact of oil and gas price crisis. With this study we hope that all stakeholders remain active doing research and exploration as attempts to National Energy Sustainbility. Lastly, we like to express our grattitude to Mr. Indro Purwaman, as Chief of Exploration Division SKK Migas, and Mrs. Shinta Damayanti as Chief of Technical G & G Department SKK Migas for all supports for this study, also to all PSC's representatives regarding all data and opinions/suggestions.

8. References

[1] Boggs, Sam, J. R., 1987, Principles of Sedimentology and Stratigraphy University of Oregon, fourth edition.

[2] Cucci, M.A., 1995, Carbonate, System Tracts of an Assymetric Miocene Build Up Near Kangean Island, East Java Sea, Proceeding of International Symposium Stratigraphy in SE Asia.

[3] Hall, R., Clements, B., Smyth, H. R., Cottam, M. A., 2007, A New Interpretation of Java Structure (2007). Proceeding of IPA Convention and Exhibition.

[4] Hall, R., 2014, Indonesian Tectonics: Subduction, Extension, Provenance and More. Proceeding of IPA Convention and Exhibition.

[5] Johansen, K. B., 2003, Depositional Geometrics and Hydrocarbon Potential within Kujung Carbonates Along The North Madura Platform, As Revealed By 3D and 2D Seismic Data. Proceeding of IPA Convention and Exhibition.

[6] Prasetyadi, C., 2007, Evolusi Tektonik Paleogen Jawa Bagian Timur, Doctoral Thesis, Institute of Teknologi Bandung. Indonesia.

[7] M.K. Boundagher-Fadel., 2013, Biostratigraphic and Geological Significance of Planktonic Foraminifera, Second Edition, University College London, London.

[8] Mudjiono, R., Pireno, G. E., 2006, Exploration of the North Madura Platform, Offshore East Java, Indonesia. Proceeding of IPA Convention and Exhibition.

[9] Satyana, A. H., 2014, New Consideration on the Cretaceous Subduction Zone of Ciletuh – Luk Ulo – Bayat – Meratus: Implications for Southeast Sundaland Petroleum Geology (2014). Proceeding of IPA Convention and Exhibition.

[10] Widarmayana, I. W. A., Anggraini, N., Stephens, N., Musgrove, F., 2014. Banyu Urip and Other Cepu Block Fields Nucleation of Early Carbonate Buildups into Isolated Platforms. Proceeding of IPA Convention and Exhibition.

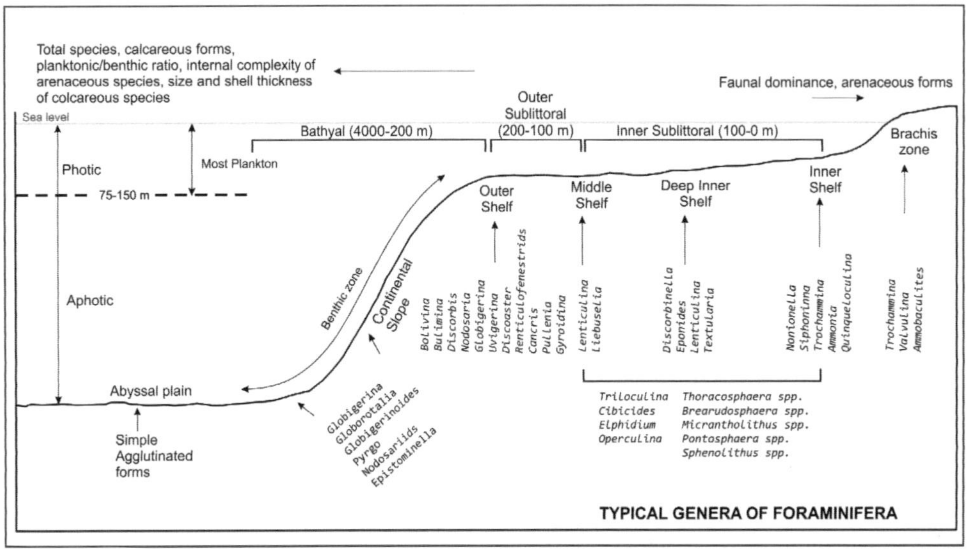

FIGURE 1, This study also made some classification modification for making simple way classification based on Hedgpeth (1957), Berggren (1978), Ingle (1980), Tipsword et.al.(1966), Brasier (1980), Bandy (1953), and Schroder et al. (1988).

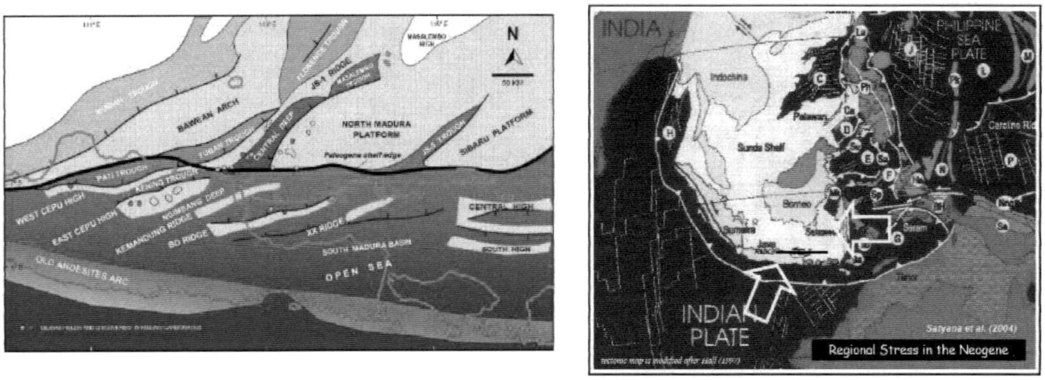

(a) (b)

FIGURE 2, Regional stress in Neogene (Satyana (2004), modified after Hall (2007) (a); Paleogene Paleogeography of East Java from Ngimbang formation (Satyana, 2005) (b).

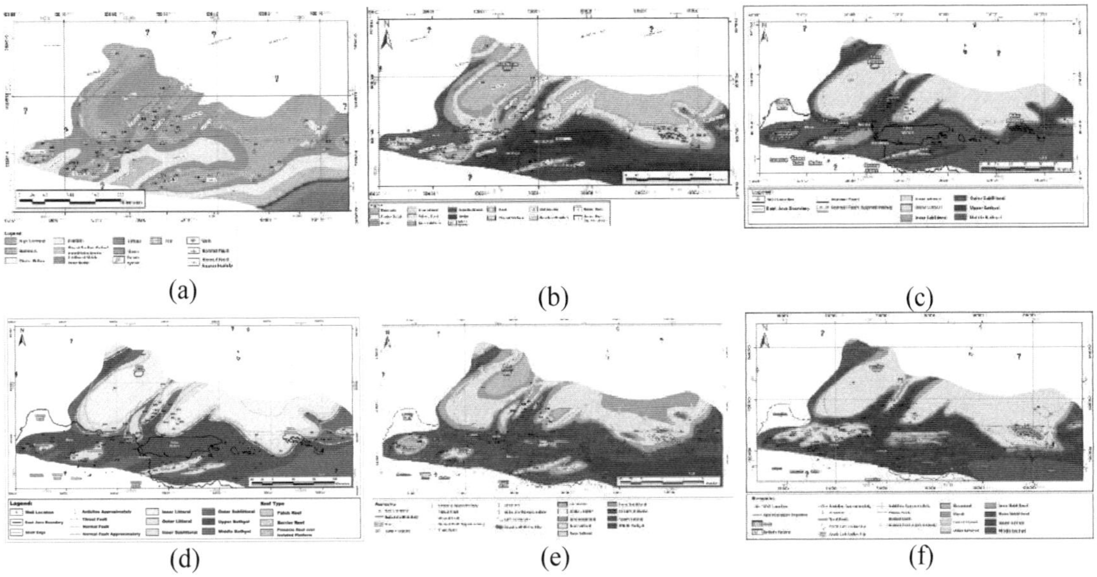

FIGURE 3, Early Tertiary Time Paleogeography map (Economic basement) (a); Middle Eocene Time Paleogeography map (Ngimbang Clastics formation) (b); Late Oligocene Time Paleogeography map (Ngimbang Carbonate and initial of Kujung formation) (c); Early Miocene time paleogeography (Kujung Formation) (d); Middle Miocene paleogeography (Ngrayong formation) (e); Pleistocene paleogeography (Paciran formation) (f).

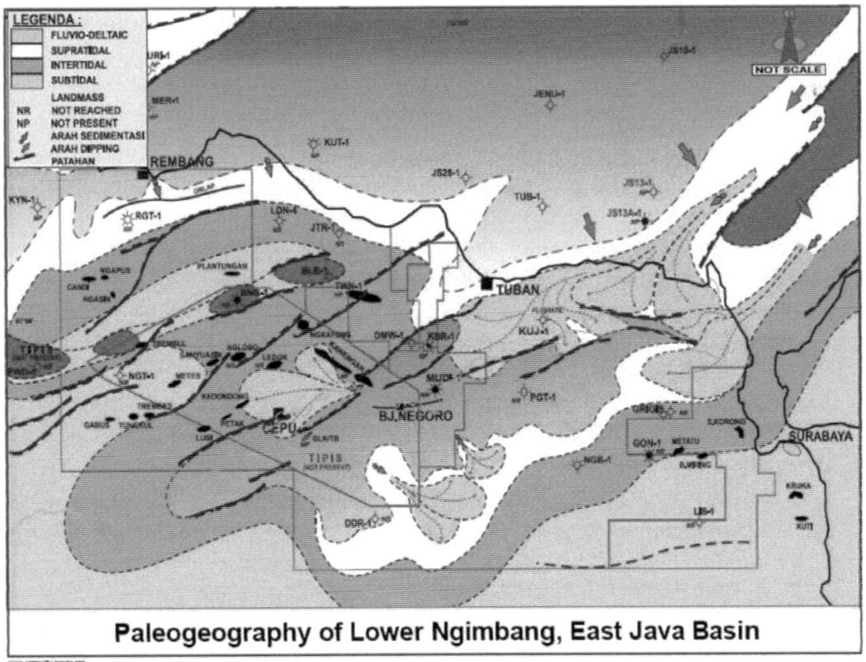

FIGURE 4, Lower part of Ngimbang Formation paleogeography map (Satyana, 2009).

Enhanced Structural Interpretation Using Multitrace Seismic Attribute For Oligo-Miocene Target at Madura Strait Offshore

Putra Pratama Wahyu Hidayat[1,a)], Antonius Hary Murti[1)], Sudarmaji[1)], Agung Shirly[2)], Bani Tiofan[2)], Shinta Damayanti[2)]

[1]Universitas Gadjah Mada, Yogyakarta, 55281, Indonesia.

[2]SKK Migas, Jakarta, 12710, Indonesia.

[a)]putrapratama08@gmail.com

Abstract. Geometry is an important parameter for the field of hydrocarbon exploration and exploitation, it has significant effect to the amount of resources or reserves, rock spreading, and risk analysis. The existence of geological structure or fault becomes one factor affecting geometry. This study is conducted as an effort to enhance seismic image quality in faults dominated area namely offshore Madura Strait. For the past 10 years, Oligo-Miocene carbonate rock has been slightly explored on Madura Strait area, the main reason because migration and trap geometry still became risks to be concern. This study tries to determine the boundary of each fault zone as subsurface image generated by converting seismic data into variance attribute. Variance attribute is a multitrace seismic attribute as the derivative result from amplitude seismic data. The result of this study shows variance section of Madura Strait area having zero (0) value for seismic continuity and one (1) value for discontinuity of seismic data. Variance section shows the boundary of RMKS fault zone with Kendeng zone distinctly. Geological structure and subsurface geometry for Oligo-Miocene carbonate rock could be identified perfectly using this method. Generally structure interpretation to identify the boundary of fault zones could be good determined by variance attribute.

1. Introduction

Exploration for Oligo-Miocene target at Madura Strait still reminds to be less activity than the others part of East Java Basin. Some petroleum system issues become obstacle, including trap geometry and vertical migration in case Ngimbang Clastic as source rock. Study area is located at Kendeng Depression which lies between RMKS Fault Zone in the north and Kendeng Thrust in the south. Fault existences become critical to answer some issue regarding hydrocarbon exploration in Madura Strait.

3D seismic as result from seismic reflection was conducted and become useful tool for evaluation process. As known before, seismic reflection has the best accuracy, resolution also the deepest penetration than the other methods.

Furthermore 3D seismic data provides information about multitrace attributes. One of the multitrace attribute used frequently is variance attribute, it can represent geological phenomenon. Variance attribute determines fault clearly and identified its trend both in large or small scale at faults dominated area.

Applying variance attribute on 3D seismic data becomes an excellent step to interpret a horizon that shows continuity or discontinuity of seismic data clearly. Mathematically variance attribute provides 1 or 0 value for the continuity of seismic data. This will enhance interpretation process, so some issues such as geometry, rocks spreading also property and resources quantification will have risks analysis in detail quantification.

2. Location and geology of study area

The study area is within the East Java Basins in offshore Madura Strait (**Figure 1**). In structural geology terminology, study area is located between strike slip RMKS fault zone fold zone and Kendeng thrust. Kendeng thrust is west-east trend fold thrust belt formed by compression tectonic deformation generating some anticlinorium (Van Bemmelen, 1949). The thrust fault is a shallow tectonic structure involving the basement (**Figure 2**; Prasetyadi, 2007). This structure zone shows compression effect as a result of Indo-Australian plate subduction beneath the Eurasian Plate.

RMKS fault zone has 15 km to 40 km wide and at least 675 km long from west to east start from Rembang area as west manifest to Kangean and Sakala Islands at eastern Madura Island. At the north part of study area, RMKS fault zone existed and divided into three parts which are west region, center region, and east region (Satyana et al., 2004). Mulhadiono (1984) identified 45 faults and 9 anticlines based on field study at Madura Island as evidence of RMKS uplifting.

The primary source rocks for oil and gas in Madura Strait area are the lacustrine shale of the Ngimbang Formation. The Ngimbang Formation is presently interpreted to be gas-mature within the main Paleogene depocentre along the northern part of the area, although it would have been oil generative during the Miocene to Early Pliocene. The charging is interpreted through the primary migration and secondary migration along fault. The primary reservoir in the study area is Oligo-Miocene carbonate or Kujung I Formation (**Figure 3**, Modified from Mudjiono and Pireno, 2002). Olige-Miocene carbonate is reefal buildups that have a thick carbonate section in local high and thinner section in the lower area. As the main traps and seals, not only Major fault of RMKS has contributed to hold hydrocarbon in reservoir, but also minor faults in this structural complex area. In addition, Mudjiono and Pireno 2002 claim that shale in the Tuban Formation provides a regionally extensive seal for Oligo-Miocene reservoir rock to trap the hydrocarbon.

Kujung play has limited data causes of less exploration activity with Kujung Carbonates as reservoir target, but several author believe that local high still possible to be a place for developing carbonate reef. Internal sequence boundary in Kujung generate porosity and permeability to preserves an amount of hydrocarbon.

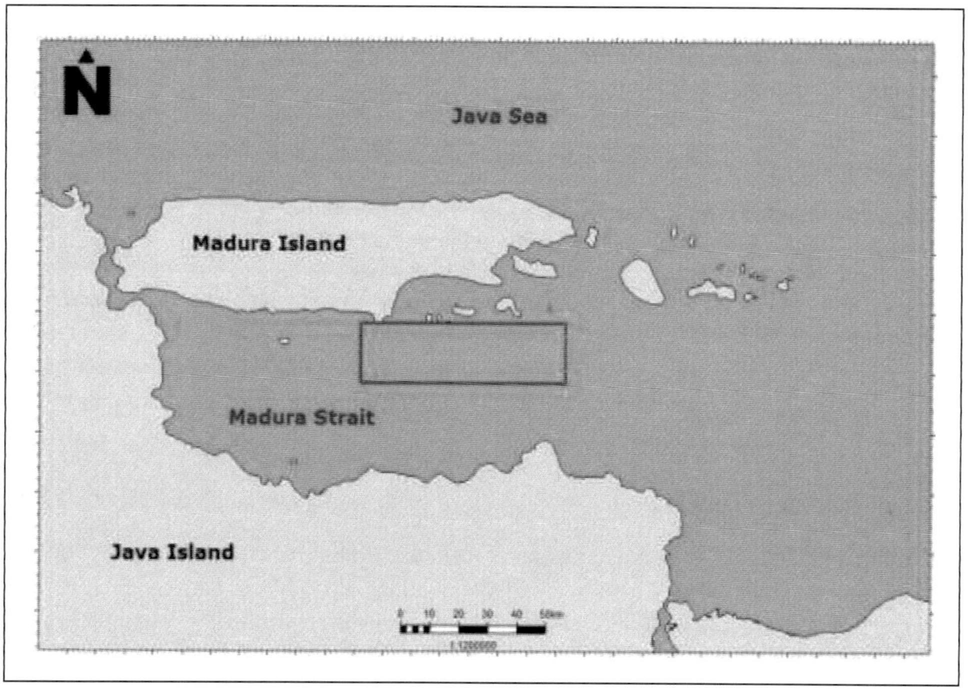

Figure 1. The location of study area

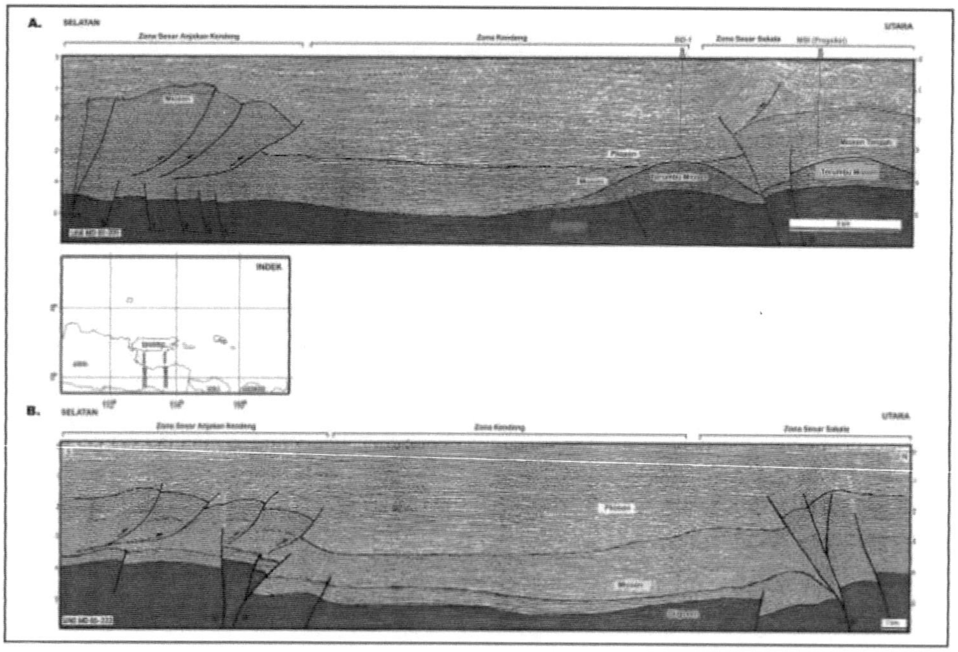

Figure 2. The geology structure of study area (Prasetyadi, 2007)

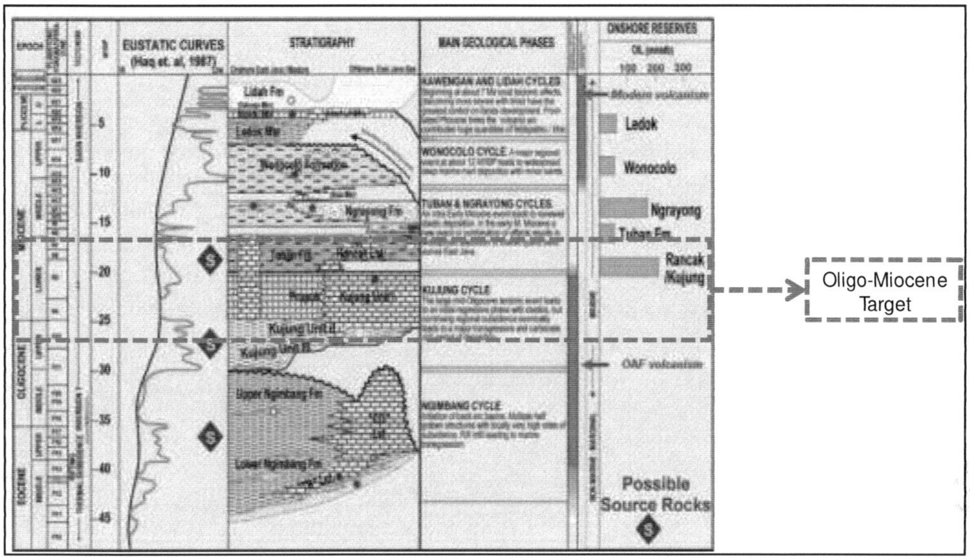

Figure 3. The stratifraphy of study area (modified from Mudjiono and Pireno, 2002)

3. Methodology

Brown (1990) in Sukmono (2000) defines attribute as the result of derivative from seismic data. All available horizons attribute and formation are not stand-alone attribute, fundamental differentiations found on data analysis from seismic data information and its display result. Time, amplitude, frequency and attenuation are the basic information from seismic data, and that information very suitable for attributes basic classification.

In Azevedo (2009), Volume attributes are computed from processed cubes or from previously computed attribute volumes which are extracted, depending on the mathematical algorithm, trace by trace or considering a group of traces (multitraces). The extraction is performed over a user-defined fixed window, where two horizontal time slice are defined as upper and lower boundaries. Multitrace seismic attributes are also computed inside a fixed vertical window with user-defined limits. In this case, besides the vertical range, the user has to define a bound in the number of traces that will be used to the attribute extraction, according to a mathematical algorithm. The output volume is the result of stacking all the time slices were the attribute computation was kept from each window position in space and time.

4. Results and discussion

Variance attribute seismic was applied to the 3D seismic data at about 1ms to 6000ms. With this range of data, the result expected to enhance seismic image by showing geological phenomenon in particularly fault from the reservoir target (Oligo-Miocene carbonate rock) in the study area. The study area is generally characterized by growth faults and its lies in the boundary of two faults zone. It is hard to identify the boundary from normal seismic data and variance attribute expected to determine this boundary. After the boundary already well defined, then we could easily determine Oligo-Miocene carbonate rock geometry.

4.1. Well correlation (Lithostratigraphy)

This study used one cube 3D seismic (1.42 x 10^4 km^2) and three (3) exploration wells (**Figure 4**). Those data have been correlated in lithostratigraphy to identify each rock formation (well correlation result is shown by **Figure 5**). The Oligo-Miocene carbonate target rock named by Kujung Lst marked by blue color.

Two wells AA-1 and NN-1 is located at a local high that since early Paleogene already exists. Carbonate reefs grow and clearly identified in both those wells. Grainstone and packestone was the main carbonate lithofacies that found from those two wells. Meanwhile CC-1, lower by position than the two other wells, is dominated by wackestone and mudstone facies and interpreted as talus from main reef body.

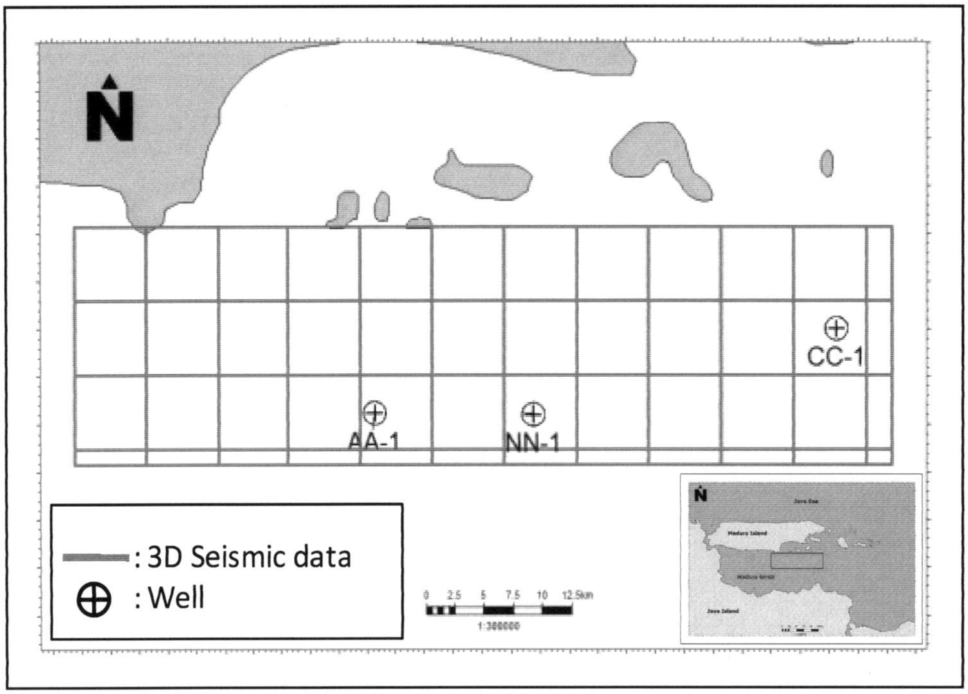

Figure 4. Base map of study area.

IOP Conf. Series: Earth and Environmental Science **132** (2018) 012007 doi:10.1088/1755-1315/132/1/012007

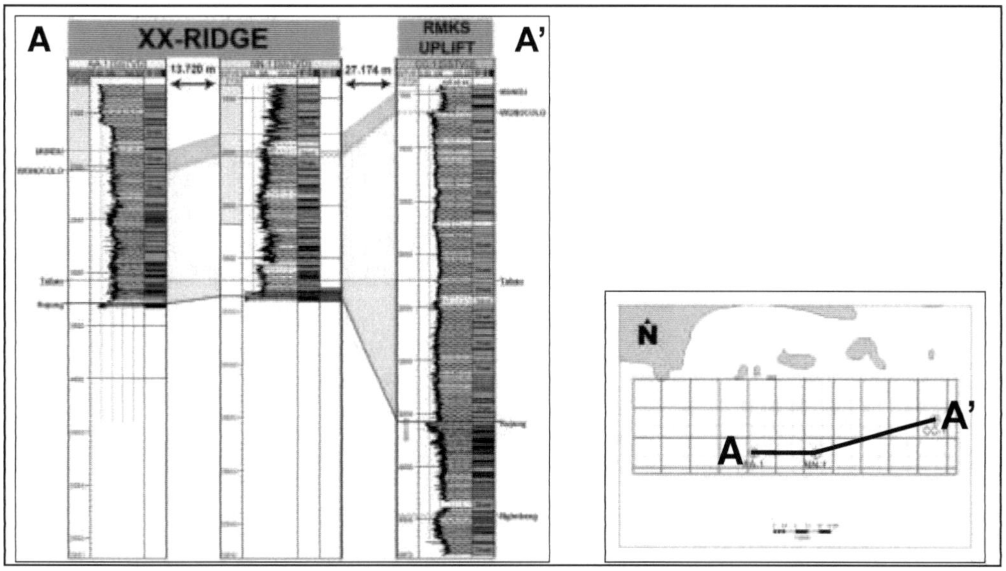

Figure 5. Lithostratigraphy correlation.

4.2. Seismic Interpretation

Well to seismic tie was done using the checkshot data from each well. Interpretation has been done for five horizons which interpreted as Basement, Ngimbang (Eo-Oligocene), Kujung Limestone (Oli-Miocene), Tuban and Wonocolo (Miocene). **Figure 6** represents arbitrary line with east west trend through all key wells. Horizons and structures interpretation was conducted to the seismic amplitude having similarity data and it's also corresponding with well data. Kujung Limestone is shown by high structure or build-up reef penetrated by AA-1 and NN-1 wells. It is distinctly recognized that AA-1 and NN-1 are the build-up body compare to structure which penetrated by CC-1.

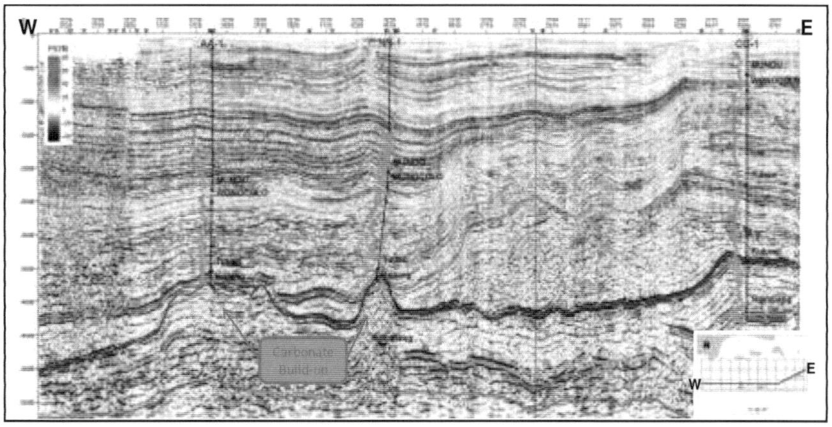

Figure 6. Horizon interpretation tied to well data

IOP Conf. Series: Earth and Environmental Science **132** (2018) 012007 doi:10.1088/1755-1315/132/1/012007

4.3. Variance Attribute Analysis and Fault Interpretation

Variance attribute was applied to the seismic data to get better geological image and help interpretation process as the main goals. Comparison between seismic section and variance attribute is shown by **Figure 7**, as north-south trending section passing through AA-1 well that shows corresponding value between continuity-discontinuity with variance value within range 0 (white) to 1 (yellow). Faults are clearly identified by variance section which have red-yellow color. High value of variance indicates intensive fault zone which well known as RMKS and Kendeng fault zone.

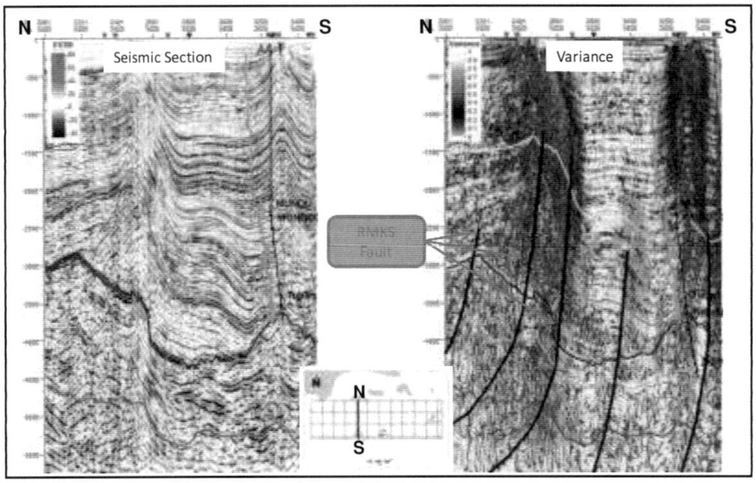

Figure 7. The comparison of seismic section and variance attribute

Figure 8 shows the existences structural setting in study area, RMKS fault zone and Paleogene structures forming XX-Ridge with the location of AA-1 and NN-1. RMKS effect could be seen in the north-east part of the seismic section as a basement involved thrust fault. RMKS fault effect is slightly located above Kujung Limestone, it caused Tuban and Wonocolo thicker than surrounding, it also caused Base of Tuban Shale as decollement. Flower structure became the distinguish boundary of up-lifted zone and Kendeng zone in southern part including XX-Ridge.

Figure 9 shows boundary between RMKS fault zone and Kendeng zone as flower structure figure. This figure shows the comparison of seismic data at time 3500 ms and variance attribute at the same time. RMKS fault zone is characterized by high value of variance (red color) represent intensive fault in those areas. Meanwhile Kendeng zone is characterized by low value of variance (grey-white color) without any significant fault shown.

The flower structure of RMKS fault passed through Kujung Limestone and spouted in Tuban Formation. Recently Kendeng zone in lower position than RMKS fault zone, although in the beginning of RMKS, Kendeng zone was interpreted in the same or maybe higher position than RMKS fault zone. This situation is shown by **Figure 10**. This figure shows the comparison of time structure map between Tuban and Kujung Limestone. Flower structure distinctly appears spouting in Tuban time structure map, it became a border between high and low area. However, in Kujung Limestone time structure map this structure passed through one area of Kujung limestone which is inverted area of southern part of RMKS fault zone.

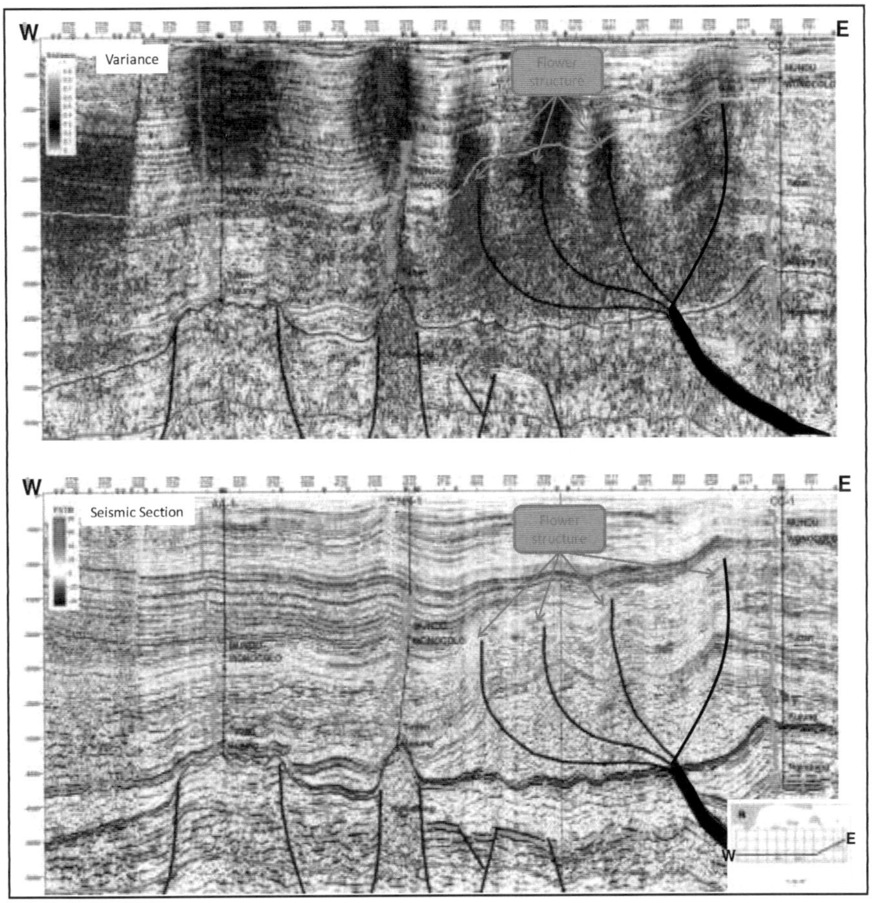

Figure 8. Fault interpretation

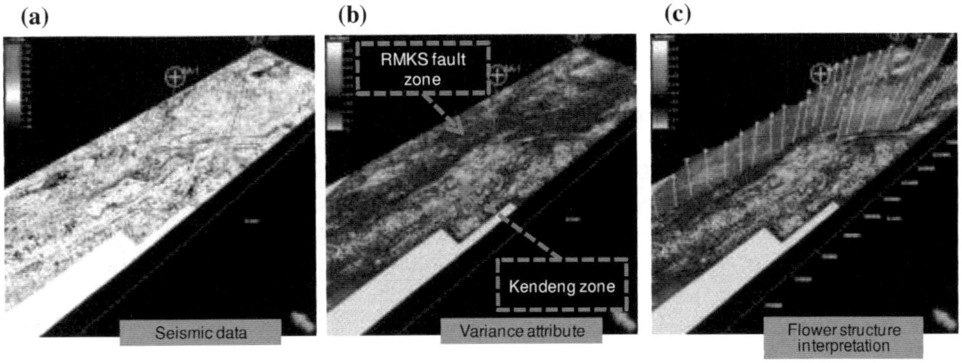

Figure 9. Time slicing at 3500 ms for (a) seismic data and (b) variance attribute, (c) flower structure interpretation to define the boundary of RMKS fault zone and Kendeng zone.

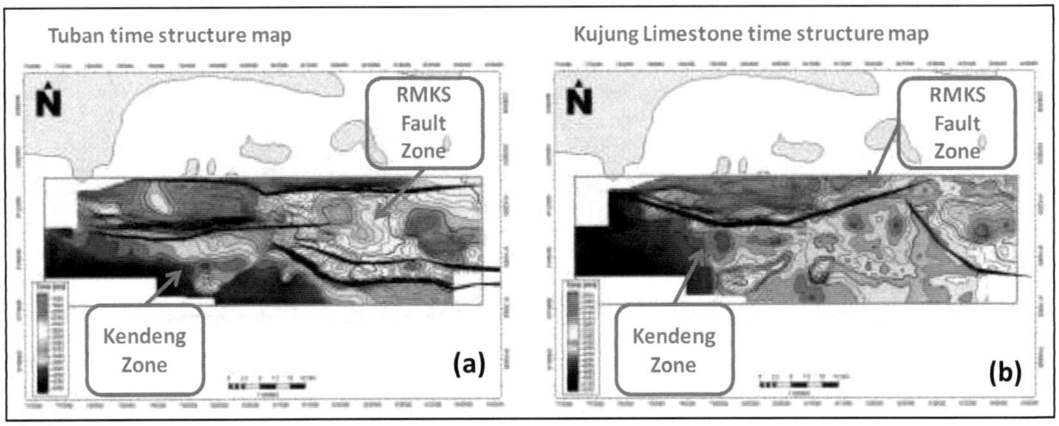

Figure 10. The comparison of the existence of RMKS flower structure in time structure map between (a) Tuban and (b) Kujung Limestone

Finally, as the result of structural as well as horizontal interpretation, play concept of hydrocarbon in Oligo-Miocene carbonate can be seen at **Figure 11**. It is clear that the htdrocarbon source rock in the study area is the shale of Ngimbang Formation, whereas hydrocarbon migrates through RMKS fault and minor fault that formed XX-ridge. Additionally, the main Cap Rock is Tuban Formation that is located on the above of XX-ridge build-up. Moreover, **Figure 5** shows that Top Kujung Formation on XX-ridge is reefal build-up, it is totally different with Top Kujung Formation from CC-1 well. This is because CC-1 well is located at RMKS uplift that is formed at the low area before it is uplifted by RMKS fault. It can be seen at **Figure 5** that the character of Top Kujung Formation in CC-1 well is siliciclastic dominated. Hence, It is clear that XX-ridge is the ideal place for accumulating hydrocarbon in the study area.

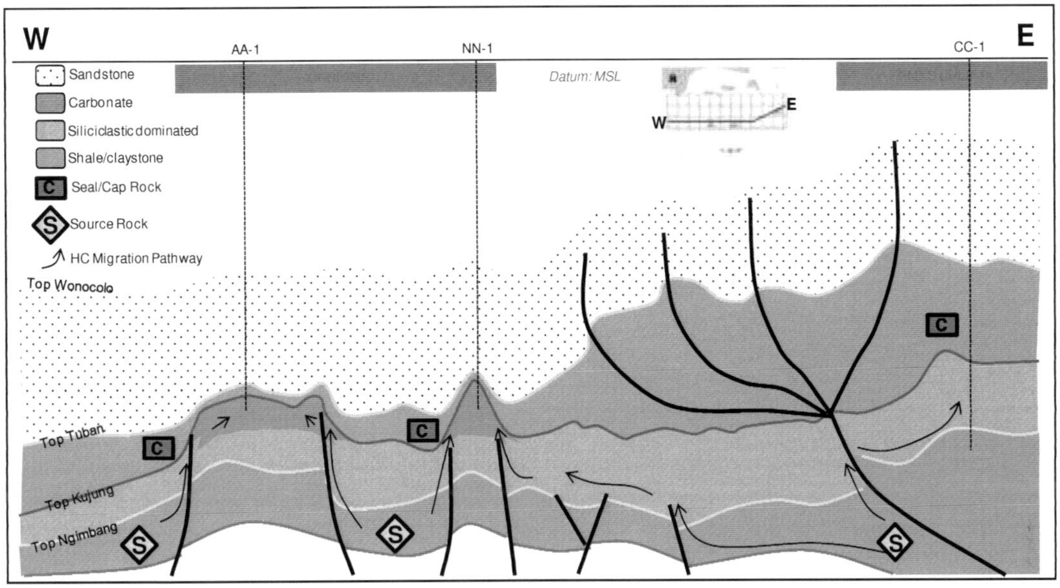

Figure 11. Oligo-Miocene carbonate Play

5. Conclusion

Interpretation of the 3D seismic data has been exercised to be more efficient by using variance attribute seismic in particularly fault interpretation. Variance attribute map has proved to be an appropriate tool to determine structural framework. Variance attribute can clearly identify the boundary of two fault zone in this case RMKS and Kendeng fault zone. With references Ngimbang formation as the main source rock in Madura Strait area, then the faults in this case RMKS fault zone become significant media for vertical migration into the Kujung carbonate reservoir. In other perspective, trap geometry cuts by faults can be well identified with this method. Therefore, application of variance attribute become very suitable tools for oil and gas exploration in faults dominated area such as study area.

6. ACKNOWLEDGEMENTS

This study initiated by SKK Migas cooperated with UPN Veteran Yogyakarta, Universitas Gadjah Mada and several PSC developing East Java Basins. This study became motivation and example for geoscientists and practitioners of oil and gas industry to keep spirit doing research and exploration in the effort of national oil and gas resources and production enhancement.

Thanks to Deputy for Planning Management Chief and Exploration Division Chief of SKK Migas for the support, also all of the PSC helping the realization of this study. Also we thank to UPP university partnership program Chevron-UGM for supporting hardware and software facility.

References

[1] Azevedo, L., *Seismic Attributes in Hydrocarbon Reservoirs Characterization* (Geoscience department of Universidade de Aveiro, 2009).

[2] Mudjiono, R., and Pireno, G. E., *Exploration Of The North Madura Platform, Offshore East Java, Indonesia* (Proceeding of Indonesia Petroleum Association, 28th Annual Convention, p. 707-726, 2002).

[3] Mulhadiono, *Stratigraphic Review and Tectonic Structure of Madura Island, East Java* (PIT-IAGI, Bandung, 1984).

[4] Prasetyadi, C., *Paleogene Tectonic Evolution of Eastern Part of Java* (Dissertation ITB, Bandung, 2007).

[5] Satyana, A. H., *New Consideration on the Cretaceous Subduction Zone of Ciletuh – LukUlo – Bayat – Meratus: Implications for Southeast Sundaland Petroleum Geology* (Proceeding of IPA Convention and Exhibition, 2004).

[6] Sukmono, S., *Stratigraphic seismic* (Engineering of Geophysics Department ITB, Bandung, 2000).

[7] Van Bemmelen, R. W., *The Geology of Indonesia* (The Hague :Govt, Netherlands, 1949).

IOP Conf. Series: Earth and Environmental Science **132** (2018) 012008 doi:10.1088/1755-1315/132/1/012008

Velocity Model Analysis Based on Integrated Well and Seismic Data of East Java Basin

Fathul Mubin[1,a)], Aviandy Widya[1], Budi Eka Nurcahya[1], Erma Nurul Mahmudah[1], IndroPurwaman[2], AryoRadityo[2], AgungShirly[2], CitraNurwani[2]
[1.] GeophysicsDepartment, Universitas Gadjah Mada, Yogyakarta, Indonesia
[2.] SKK Migas, Jakarta, Indonesia

a). mubinugm@gmail.com

Abstract. Time to depth conversion is an important processof seismic interpretationtoidentify hydrocarbonprospectivity. Main objectives of this research are to minimize the risk of error in geometry and time to depth conversion. Since it's using a large amount of data and had been doing in the large scale of research areas, this research can be classified as a regional scale research.The research was focused on three horizons time interpretation: Top Kujung I, Top Ngimbang and Basement which located in the offshore and onshore areas of east Java basin. These three horizons was selected because they were assumed to be equivalent to the rock formation, which is it has always been the main objective of oil and gas exploration in the East Java Basin. As additional value, there was no previous works on velocity modeling for regional scale using geological parameters in East Java basin. Lithology and interval thickness were identified as geological factors that effected the velocity distribution in East Java Basin. Therefore, a three layer geological model was generated, which was defined by the type of lithology; carbonate (layer 1: Top Kujung I), shale (layer 2: Top Ngimbang) and Basement. A statistical method using three horizons is able to predict the velocity distribution on sparse well data in a regional scale. The average velocity range for Top Kujung I is 400 m/s – 6000 m/s, Top Ngimbang is 500 m/s – 8200 m/s and Basement is 600 m/s – 8000 m/s. Some velocity anomalies found in Madura sub-basin area, caused by geological factor which identified as thick shale deposit and high density values on shale.Result of velocity and depth modeling analysis can be used to define the volume range deterministically and to make geological models to prospect generation in details by geological concept.
Keywords: Depth Conversion, Velocity Modeling, East Java

1. Introduction

1.1 Background
The world needs on oil and natural gas still on high demand, makes the exploitation and exploration really needed to cover the less back up in the future. As we know, the existence of hydrocarbons prospect on oil and natural gas cannot be easily found on the surface, so we need indirectly methods such as with geology and geophysical interpretation. The knowledge of subsurface regional geology and tomography technology insist that seismic method and logging are really absolutely needed to hydrocarbon

prospectivity identification in an accurately and small risk on value in the next stages. Geophysical methods that play a role in the identification of hydrocarbon prospectivity is the seismic interpretation. One of the most important part on seismic interpretation is conversion from time domain to depth domain on the structure contour map. Lot of data and large research area delivered this research to regional scale research in East Java basin.

1.2. Challanges

There are many conversion study which are do in this East Java basin but none of them belongs to regional scalecommonly limited by block. This study would be very supportive and helpful for the next studies on the regional scale to look at all the concepts of regional geology.

In simple interpolation method, velocity model would be depending on calibration wells/references. External Drift Kriging (Kriging with External Drift/KED) is a statistic method that used in this study to populating the well data as a calibration factor which is using contour map of time as a secondary parameter (Hengl et al., 2003). KED was considered as method on this regional scale study because it's able to interpolate in a good way rather than others statistic methods. Geological analysis in order to velocity model analysis also required as a validation for statistic method.KED able to provide the image of velocity model which is fit with geological research area, this would be act as added value than other statistical methods,

Although many wells used in this study, limited checkshot data reminds as challenges to further evaluation. Checkshot number and distribution highly influenced the result of velocity model. In regional scale, if checkshot number is not well distributed, then horizons as second parameter data can be optimalized with this geostatistical method (KED).

1.3. Objective

Regarding to the background and problems, there are some questions that will be answered through this study:

- Which geological factors affecting the anomalous velocity model in the study area?
- Are the use of data which is undistributed well or sparse in each block can be used to predict velocity models on regional scale optimally?
- Is the Kriging geostatistic with KED capable to generate velocity model and depth conversion which fit to the geological model?

2. Geology of Research Area

The East Java Basin bordered covers East Java, Madura Island, Java Sea to Kangean Island (BPMIGAS 2008). In geological structure, the research area divided into 8 (eight) main structures, There are: Bawean Arch, Muriah Trough, JS-1 Ridge, Tuban-camar Trough, Central Depression, North Madura Platform, RMKS zone dan Kendeng zone.

In the offshore, East Java basin subdivide as two zones of structure, central part and eastern part of Java Sea. The central part of Java Sea is dominated by structured which extend toward southeast-northwest (SE-NW) direction. Some varieties in morphology within this present Java Sea are Karimunjawa Arch, Bawean Arch, Meratus Arch, danPulauLaut Arch. KarimunJawa Arch are paleo high which have broad geometry and metasediment outcrop in Pra-Tersier age (Koesoemadinata and Pulunggono, 1975). Based on previous geophysics research, this paleo high had granite like in Bangka-Belitung Islands and granite-batholith as basement for this paleo high (Guntoro, 1996). Bawean Arch is composed by vulcanic Tertiary rock, while Meratus Arch and Pulau Laut Arch is developed by mélange Cretaceous and appear in Meratus Mountain Range. Pulau Laut extend to the south until reach northern part of Java Island called JS-Ridge (north Tuban for exact location).

Despite the paleo high, some paleo low (trough) also found from west to east, Muria Trough, Tuban Trough, and Central Deep. Muria Trough is the half-graben structure developed in west Karimun Jawa Island which spread relative toward NE-SW direction till reach north Muria Mountain. Seismic section toward NW-SE along Karimun Jawa Arch showed a developed basin from normal fault NE-SW trending direction. Tuban Trough from tectonic point of view, is respectively a half-graben structure similar with Muria Trough. The southern part of Tuban Trough is bounded by fault system toward E-W direction and gradually vanishes toward northeast (NE) direction in the boundary of Bawean Arch and JS-1 Ridge. In the far east, Central Deep also developed half-graben structure type toward NE-SW direction. Central Deep is bounded by RMKS (Rembang-Madura-Kangean-Sakala) fault system toward E-W direction in southwest. Though, in the east, Central Deep is bounded by a paleo high called Northern Madura Platform.

Northern Madura Platform is broad carbonate sedimentation area during Miocene to Pliocene (Prasetyadi, 2007). In south is bounded by RMKS fault zone toward E-W direction and created mophology which has lower elevation than Madura Island due to compression along RMKS fault zone. In the eastern part of East Java basin is developed Sibaru Platform which is elongation from Pasternoster Platform in north. Sibaru Platform is a reefal platform which is bordered by reefal complex on the east and south of this area (Prasetyadi, 2007).The western and central part of Java Sea is commonly controlled by structure which elongate NE-SW, whereas in the eastern part of Java Sea developed structure toward W-E direction. It represent with the presence of Kangean and Lombok paleo high. Kangean paleo high is the most eastern part of RMKS fault zone which is plunging toward W-E, while Lombok paleo high located in southwest. Structure type in the eastern Java Sea is the extension of two major structures in present onshore East Java, Rembang and Kendeng zone, which elongate toward W-E direction.

3. Theoritical Framework

3.1. VelocityModels

There are many kinds of seismic wave of velocity, the different mathematic calculation in measuring seismic velocity also affect the definition of the velocity itself. These are some kinds of velocity:

3.1.1. Average Velocity (V_{avg})

In horizontal layer formations, average velocity from n's layer defined as:

$$V_a = \frac{1}{T_0} \sum_{k=1}^{n} V_k . t_k$$

V_k = velocity from k's layer, t_k = Two way time (TWT) from k's layer and
T_0 = Normal time (Two way normal incident) that can be formulated by:

$$T_0 = \sum_{k}^{n} t_k = 2 \sum_{k=1}^{n} \frac{h_k}{v_k}$$

h_k= the thickness of k's layer.

3.1.2. RMS Velocity (Root Mean Square)

RMS velocity is the total velocity from horizontal layers system in the form of average square roots. If considered Δt_1, Δt_2, Δt_3,…,Δt_n and velocity in each layers V_1, V_2, V_3,....V_n, then V_{RMS} for the number of "*n*" can be formulated :

$$V_{RMS} = \sqrt{\frac{\sum_{k=1}^{n} V_k^2 . \Delta t_k}{\sum_{k}^{n} \Delta t_k}}$$

The RMS velocity is not the result of velocity measuring, but the result from mathematic calculationprocess.

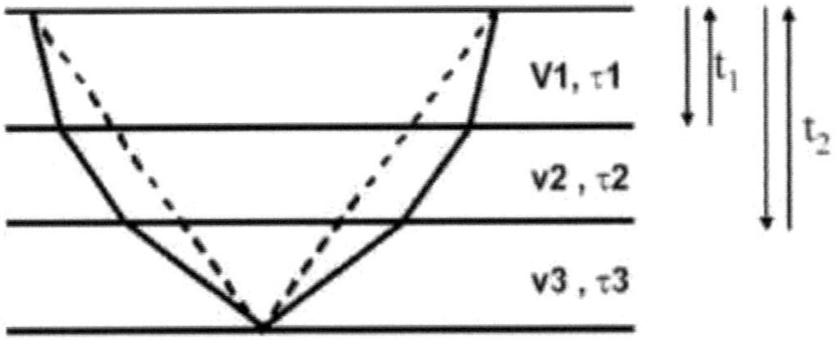

Figure 1. *RMS* velocity in each layers

3.1.3. Interval Velocity

Interval Velocity *Vint (Interval Vint)* is the average speed between 2 points measured perpendicular toward velocity layers that considered equals, Interval velocity considered as the velocity in each layers that can be formulatedby:

$$V_{\text{int}} j = \frac{\Delta z_j}{\Delta t_j}$$

The velocity definition can be understood easily by:

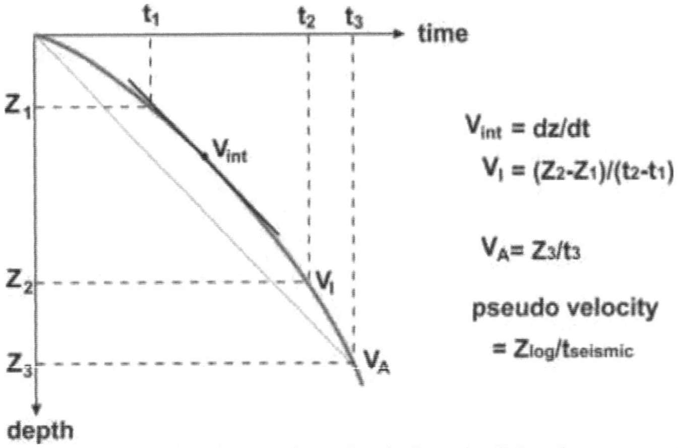

Figure 2. The velocity definition on time-depth domain (Marsden, course handout)

3.2. Geostatistical Methods

Geostatistical methods has been used in many geophysical exploration field,thus it can integrate various type of data for modeling.The Geological model usually based on log data that really depend on the number and distribution of wells. Seismic provide spatial geometry information, but has certain of limitations such as resolutions and noises. Those limitations can be minimalized with integrating between 2 types of those data to produce a good model with the existing geological concepts.

3.3. Kriging with External Drift (KED) Methods

KED method in estimating a new prediction of point value, withequation:

$$\hat{z}_{\mathrm{KED}}(\mathbf{s}_0) = \sum_{i=1}^{n} w_i^{\mathrm{KED}}(\mathbf{s}_0) \cdot z(\mathbf{s}_i)$$

for

$$\sum_{i=1}^{n} w_i^{\mathrm{KED}}(\mathbf{s}_0) \cdot q_k(\mathbf{s}_i) = q_k(\mathbf{s}_0)$$

With $k = 1, \ldots, p$or at matrix notation:

$$\hat{z}_{\mathrm{KED}}(\mathbf{s}_0) = \delta_{\mathbf{0}}^{\mathbf{T}} \cdot \mathbf{z}$$

With z : variable target, q_k : prediction variable (prediction value),$(\mathbf{s}_0)$, $\delta_{\mathbf{0}}$: vector from KED (w_i^{KED}), p: amount that predictedand $\mathbf{z}$: vector of n value. KED uses extended matric, equation:

$$\lambda_{\mathbf{0}}^{\mathrm{KED}} = \left\{ w_1^{\mathrm{KED}}(\mathbf{s}_0), \ldots, w_n^{\mathrm{KED}}(\mathbf{s}_0), \varphi_0(\mathbf{s}_0), \ldots, \varphi_p(\mathbf{s}_0) \right\}^{\mathbf{T}} = \mathbf{C}^{\mathrm{KED}-1} \cdot \mathbf{c}_{\mathbf{0}}^{\mathrm{KED}}$$

Matrix system of KED are(Webster and Oliver, 2007):

$$\mathbf{C}^{\mathrm{KED}} = \begin{bmatrix} C(\mathbf{s}_1,\mathbf{s}_1) & \cdots & C(\mathbf{s}_1,\mathbf{s}_n) & 1 & q_1(\mathbf{s}_1) & \cdots & q_p(\mathbf{s}_1) \\ \vdots & & \vdots & \vdots & \vdots & & \vdots \\ C(\mathbf{s}_n,\mathbf{s}_1) & \cdots & C(\mathbf{s}_n,\mathbf{s}_n) & 1 & q_1(\mathbf{s}_n) & \cdots & q_p(\mathbf{s}_n) \\ 1 & \cdots & 1 & 0 & 0 & \cdots & 0 \\ q_1(\mathbf{s}_1) & \cdots & q_1(\mathbf{s}_n) & 0 & 0 & \cdots & 0 \\ \vdots & & \vdots & 0 & \vdots & & \vdots \\ q_p(\mathbf{s}_1) & \cdots & q_p(\mathbf{s}_n) & 0 & 0 & \cdots & 0 \end{bmatrix}$$

So $\mathbf{c}_{\mathbf{0}}^{\mathrm{KED}}$:

$$\mathbf{c}_{\mathbf{0}}^{\mathrm{KED}} = \{ C(\mathbf{s}_0,\mathbf{s}_1), \ldots, C(\mathbf{s}_0,\mathbf{s}_n), q_0(\mathbf{s}_0), q_1(\mathbf{s}_0), \ldots, q_p(\mathbf{s}_0) \}^{\mathbf{T}}; q_0(\mathbf{s}_0) = 1$$

Drift (geo-regresion) is the results of the KED operation, which is a function of the variogram that predict average regression against linear regression functions. This regression has proportional value from the impact on the entire data.

4. Methodology

In this study, the types of data that used by the researcher are 2D seismic data (802 tracks), 3D seismic (8 fields), wells logging data (106 wells), Checkshot (25 wells).

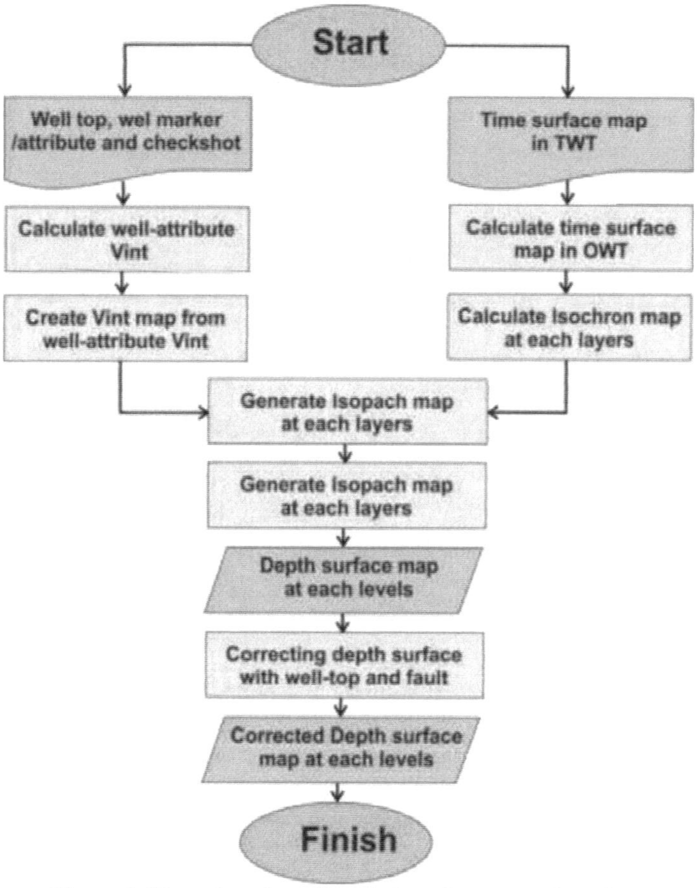

Figure3. Time domain to Depth domain conversion plot.

The first step to do time-depth conversion is tie the wireline log data from well and seismic data in the exact marker for each horizons/ formations, then used a line from 2D seismic survey as reference to standardize the reflector in each line both 2D and 3D seismic data. However, before converting to depth domain, we used well-seismic tie from previous process to pick the correct reflector for each formations and resulting seismic interpretation in time domain. Availability of 2D and 3D seismic data also well data for this research are shown in Figure 4.

At this moment, all resulted horizons from seismic interpretation remain in time domain and should be converted to depth domain. To acknowledge that situation, velocity map for each horizon has created. The velocity maps made from interval velocity which is generated from checkshot data in all wells. Missing checkshot data can be utilized by taking checkshot from the closest well. After created interval velocity in

IOP Conf. Series: Earth and Environmental Science **132** (2018) 012008 doi:10.1088/1755-1315/132/1/012008

each well using checkshot data, a grid was generated from each point to build velocity interval maps for every horizons.

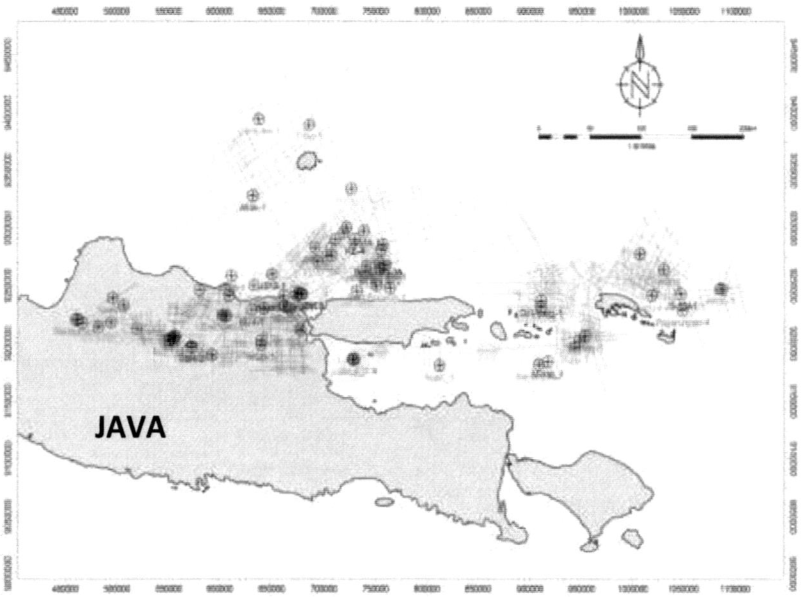

Figure 4. 2D and 3D Seismic data distribution and wells data on the research area.

5. Result and Discussion

5.1. Time Surface Map

(a)

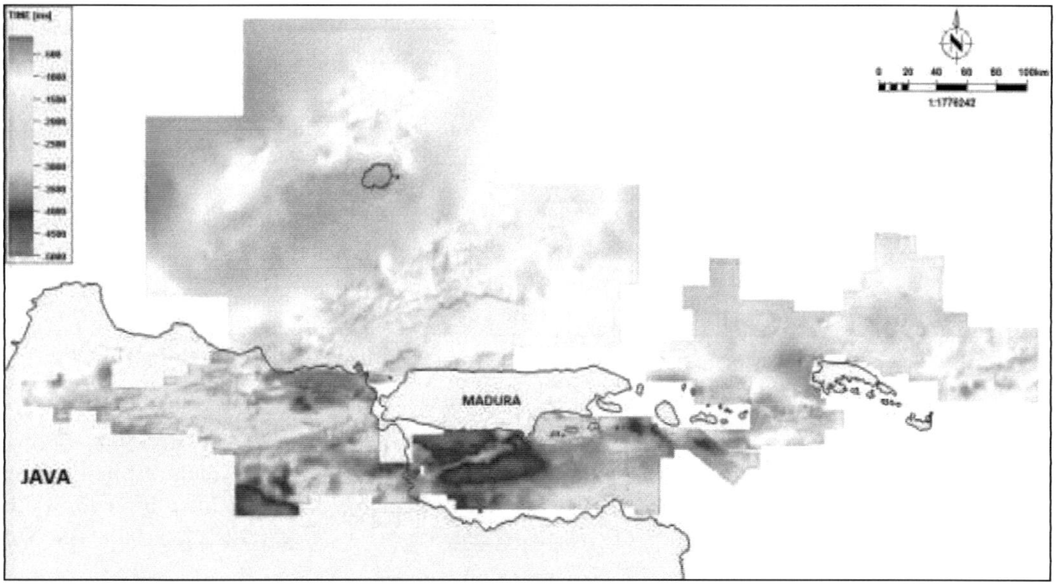

IOP Conf. Series: Earth and Environmental Science **132** (2018) 012008 doi:10.1088/1755-1315/132/1/012008

(b)

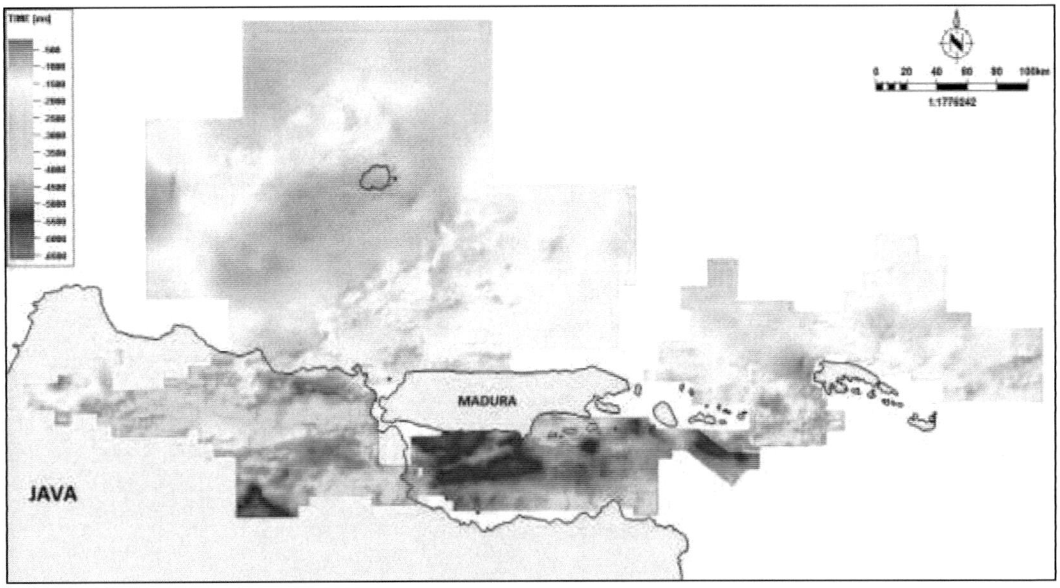

(c)

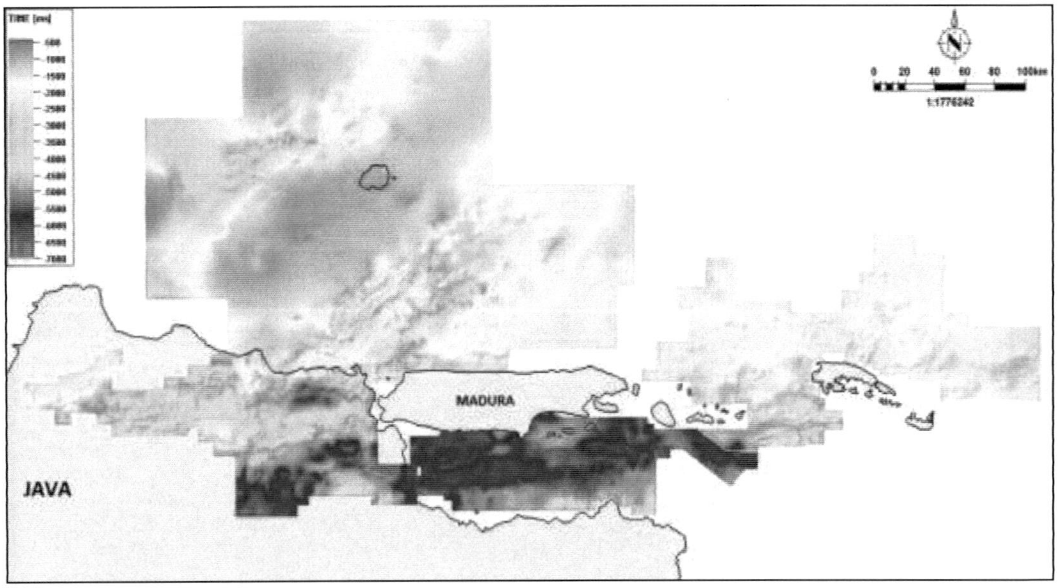

Figure 5. Time Surface map on the horizon a) Top Kujung 1, (b) Top Ngimbang Klastik, (c) Basement.

From the time maps result in those horizons (Figure 5), the study area relatively have value 500 – 5000 ms with the deepest part represent as dark color (purple-dark blue) is located in the south of Madura island (Madura subbasin) reach almost 3500 ms till 5000 ms due to depression phase which differ from other place in this basin, while the shallowest part denote as bright color (light blue-red) is located in the North

East Java sea extent from 100 ms till 3500 ms. In fact, morphological of every maps is quite similar due to basement involved in each sedimentology process for younger formations.

5.2. Velocity Model

(a)

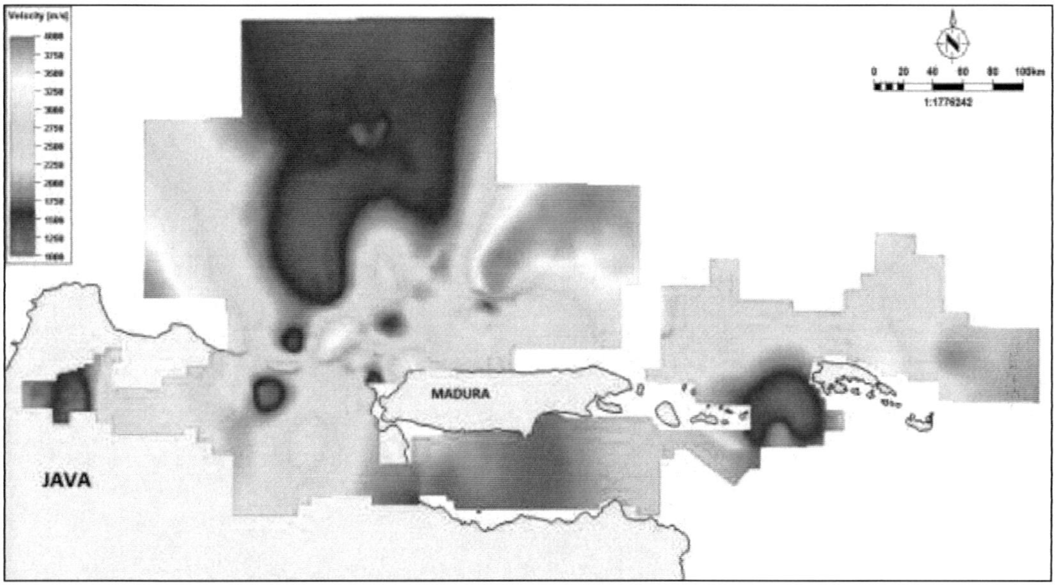

(b)

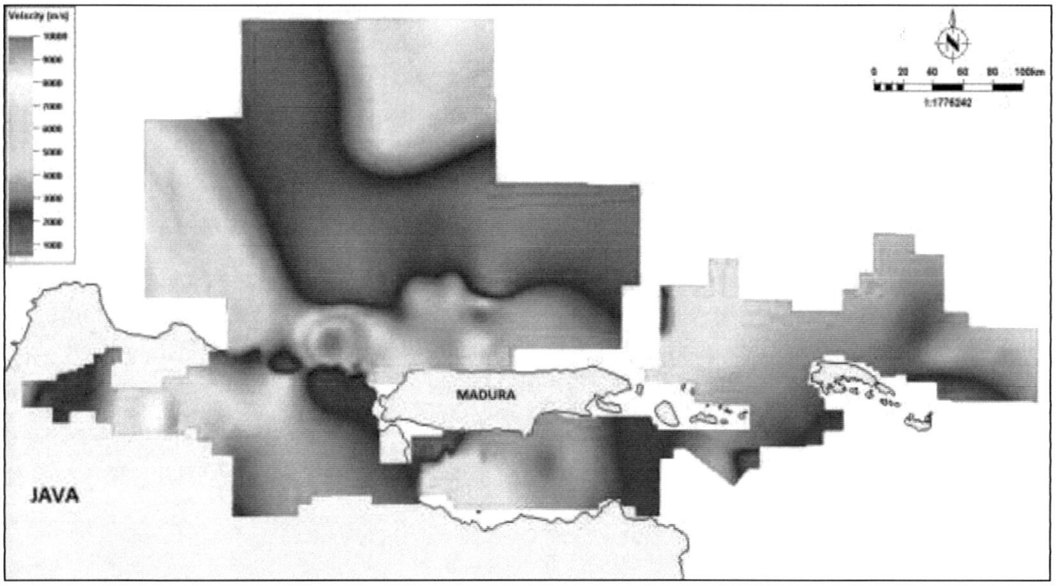

(c)

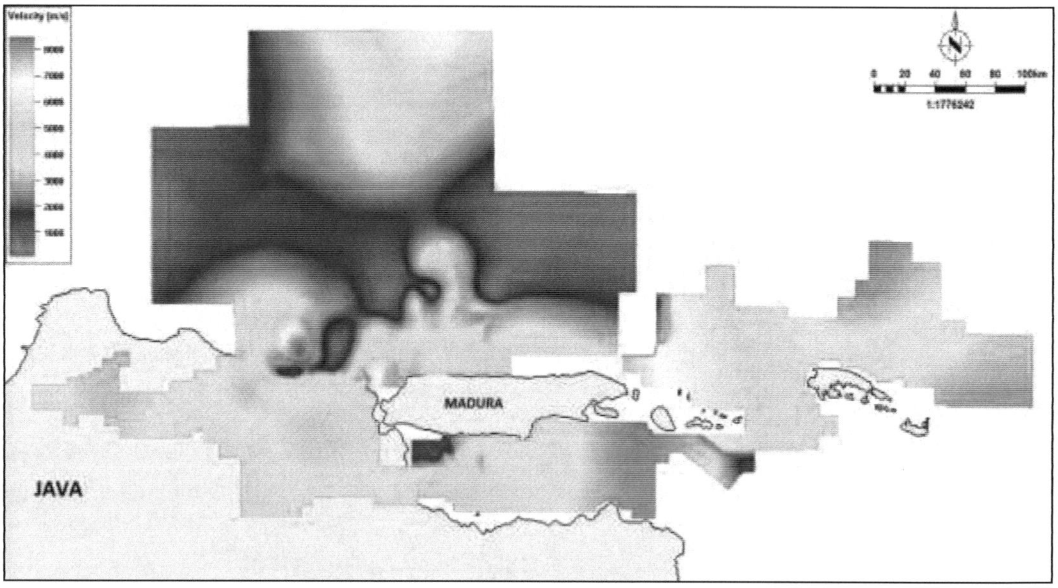

Figure 6. Interval Velocity model (Vint) on horizon (a) Top Kujung 1, (b) Top Ngimbang Klastik, (c) Top Basement.

Figure 6. shows interval velocity model in every single horizon based on checkshot data. Some difficulties come from the complexity in structural variation within study area and lack ofcheckshot data. Lackof checkshot was corrected by using the nearest well which hascheckshot. Some pseudo wells was created not only to solve the gap between one to another well but also to accommodate some areas with complex structure, for example in the southern of Madura Island.The pseudo well is developed by using checkshot from nearby well then calculating interval velocity in particular area. Other parameter is secondary parameter from interpreted horizon to calculate interval velocity model for every horizon. Based on lithology and formation thickness, interval velocity will calculate high value when its lithology had high density and thick.

Top Ngimbang Clastic Map (figure 6.b.) in Madura subbasin have high velocity value (more than 3000 m/s) than its surrounding area due to low area and thick sediment deposit about 14000 ft thick filled with mostly shale in lithology. However, this thick sediment accumulation interpreted as source for hydrocarbon source for this area.Thick shale in this area also is the main factor for increase interval velocity.

IOP Conf. Series: Earth and Environmental Science **132** (2018) 012008 doi:10.1088/1755-1315/132/1/012008

5.3. Depth Surface Map of Conversion Results

(a)

(b)

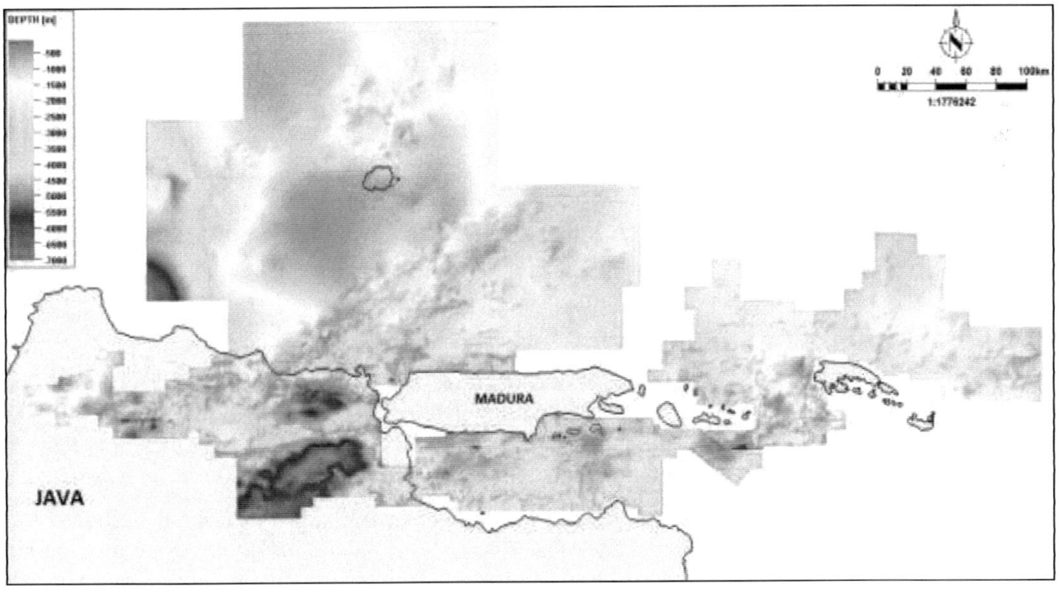

(c)

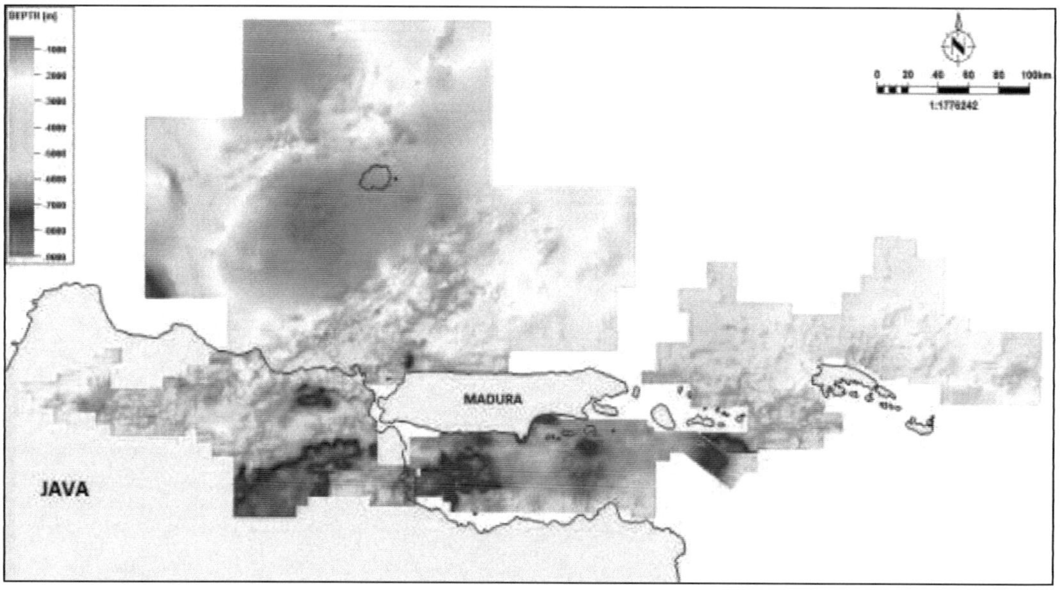

Figure7. Depth surface map of conversion results on horizon (a) Top Kujung 1, (b) Top Ngimbang Klastik, (c) Basement.

Depth contour map in Figure 7 relatively has the same structure pattern with time contour map. It became the early control to measure whether the process of making velocity model is correct or not.

6. Conclusion

Lithology and thickness of each formation are the main geological factor controlling the velocity model in East Java Basin. Depression area known as Madura Sub Basin dominated by thicker clay compare to surrounding area and indicating velocity anomalies. Kriging with External Drift (KED) method in geostatistical is able to make a depth conversion in regional scale pretty well. The main obstacle that needs to be completed for the further research is the availability and distribution of seismic and checkshot for cover regional area.

Acknowledgements

This study initiated by SKK Migas cooperated with UPN Veteran Yogyakarta, Universitas Gadjah Mada and several PSC (production sharing contract) companies that developing East Java Basins. This study became motivation and example for geoscientists and practitioners of oil and gas industry to keep spirit doing research and exploration in the effort of national oil and gas resources and production enhancement.

Thanks to Deputy for Planning Management Chief and Exploration Division Chief of SKK Migas for the support, also all of the PSC helping the realization of this study. Data and result have a permission from SKK Migas and PSC to be published. Also we thank to UPP university partnership program Chevron-UGM for supporting hardware and software facility.

References

[1]. Chambers, R. L., Yarus, J. M, Hird, K.B, 2000, *Petroleum geostatistics for non-geostatisticians*, The Leading Edge, May 2000 (Part 1), June 2000 (Part 2).
[2]. Hengl, T., Heuvelink, G.B.M., and Stein, Alfred, 2003, *Comparison of Kriging with External Drift and Regression-Kriging*. Technical note, DepartmentofEarthSystemsAnalysis,InternationalInstituteforGeo-informationScienceandEarthObservation(ITC), Netherlands.
[3]. Kaufman, H., 1953, *Velocity functions in seismic prospecting*, Geophysics, 18, 289-297.
[4]. Prasetyadi, C., *Paleogene Tectonic Evolution of Eastern Part of Java* (Dissertation ITB, Bandung, 2007).
[5]. Reilly, M., 1993, *Integration of well and seismic data for 3D velocity model building*, First Break, 11, no. 6, 247-260.
[6]. Satyana, A. H., *New Consideration on the Cretaceous Subduction Zone of Ciletuh – LukUlo – Bayat – Meratus: Implications for Southeast Sundaland Petroleum Geology* (Proceeding of IPA Convention and Exhibition, 2004).
[7]. Schlumberger, 2010

An Extreme Meteorological Events Analysis For Nuclear Power Plant (NPP) Siting Project at Bangka Island, Indonesia

Deni Septiadi [1,2], **Yarianto Sugeng B. S**[3], **Sriyana**[3], **Kurnia Anzhar**[3], **Hadi Suntoko**[3]

[1] Indonesian School of Meteorology Climatology and Geophysics (STMKG)
[2] Indonesian Agency for Meteorology, Climatology and Geophysics (BMKG)
[3] National Nuclear Energy Agency of Indonesia (Batan)

deni.septiadi@stmkg.ac.id, zeptiadi@yahoo.co.id

Abstract. The potential sources of meteorological phenomena in Nuclear Power Plant (NPP) area of interest are identified and the extreme values of the possible resulting hazards associated which such phenomena are evaluated to derive the appropriate design bases for the NPP. The appropriate design bases shall be determined according to the Nuclear Energy Regulatory Agency (Bapeten) applicable regulations, which presently do not indicate quantitative criteria for purposes of determining the design bases for meteorological hazards. These meteorological investigations are also carried out to evaluate the regional and site specific meteorological parameters which affect the transport and dispersion of radioactive effluents on the environment of the region around the NPP site. The meteorological hazards are to be monitored and assessed periodically over the lifetime of the plant to ensure that consistency with the design assumptions is maintained throughout the full lifetime of the facility.

Keywords : meteorological, nuclear power plant, hazards.

1. Introduction

Nuclear installations must be operated safely, securely and reliably to meet the requirements and safety criteria since the early stages of siting, design, construction, and even activities in the previous stage to the operational stage and also safety when handling and during decommissioning and radioactive waste management. According to the guidelines on siting guide: site selection and evaluation criteria for an early site permit application, that the siting, design, construction, operation and decommissioning are the five major stages that need to be considered in the installation of nuclear power plants. The site evaluation program consists of provisions and activities to be implemented for the evaluation and characterization of the site to ensure the fulfillment of all the requirements. Furthermore, every nuclear power plant is also necessary to formulate a comprehensive emergency plan that will help ensure public safety. This plan should provide appropriate measures, by way of security measures, for implementation at the specified time so that the radiation exposure to members of the general public will remain in the level of intervention (Bapeten , 2014; Batan, 2014; Bishnoi et al., 2005; and EPRI, 2002).

Meteorological phenomena may cause danger either singly or in combination, which may affect the safety of nuclear installations. Thus, evaluation of the nuclear installation site for the meteorological aspects referred to in Bapeten (2014) about the site evaluation for the nuclear installation meteorological aspects, which include: (a) monitoring and data collection as well as meteorological information; (B) Evaluation of meteorological hazards; and (c) determination of the basic parameters of the design for the meteorological aspects. Meteorological measurements in the vicinity of a nuclear power plant are needed to assess the dispersion of the releases of radioactive materials in the atmosphere. One purpose of this paper is to examine the potential of meteorological hazards that can hinder the process of installation and construction of nuclear power plants. Monitoring and data collection as well as meteorological information referred to in points (a) includes: a. The collection of data and meteorological information from outside of site locations (off site); and b. Monitoring and gathering meteorological data at the site locations (on site). Meteorological hazard evaluation referred to in points (b) includes: a. Value extreme meteorological parameters; and b. Extreme weather phenomena. Meanwhile, the determination of the basic parameters of the design for the meteorological aspects referred to in points (c) includes: a. Value extreme meteorological parameters; and b. Extreme weather phenomena (see Bapeten , 2014; CNSC, 2008 and Roshan et.al., 2006).
There were four meteorological variables which need to be addressed for meteorological hazard. Those four variables are air temperature, wind speed, precipitation (liquid equivalent) and snowpack. It also needs to analyse hazardous, rarely occurring meteorological phenomena such as lightning, tropical cyclones, typhoons, hurricanes, tornadoes and waterspout. Lightning was a common event in the Indonesian region, including Bangka, due to high convective activity in this region, but typhoons, hurricanes, and tornadoes almost never happen in this region. No record has ever been found about the occurrence of such phenomena. Water spouts is also a rare phenomenon, but it was likely to occur in this region (IAEA, 2011). The objective of this paper is to present all results of the studies, assessments and investigations in relation to the meteorological hazards assessment and dispersion characteristics. Furthermore, the scope of this paper covers the description of the tasks and the results obtained in relation to the meteorological aspects regarding to the assessment of meteorological hazards for the design of the Nuclear Power Plant at the West Bangka area of interest.

2. Methodology

The study area located in West Bangka area of interest in Indonesia (see Figure 1). For assessing the extreme values of meteorological variables and rarely occurring hazardous meteorological phenomena as well as for determining the specific characteristics of the site that affect the transport and dispersion of radioactive effluents to the environment, specific and detailed information should be collected. The meteorological instruments (on site) were already installed on the tower and on meteorological instrument park and full operation since 1 January 2012. In detail, the meteorological instruments and the sensors arrangement are presented in Table 1. Meteorological data available from the site consist of wind (speed and direction) and air temperature at four heights, i.e. 10, 40, 60, and 80 m; air pressure and relative humidity at one height i.e. 10 m. Other data available are precipitation and solar radiation. All data are stored in a data logger and transmitted to a personal computer every one hour where the data are stored.

An off site meteorological data obtained by formal and non-formal meteorological stations. Indonesian Agency for Meteorology Climatology and Geophysics (BMKG) is the formal meteorological stations associated to the World Meteorological Organization (WMO) and reporting all meteorological data such as precipitation, air temperature, air pressure, wind, relative humidity etc. (BMKG, 2010). Non-formal meteorological stations operated by Agriculture Department and only reporting precipitation data. These meteorological stations used for the baseline analysis are listed in Table 2 along with their distance from West Bangka (WB) area of interest, time period and consistency of data record. Meteorological hazards are associated with extreme meteorological conditions and with rarely occurring hazardous meteorological phenomena.

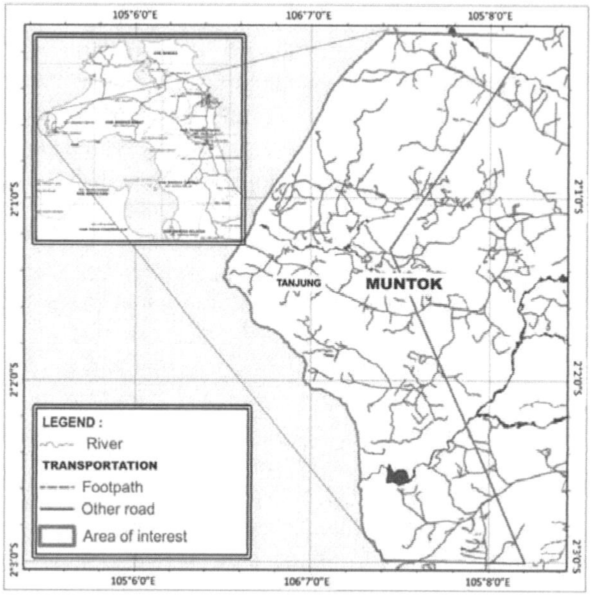

Figure 1. Location of West Bangka Area of Interest.

The meteorological hazards as well as treated in the IAEA Safety Guide are those caused by external events i.e. unconnected events with the operation of a facility or the conduct of an activity that could have an effect on the safety of the facility or activity. Methods for the assessment of hazards are often divided into two broad approaches: deterministic methods and probabilistic methods. When a statistical analysis is performed, it is typically based on time series analysis and synthesis. It is assumed that the series represents both deterministic components and an unknown number of random components, and that the random components are reasonably independent. Using these methods, gaps and missing data and outliers of the available data set should be adequately taken into account. Data analysis including tabulation of the primary and secondary meteorological data, short and long-term, both of on sites or even off sites. These tabulations are made into yearly, monthly, daily and hourly. Missing data analysis being concerned since in-situ measurements or reanalysis and satellite data provides inconsistency.

Table 1. List of the instruments installed on the West Bangka area of interest.

No.	Name of Instruments	Placement	Utility	Qty. (pc)	Status
1	Anemometer	10, 40, 60, 80 m	Wind speed and direction measurement	4	ok
2	Wind Vane	10, 40, 60, 80 m	Wind speed and direction measurement	4	ok
3	Humidity Temperature sensor with weather & Radiation S.	10 m	Multi measurement	1	ok
4	Anemometer Young propeller	10, 40, 60, 80m	Wind speed and direction measurement	4	ok
5	Solar Panel ET Module	1.5 m	Generate solar power	1	ok
6	Solar Powered Warning (lamp)	80 m	Power indicator	1	ok
7	Lighting Arrester	83 m	Prevent power line interruptions by lightning	1	ok
8	Tipping Bucket Rain Gauge	1 m	Precipitation measurement	1	ok
9	Pyrometer	1.5 m	Radiation measurement	1	ok
10	Data logger	4 m	Record data	1	ok
11	Temperature Sensor	10 m	Temperature measurement	1	ok
12	Barometric Pressure sensor	4 m (in cabinet)	Pressure measurement	1	ok
13	Net radiometer	1.5 m	Radiation measurement	1	ok
14	Lighting sensor	1.5 m	Lightning Measurement (Cloud to Ground)	1	ok
15	Air pressure sensor	1.5 m	Pressure measurement	1	ok
16	Solar Panel ET Module	1.5 m	Generate solar power	1	ok
17	Tipping Bucket Rain Gauge	1 m	Precipitation measurement	1	ok
18	Pyranometer	1.5 m	Radiation measurement	1	ok
19	Net radiometer	1.5 m	Radiation measurement	1	ok

Table 2. Summary of meteorological stations in the regional of the NPP study area.

Station name	Coordinates		Distance (km) from WB	Period	Consistency**
	Lat.	Lon.			
Pangkalpinang	2°10'0.1"S	106°7'59.9"E	114	1980-2012	Good
Palembang	2°54'0.0"S	104°42'0.0"E	109	2002-2012	Good
Klapa*	1°53'17.70"S	105°38'47.61"E	60	1971-2000	Poor
Muntok*	2°01'23.03"S	105°14'26.94"E	14	1971-2000	Poor
Pemali*	1°55'50.16"S	106°00'20.48"E	99	1971-2000	Poor
Sungai Liat*	1°52'32.91"S	106°05'40.03"E	110	1971-2000	Poor
Belinyu*	1°38'13.60"S	105°49'48.88"E	89	1971-2000	Poor

*Non-formal station
** Calculated by total data availability

3. Results and Discussion

3.1. Extreme Meteorological Events

Extreme winds, air temperature and precipitation were studied. A minimum of a 30-year data set of on-site measurements is necessary to perform an extreme value analysis. Where a 30-year dataset was not available, it was possible to interpolating data using regression analysis in order to create a minimum 30-year dataset. Given that the West Bangka areas of interest measurements do not cover a 30-year period, this became a necessary calculation before being able to perform the extreme value analysis. Pangkalpinang was used as nearby meteorological station data to develop a long-term analytical data set for the West Bangka areas of interest. The values presented in this section are independent of the possible variations due to climate change.

3.1.1 Wind speed. A 30-year data set was created by correlating the existing of data on site to 30-year data from Pangkalpinang station, the nearest meteorological station with a similar climatology that recorded sustained wind speeds for that period. An extreme value analysis was then performed on this new 30-year data set in the West Bangka areas of interest. The results of this statistical analysis are predictions of maximum values for a set period in the future. A return period of 2, 5, 10, 25, 50, and 100 years is the most interest as it encompasses the entire lifecycle of the plant. Standard deviation is quoted as a measure of uncertainty in the calculation and all subsequent extreme value analysis. This exercise was done for the10 m heights (see Table 3) and also summarized of maximum wind speed (see Table 4) :

Table 3. Extreme Value Predictions for 10 m of Wind speed at the West Bangka areas of interest.

No.	Return Period (years)	Wind speed (m/s)
1	2	28
2	5	30
3	10	32
4	25	33
5	100	35

Table 4. Summary of Maximum of Wind Speed at the West Bangka areas of interest.

No.	Date	Wind (m/s)
1	12 Januari 2012	16.9
2	18 Februari 2012	15.8
3	20 March 2012	17.1
4	01 April 2012	15.5
5	08 May 2012	16.3
6	09 June 2012	19.2
7	26 July 2012	12.5
8	05 August 2012	12.5
9	02 September 2012	12.4

3.1.2. Air temperature. Similar to the wind speed data, a few of temperature measurements are only available at the South Bangka areas of interest. A regression analysis was done with the Pangkalpinang station, the nearest meteorological station with a similar climatology that recorded sustained air temperature for that period.These analysis was done as seen in Table 5 below. For the analysis of maximum temperature >32.2 °C and summary of maximum air temperature described in Figure 2 and Table 6 respectively.

Table 5. Extreme Value Predictions for Air Temperature at the West Bangka areas of interest.

No.	Return Period (years)	Air Temperature (^{0}C)
1	2	32.2
2	5	32.3
3	10	32.4
4	25	32.8
5	100	34.6

Table 6. Summary of Maximum of Air Temperature at the West Bangka areas of interest.

No.	Date	Temperature (°C)
1	15 January 2012	30.7
2	25 February 2012	30.5
3	10 March 2012	30.5
4	23 April 2012	32.6
5	17 May 2012	33.0
6	04 June 2012	32.6
7	21 July 2012	32.0
8	12 August 2012	32.7
9	23 September 2012	33.0

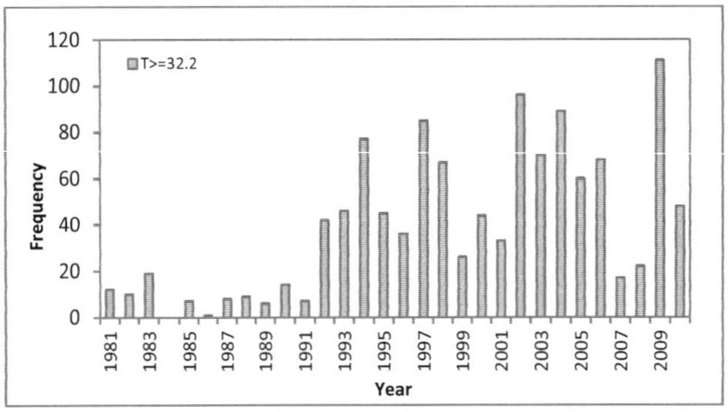

Figure 2. A 30-year of maximum temperature with a threshold 32.2 °C.

3.1.3. Precipitation.. While a few of precipitation measurements available at the West Bangka areas of interest, Pangkalpinang station are the nearest meteorological station with a similar climatology that recorded sustained precipitation for that period.

Table 7. Extreme Value Predictions for Precipitation at the West Bangka areas of interest

No.	Return Period (years)	Precipitation (mm/hr.)
1	2	104
2	5	132
3	10	150
4	25	171
5	100	210

Figure 3 below show the frequency of the precipitation with a >50 mm of threshold, with a total frequency during the years 1981-2010 are 263 and exist for every year. From the total of frequency as a cumulative, the probability occurrence of precipitation>50 mm/day calculated and shown as well as in Figure 4 which is more than 14% lie on January and April; at least 10% on November, December and March. Summary of maximum precipitation described in Table 8 below.

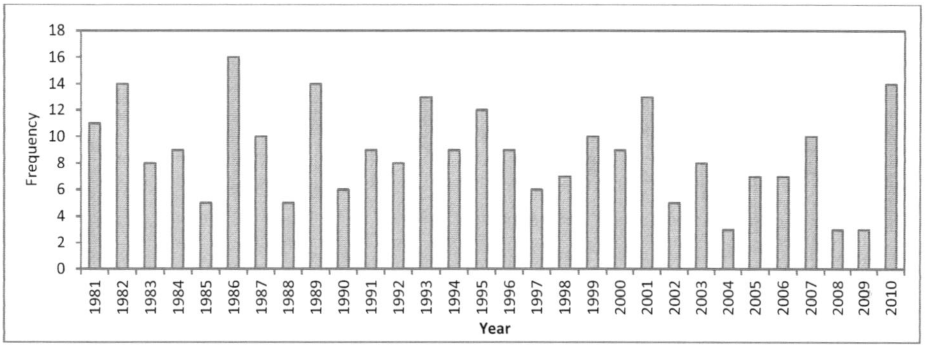

Figure 3. 30-year frequency of precipitation with a >50 mm/day of threshold.

Table 8. Summary of Maximum of Precipitation at the West Bangka areas of interest.

No.	Date	Precipitation
1	20 January 2012	28.8
2	26 February 2012	76.1
3	15 March 2012	93
4	24 April 2012	16.9
5	23 May 2012	31.5
6	16 July 2012	15.1

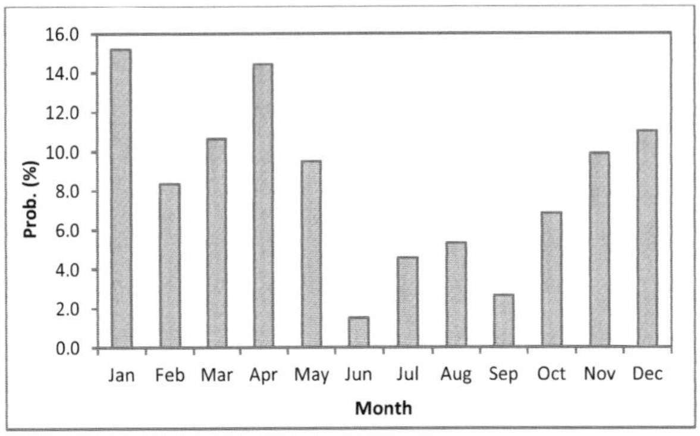

Figure 4. The probability occurrences of precipitation with a >50 mm/day of threshold.

3.2. Interpretation of Design Basis Parameter

The proposed design basis value for meteorological parameters with an interim interpretation at West Bangka area of interest are established as follows, if a return period of 100 years is adopted:

Wind Speed at H = 10 m	35.0 m/s
Air Temperature at H = 10 m	34.6 ^{0}C
Precipitation	210 mm/day.

Concerning the proposed design basis value, the following point should be noted:

The proposed design basis value is established by applying of Pangkalpinang meteorological station as data reference, because meteorological data at West Bangka area of interest is only one (1) year at this moment. it is however recommended that the design basis values for West Bangka area of interest should finally be re-evaluated and established based on the long-term meteorological data at the site itself.

4. Conclusions

The potential sources of meteorological phenomena in West Bangka area of interest are identified and the extreme values of the possible resulting hazards associated which such phenomena are evaluated to derive the appropriate design basis for the nuclear power plant. The appropriate design bases shall be determined according to the Bapeten applicable regulations, which presently do not indicate quantitative criteria for purposes of determining the design bases for meteorological hazards. These meteorological investigations are also carried out to evaluate the regional and site specific meteorological parameters which affect the transport and dispersion of radioactive effluents in the environment of the region around the nuclear power plant site. The meteorological hazards are to be monitored and assessed periodically over the lifetime of the plant to ensure that consistency with the design assumptions is maintained throughout the full lifetime of the facility. For all such purposes indicated above, the recommendations and guidelines from IAEA Safety Guide SSG-18 "Meteorology and Hydrology Hazards in Site Evaluation for Nuclear Installation", were applied. The proposed design basis value for meteorological parameters with an interim interpretation at Bangka area of interest are established as follows, if a return period of 100 years is adopted: wind speed at height 10 m are 35.0 m/s; air temperature at height 2 m are 34.6 °C; and rainfall are 210 mm/day. Concerning the proposed design basis value, the following point should be noted: The proposed design basis value is established by applying Pangkalpinang meteorological station as data of reference, because meteorological data at site area of interest less than 2 years at this moment. it is however recommended that the design basis values for site area of interest should finally be re-evaluated and established based on the long-term meteorological data at the site itself.

References

[1] Bapeten, 2014. Evaluasi tapak instalasi nuklir untuk aspek meteorologi dan hidrologi, *Peraturan Kepala Bapeten Nomor 6 Tahun 2014,* Jakarta.

[2] Batan, 2014. Program Evaluasi Tapak (PET) Reaktor Daya Eksperimetal di kawasan Puspiptek Serpong, Kota Tangerang Selatan.

[3] Bishnoi, L.R. and Prabir C. Basu, 2005. Siting of nuclear installations", *Nuclear India*, Vol 38, No.7-8.

[4] BMKG, 2010. Prosedur standar operasional pelaksanaan peringatan dini, pelaporan, dan diseminasi informasi cuaca ekstrim, *Peraturan Kepala Badan Meteorologi, Klimatologi, dan Geofisika Nomor : Kep. 009 tahun 2010,* Jakarta.

[5] EPRI, 2002. Siting Guide: site selection and evaluation criteria for an early site permit application. Electric power research institute, *Technical Report*, California.

[6] IAEA, 2011. Safety Guide on Meteorological and Hidrological Hazards, *Safety Standards Series No. SSG-18,* IAEA Vienna.

[7] Roshan A.D, Shylamoni P., and Sourav A., 2006. Monograph on siting of nuclear power plants, Civil and structural engineering division, *Atomic Energy Reulatory Board*, Niyamak Bhavan Anushakti Nagar, Mumbai-400094, India.

[8] WMO, 2008. *Guide to Meteorological Instruments and Methods of Observation*, Guide to Meteorological Instruments and Methods of Observation, WMO-No. 8, Seventh edition, Geneva 2, Switzerland.

[9] ________*U.S Nuclear Regulatory Commission*, 2007. Revision 1 regulatory guide office of nuclear regulatory research regulatory guide 1.76, design-basis tornado and tornado missiles for nuclear power plants.

[10] ________*Canadian Nuclear Safety Commission*, 2008. Regulatory Document Rd-346, Site Evaluation For New Nuclear Power Plants, Canadian Nuclear Safety Commission.

Acoustic Impedance Inversion To Identify Oligo-Miocene Carbonate Facies As Reservoir At Kangean Offshore Area

Arif Zuli Purnama[1,a], Pritta Ariyani Machmud[1], Budi Eka Nurcahya[1], Miftahul Yusro[2], Agung Gunawan[2], Dicky Rahmadi[2]

[1] Universitas Gadjah Mada, Yogyakarta, 55281, Indonesia
[2] SKK Migas, Jakarta, 12710, Indonesia

[a] arif_zuli@mail.ugm.ac.id

Abstract. Model based inversion was applied to inversion process of 2D seismic data in Kangean Offshore Area. Integration acoustic impedance from wells and seismic data was expected showing physical property, facies separation and reservoir quality of carbonate rock, particularly in Kangean Offshore Area. Quantitative and qualitative analysis has been conducted on the inversion results to characterize the carbonate reservoir part of Kujung and correlate it to depositional facies type. Main target exploration in Kangean Offshore Area is Kujung Formation (Oligo-Miocene Carbonate). The type of reservoir in this area generate from reef growing on the platform. Carbonate rock is a reservoir which has various type and scale of porosity. Facies determination is required to to predict reservoir quality, because each facies has its own porosity value. Acoustic impedance is used to identify and characterize Kujung carbonate facies, also could be used to predict the distribution of porosity. Low acoustic impedance correlated with packstone facies that has acoustic impedance value below 7400 gr/cc*m/s. In other situation, high acoustic impedance characterized by wackestone facies above 7400 gr/cc*m/s. The interpretation result indicated that Kujung carbonate rock dominated by packstone facies in the upper part of build-up and it has ideal porosity for hydrocarbon reservoir.

1. Introduction

In general, there are three main steps of seismic exploration method used to find hydrocarbon reserves i.e. acquisition, processing, and interpretation of seismic data. Seismic interpretation is used to determine and predict geological phenomenon from seismic data generated from processing (Sheriff and Geldart, 1995). Seismic interpretation has many types, one of them is reservoir characterization. Reservoir characterization is a process combining all existing data to define reservoir geometry and physical parameter distribution of hydrocarbon reservoir. The result of reservoir characterization is expected to generate subsurface model.

One method used frequently to define subsurface image is acoustic impedance seismic inversion. This inversion is applied to convert seismic wiggle into acoustic impedance value. Acoustic impedance is physical rock property having relation with porosity, density, even the thickness of reservoir rock target. Acoustic impedance inversion is divided into three types, one of them is model based acoustic impedance inversion. Model based generates subsurface model more accurate than the other type.

Kangean offshore was selected as study area. In general, carbonate rock reservoir has varies type and scale of porosity. Its porosity is highly related to facies type. So that in this study, model based acoustic impedance inversion defined physical property of rock is expected to separate varying types of facies from carbonate rock reservoir (Kujung Limestone).This is the reason that model based is selected in this study area.

2. Location and geology of study area

The study area is located in offshore area ± 50 km west from Kangean Island, this is within the Northeast Java Basins. Figure-1 shows the study area marked by Green polygon.

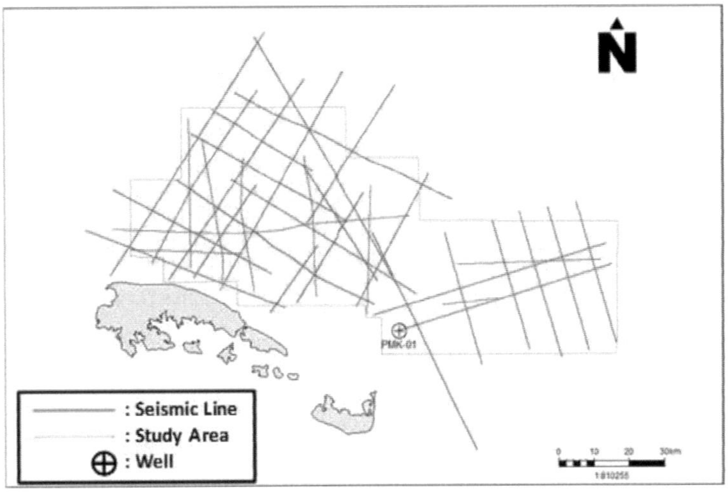

Figure 1. Base map of study area

The study area is controlled by one major fault, which is RMKS fault (Rembang-Madura-Kangean-Sakala). RMKS fault zone has wide in between 15 km to 40 km and length about 675 km from Rembang area on west to Tuban area on east, until Kangean Islands on the east of Madura Island. Inversion structure could be seen in this zone as a result of horizontal fault deformation i.e. folding, thrust fault. Depending on the diversity of its structure, RMKS fault zone is divided into three parts i.e. west region, center region, and east region (Satyana et al., 2004). The study area is included the east region of RMKS fault zone. It covered from Kangean islands to Flores sea area. The exposed rock on Kangean Island is older than Maduras rock, it proved the effect of RMKS fault in its zone that was bigger than Madura and Rembang zone (Satyana et al., 2004). RMKS fault inverted Paleogene basin formed by normal faults trending west-east. The structure of RMKS fault is shown by Figure-2. This figure shows Sepanjang in the south and Sakala in the north controlling the study area.

Mudjiono and Pireno (2001) divided the Northeast Java Basins into two parts i.e. offshore and onshore East Java/Madura, in addition they defined several sedimentary cycle i.e. Ngimbang, Kujung, Tuban and Ngrayong, Wonocolo, and Kawengan and Lidah cycle. This is shown by Figure-3. The oldest sediment is Pra-Ngimbang formation, this is terrestrial clastic sediment consisting sandstone, claystone, and intercalation of coal. Pra-Ngimbang formation was deposited in graben area at Pra-Tersier to Early Tersier by lacustrine and deltaic environment, so that this can be source rock of Northeast Java Basins. Ngimbang formation was deposited at Ngimbang Cycle (Eocene), It had half grabben filled by syn-rift sediment dominated by thick claystone. Kujung Formation is divided into three parts i.e. Kujung Limestone dominated by reef build-up, Kujung Shale dominated by limy shale and thin carbonate, Kujung Sandstone dominated by clastic sediment. Kujung Limestone Formation was believed as ideal hydrocarbon reservoir. The various facies of Kujung Limestone enable to stored hydrocarbon. Tuban Formation, in the study area, is dominated by shale. It is a massif sediment, unstructured, rich of green shale and foraminifera plankton. Tuban Formation could be caprock of Kujung Limestone reservoir.

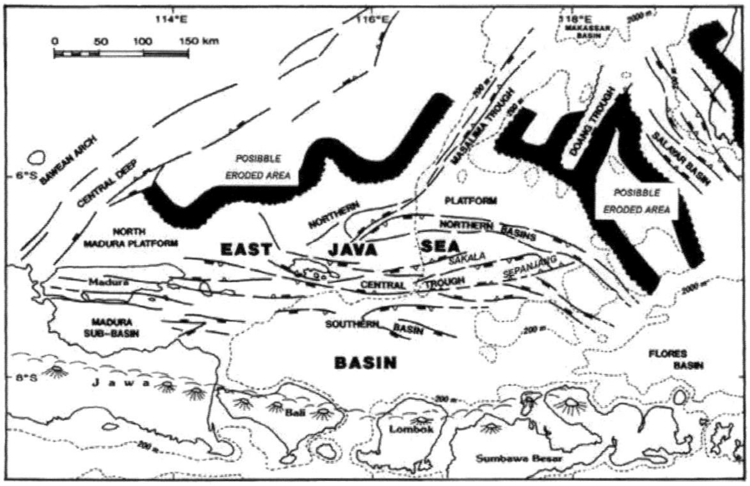

Figure 2. Tectonic structure of eastern region of RMKS fault zone (Bransden and Matthews., 1992)

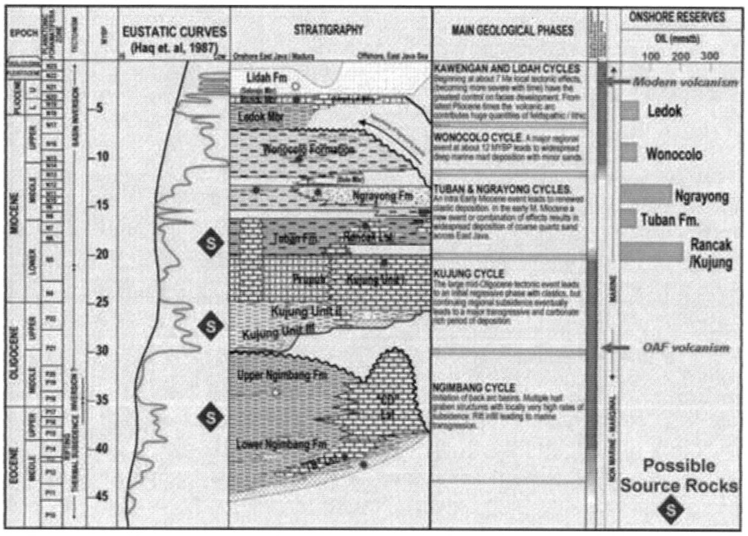

Figure 3. Stratigraphy and petroleum system of Northeast Java Basins (modify of Mudjiono and Pireno., 2001)

3. Theoretical background

Acoustic impedance is defined as the rock capability allowed seismic wave traveling through it. Physically, acoustic impedance is the product between medium density with seismic wave velocity (equation-1).

$$Z_i = \rho_i . V_i \quad (1)$$

where is the acoustic impedance in i medium, is i medium density, and is seismic wave velocity in i medium.

Acoustic impedance is the rock property related to lithology, porosity, fluid content, depth, pressure, and temperature. This means that acoustic impedance is not surface property of two rock layers but it is rock property. Antsey, in Sukmono (2000), determined the relation between hardness of rock with acoustic impedance. The harder rock will be more easily traveled by acoustic wave than the softer one.

Seismic inversion is a technique to generate subsurface model using seismic data as an input and well data as a control (sukmono, 2000). Based on the definition, seismic inversion is an reverse technique from forward modeling. The result of seismic inversion is synthetic seismic section based on earth model.

In order to get maximum result, seismic inversion is applied only to the post-stack seismic data and acoustic impedance is extracted from well data. Acoustic impedance is the product between density with velocity log of well data. It gives high quality vertical resolution but low quality in lateral. Otherwise, acoustic impedance seismic gives high quality lateral resolution but low in vertical. So that, the integration between seismic and well will generate an effective and efficient way to characterize reservoir.

Model based inversion is a type of seismic inversion generating blocky acoustic impedance model from well data. The first step of model based inversion is building an acoustic impedance model. Then wavelet convolution will be applied on thus model. The result of convolution is synthetic seismic that will be compared to the original seismic data. The comparison of the model (convolution result) with the original seismic using iteration will be ended when it has small error value. In order to get the smallest error value, the model must be updated continuously. Updating the model is used generalized linear inversion method. This is a method using the most suitable acoustic model derivative to the original seismic data, it called Marquardt-Levenburg method (Russell, 1996). Model based inversion is based on convolution model shown by equation-2.

$$seismic = wavelet * reflectivity + noise \quad (2)$$

With the assumption that seismic trace and wavelet have been known, also the noise is not interconnected and random. Model based inversion is blocky model, it consists of small blocks representing inversion image resolution. The smaller the block, the longer inversion process, but the higher resolution will be obtained. With the high resolution of seismic data, it is highly useful for accurately interpretation (chopra et al., 2006).

4. Results and discussion

The data given in this study area are 35 seismic lines and a PMK-01 well. The base map of study area including seismic and well data is shown by Figure-1. Acoustic impedance inversion was applied to seismic data controlled by well. This is expected to determine characteristics of target reservoir. The target reservoir is Oligo-Miocene carbonate rock called as Kujung Limestone, its facies become characteristics determined by acoustic impedance inversion. The inversion is processed from time -100ms Kujung Limestone to +100ms Ngimbang Horizon.

PMK-01 well is located at deep marine, then it became high area as the result of RMKS inversion. Sedimentation supply came from Lombok High, so that Lombok Ridge highly affected its sedimentation. The Oligo-Miocene carbonate rock in this well is dominated by wackestone in the lower part, marl in the middle part, and interbedded of marl and packstone in the upper part. The existence of marl and wackestone is caused by Fore Reef environment. Fore Reef is characterized by lithology i.e. shale, skeletal wackestone, mudstone, marl, rudstone/floatstone, and dolomite (mixing zone) (Hadnford and Loucks .,1993).

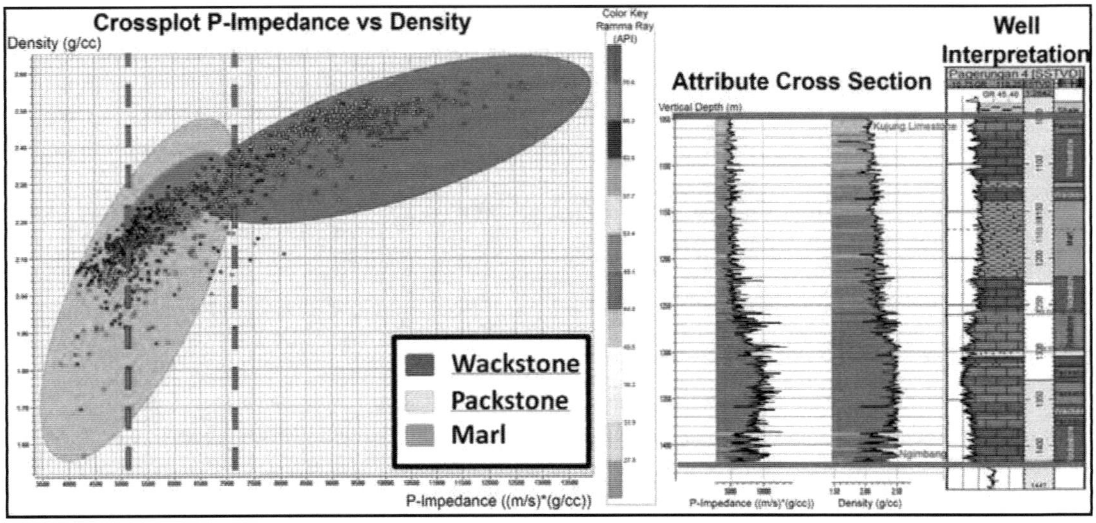

Figure 4. Crossplot analysis of PMK-01 Well

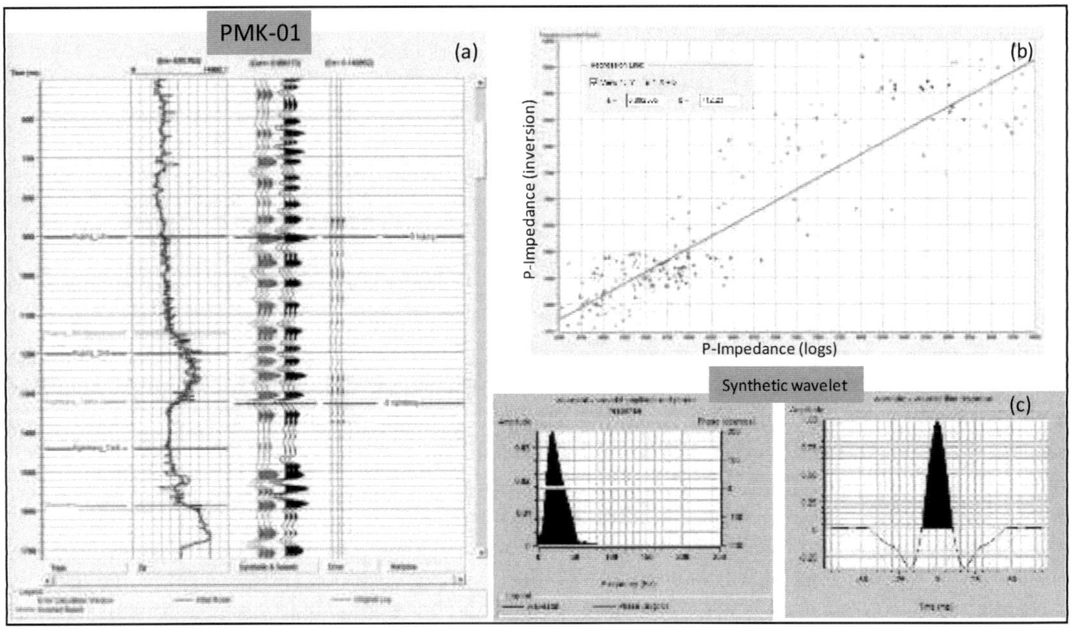

Figure 5. The inversion parameter i.e. (a) comparison log, (b) crossplot acoustic impedance log between inversion result and original log, (c) the inversion wavelet.

The crossplot analysis is conducted as the first step in acoustic impedance inversion. The crossplot analysis is shown by Figure-4, it describes the relation of acoustic impedance to density with gamma ray color key within Kujung Limestone. Kujung Limestone is divided into three facies with various acoustic impedance values. There are two cut off value of acoustic impedance separating its facies i.e. 5200 gr/cc*m/s for packstone-marl and 7400 gr/cc*m/s for marl-wackestone. Interbedded of wackestone and packstone (well interpretation) was hard to defined by acoustic impedance, it possibly is caused by thin section of packstone-wackestone. It is smaller than minimum range of crossplot picked data, so that acoustic impedance does not separate packstone-wackestone interbedded.

IOP Conf. Series: Earth and Environmental Science **132** (2018) 012010 doi:10.1088/1755-1315/132/1/012010

The tying technique between seismic data to well log called as well to seismic tie is the first process of inversion. Well to seismic tie correlates seismic data (time domain) to the well log (depth domain). It is used synthetic seismogram generated by velocity survey. Velocity survey is convolution process from wavelet and coefficient reflection. Coefficient reflection is the product of sonic and density log from the well data, it also represented acoustic impedance log.

Inversion analysis was conducted to define the best parameter for inversion. It compares the initial model (product of applying acoustic impedance log to seismic) to initial log. This method is shown by Figure-5. The comparison between initial log (blue), initial model (black), and inversion result (red) shown by Figure-5 (a) has 0.99 correlation value and 0.14 of error. Inversion analysis was applied to the target area which is Oligo-Miocene carbonate rock called Kujung Limestone, so that the correlation and error value were calculated from Kujung to Ngimbang. The synthetic seismogram is identically the same as seismic data. It describes that inversion result will be the same relatively as acoustic impedance log generated before. Figure-5(b) shows the crossplot of acoustic impedance log from inversion result and original log. The distribution of log data have linear relation, it describes the similarity of inversion result to original log. Figure-5(c) shows the wavelet generated by statistically calculation from seismic. It is the same used in the well to seismic tie process.

There are two horizons picked in this study which are Top Kujung Limestone and Top Ngimbang (as Base Kujung). Interpretation of thus horizons was controlled by marker data from well and seismic stratigraphy. Build-up form is located below Tuban-Kujung unconformity. It has different composition of seismic amplitude than the other surrounding it. It means the lithology within build-up also different. Meanwhile Ngimbang was picked by the continuity from well marker to seismic.

The seismic interpretation is shown by Figure-6 (a). The result of acoustic impedance inversion is shown by Figure-6 (b). The acoustic impedance inversion could separate carbonate facies, but not for thin interbedded facies. Thin interbedded of wackestone-packestone in the lower part became fully wackstone facies. Also thin interbedded of wackestone-packstone in the upper part became packstone facies with thin marl within it. In the other hand, wackestone facies is represented by blue-purple color, it is above 7400 gr/cc*m/s acoustic impedance value. Packstone facies is represented by cyan-green color having below 7400 gr/cc*m/s acoustic impedance value.

Oligo-Miocene carbonate build-up in this study was the result of fore reef environment. It made wackstone and marl dominating its build-up. The few amount of packstone in the upper part of build-up was the result of changing environment to reef core. The good reservoir is marked by packstone facies in the upper part of build-up Kujung Limestone represented by light green color of acoustic impedance.

Figure-7 shows acoustic impedance map overlay with Kujung Limestone time structure map. It explains the dominant facies of the study area which is packstone facies. On the southern part of the study area has build-up with dominant wackestone facies. This build-up possibly was formed in the lower area, so that wackestone facies grew perfectly. On the northern part of the study area has build-up with dominant packstone facies. It possibly was formed in the platform area with steady position of Kujung Limestone.

The prospect area for the next development of oil and gas exploration is northern part of the study area. It has low acoustic impedance indicating packstone facies and goo porosity. It las so has build-up in the several location with constantly the thickness of Kujung Limestone.

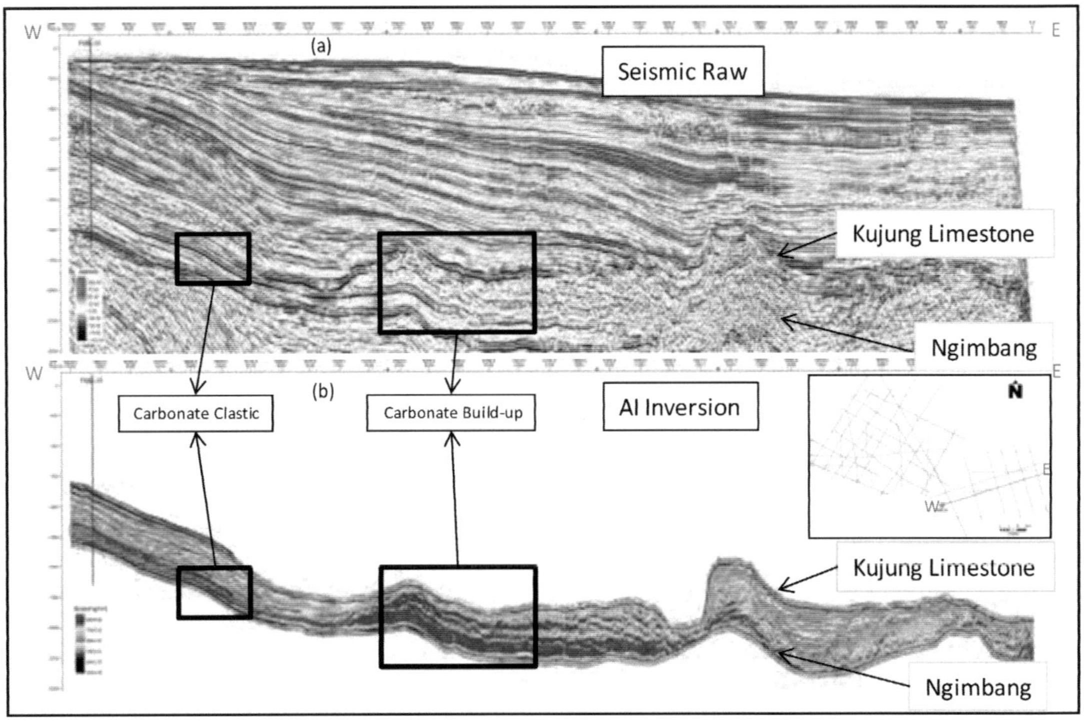

Figure 6. The section of (a) seismic interpretation and (b) inversion result.

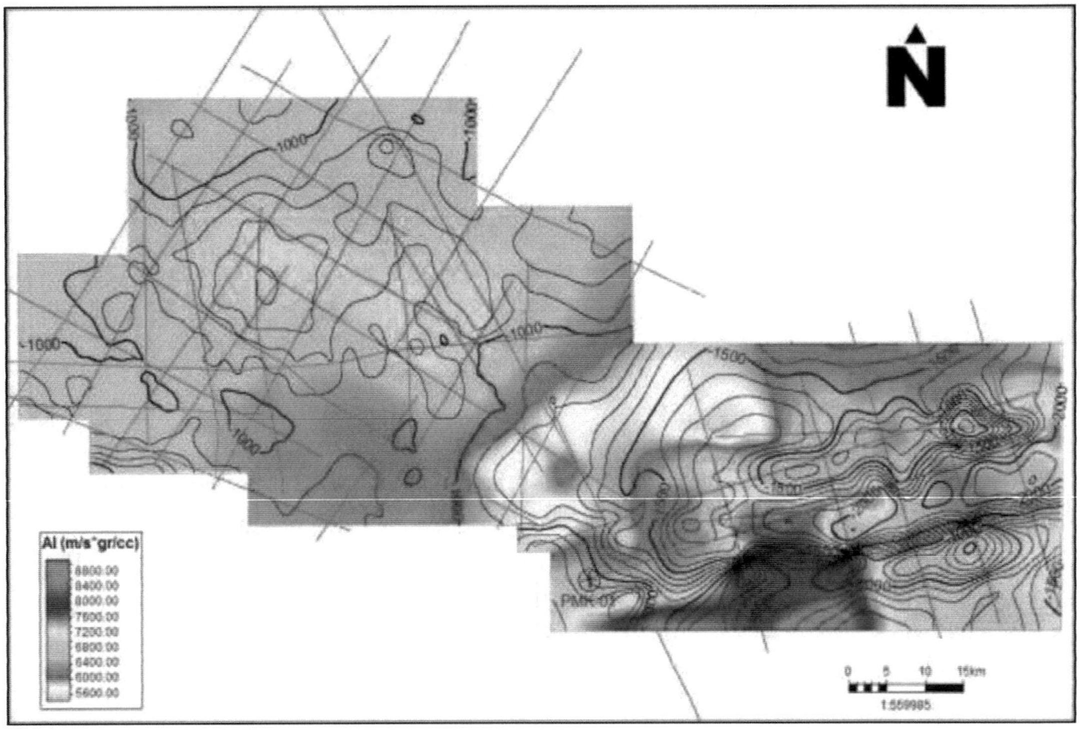

Figure 7. The accoustic impedance map overlay Kujung Limestone time structure map.

5. Conclusion

Based on well log sensitivity test, Oligo-miocene carbonate facies of Kujung can be well separated by it's acoustic impedance and density value. Packstone has low acoustic impedance below 7400 gr/cc*m/s, meanwhile wackestone has higher acoustic impedance. Acoustic impedance inversion was done using a model based method. Oligo-miocene carbonates of Kujung built up are dominated by wackestone and marl facies, where the packstone facies is found at the top of built up. This built-up succession shows the deposition environment of the fore reef. The acoustic impedance map overlaid by structural maps provides an overview of the built-up and clastic carbonate distribution, where built up carbonates are found in the northern and southern areas. The northern area was found abundant of good porosity packstone, while the southern part was abundant of poor porosity wackestone. This corresponds to the regional geology of fore reef in study area, where the northern part was on the platform environment while the southern part was deeper. The northern area are a potential area for hydrocarbon reservoir.

References

[1] Chopra, S., Castagna, J., Portniauguine, O., 2006.Seismic resolution and thin-bed reflectivity inversion.CSEG Recorder.
[2] Handford, C. R., and R. G. Loucks, 1993. Carbonate depositional sequences and systems tracts-responses of carbonate platforms to relative sea-level change. Recent Advances and Applications: American Association of Petroleum Geologists Memoir 57, p. 3-41.
[3] Mudjiono, R., and Pireno, G. E., 2006.Exploration of the North Madura Platform, Offshore East Java, Indonesia.Proceeding of IPA Convention and Exhibition.
[4] Russell, B. H., 1996. Strata Workshop.Hampson-Russell Software Services Ltd.
[5] Satyana, A. H., 2014, New Consideration on the Cretaceous Subduction Zone of Ciletuh LukUlo – Bayat – Meratus: Implications for Southeast Sundaland Petroleum Geology. Proceeding of IPA Convention and Exhibition.
[6] Sheriff, R.E. andGeldart, L.P., 1995.Exploration Seismology. Cambridge University Press, Second Edition.
[7] Sukmono, S., 2000.Seismic Inversion for Reservoir Characterization. Laboratorium Teknik Geofisika, TeknikGeofisika ITB, Bandung.
[8] Wisatria, B., 2009. Hydrocarbon Prospect Zone Determination Using Attribute Analysis and Seismic Inversion of "INDAH" Field.Geophysics Laboratory, Physics Department, Mathematic and Natural Science Faculty, UniversitasGadjahMada, Yogyakarta.

Acknowledgments

This study initiated by SKK Migas cooperated with UPN Veteran Yogyakarta, Universitas Gadjah Mada and several PSC developing East Java Basins. This study became motivation and example for geoscientists and practitioners of oil and gas industry to keep spirit doing research and exploration in the effort of national oil and gas resources and production enhancement. Thanks to Deputy for Planning Management Chief and Exploration Division Chief of SKK Migas for the support, also all of the PSC helping the realization of this study. Also we thank to UPP university partnership program Chevron-UGM for supporting hardware and software facility.

IOP Conf. Series: Earth and Environmental Science **132** (2018) 012011 doi:10.1088/1755-1315/132/1/012011

Near-Surface Modeling Using First Arrival Tomography: A Case Study on High Variation Topography and High Surface Energy Attenuation Area

Teguh SUROSO[a] and Aurum DATAMETRIANA

Pertamina – Upstream Technology Center, Jakarta, Indonesia

[a)]teguh.suroso@pertamina.com

Abstract. This paper describes a method to build a near-surface velocity model. The purpose of this work is to have a better understanding of the near-surface condition below area with high variation topography and high surface energy attenuation. The velocity model built then was used as input for statics calculation. The method applied to this work is first arrival tomography. This method was tested using two datasets with different acquisition geometries. And to verify the result, we compare the calculated statics correction with a conventional method. The results of this work is two similar near-surface models with detail model and the statics applied on the seismic data showed better correction compare to the conventional one. We conclude that the method is applicable accurately on the area with high variation of topography and high surface energy attenuation.

1. Introduction

Some near surface problems can affect seismic data quality. Near surface problem such as extreme topography and high attenuation at near surface can make reflector looks discontinuity. The model velocity that build from travel time tomography method can be used to near surface

Subsurface condition can be determined from velocity model that build from traveltime tomography. Furthermore, that model velocity can be used for static correction calculation of the seismic data (Zhang, 2015). Static correction with this velocity model expected better than static correction with elevation static method.

The case study of this work use two kinds of acquisition method to model velocity building control. The first type of acquisition method use the conventional method with buried source beneath the surface and receiver at the surface. The second acquisition method is almost the same as first parameter except for receiver depth. The receiver for second acquisition method is buried beneath the surface. Location of the source and receivers for second acquisition method the same as first acquisition method.

2. Methodology

2.1. Fermat's Principle

Important property of seismic raypath is given by fermat's principle. Raypath is spatial curve along the medium which traveltime with minimum slowness is stationary. After all curve are connecting, minimum traveltime between source and receiver can be found (Moser, 1991). Traveltime between source and receiver can be calculated by summing traveltime of each curve.

2.2. Refraction Traveltime Tomography Method

A number of refraction traveltime tomography methods have been developed. White (1989) described a refraction traveltime tomography method that applies a two point raytracing algorithm and solves a damped least-squares problem for both velocities and refractor depths. Zhu and McMechan (1989) performed refraction tomography using an analytic traveltime solution and requires the initial model to have positive velocity gradients. Ammon and Vidale developed a refraction traveltime tomography method that regularizes the inverse problem with second-order model derivatives in a jumping fashion. The use of the traditional two-point ray-tracing algorithms limits the accuracy of tomography inversion. The ray methods suffer from the problem of converging to a local minimum traveltime path and occasionally missing the global minimum (Moser, 1991). An ill-posed inverse problem must be properly regularized in order to obtain a stable solution, which is independent of the model discretization (Delprat-Jannaud and Lailly, 1993). Rodi (1989) and Zhang et al. (1996) showed that regularizing the model stepsize (creeping) rather than the model itself (jumping) cannot ultimately avoid solving an illposed inverse problem. Further, not all the regularization approaches perform well, and an appropriate criterion for refraction traveltime tomography must be defined.

2.3. Shortest Path Ray Tracing Method

To perform a tomography study, we need ray tracing technique to travel time and raypath calculation. There are some solutions to the wavefront ray tracing calculation. One of the solution is shortest path ray tracing (SPR). On the SPR method use graph theory expand a wavefront by finding the shortest path. Seismic raypath can be found by the shortest traveltime paths through a network that represents the earth. The network consists of nodes, and the nodes connection is based on a graph template. These node functioning as new source on the wavefront calculation (Zhang dan Toksöz, 1998).

The SPR method include three steps. First, timing nodes along an expanding wavefront from its original source or secondary source. Second, finding the minimum traveltime point along the wavefront and taking this point as a new secondary source. Third, expanding the wavefront from this minimum time point. These steps are repeated until the whole model is traced. (Zhang dan Toksöz, 1998). With a small number of ray legs in a graph template, SPR usually yields zig-zag raypaths in homogeneous or smooth velocity zones in the model, and produces longer raypaths (Fischer and Lees, 1993).

3. Acquisition Parameter

There are two parameter acquisition at this work. Both of acquisition parameter, source and receiver at the same location. First parameter, receiver is buried at surface (conventional method). Second parameter, receiver is buried 6 meters beneath the surface. The source in buried about 40 meters beneath the surface.

4. Case Study

The survey area of this case study contain an extremely topography (fig. 1). The extremely topography make discontinuity reflector at near surface. The receiver elevation is about 79.3 to -6.4 meter.

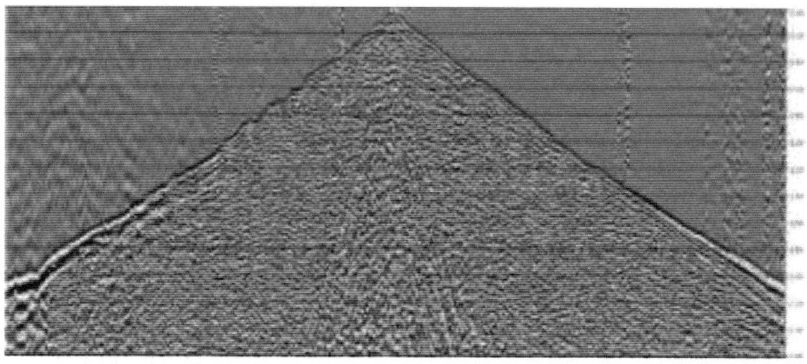

Fig. 1. Shot gather with conventional acquisition method.

Aside from topography problem, in this case there is strong surface attenuation, so energy from source can't be well propagated. Shot gather from surface source contained low frequency velocity and make reflector aliasing with low frequency noise. Fig. 2 shows shot gather obtained from conventional acquisition with strong surface attenuation. This attenuation is caused by weathering layer with velocity range from 880-1100 m/s. fig. 3 shows velocity model with surface receiver.

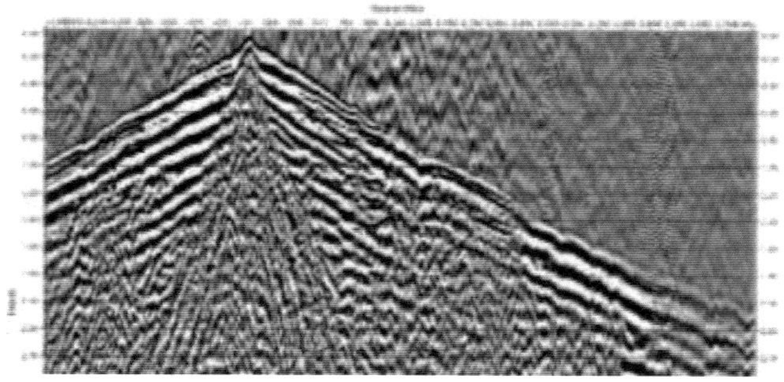

Fig 2. Shot gather obtained from conventional acquisition and high attenuation.

In this study, we use 2 different acquisition parameter as comparison. First acquisition parameter is conventional acquisition method with source and receiver buried 40 m beneath surface. Second acquisition parameter is also conventional acquisition method with source buried 40 m and receiver 6 m beneath surface. Velocity model from both acquisition relatively identical. Fig 3 shows velocity model with surface receiver. Fig 4 shows velocity model with buried receiver 6 m beneath surface. Fig 5 shows velocity model different between 2 velocity model.

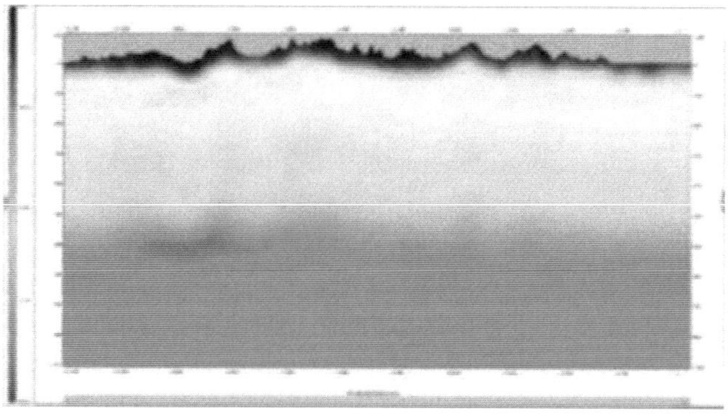

Fig. 3. Velocity model with surface receiver.

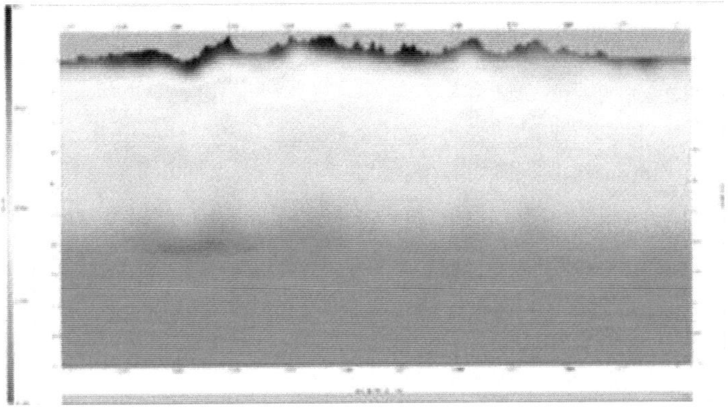

Fig. 4. Velocity model with buried receiver.

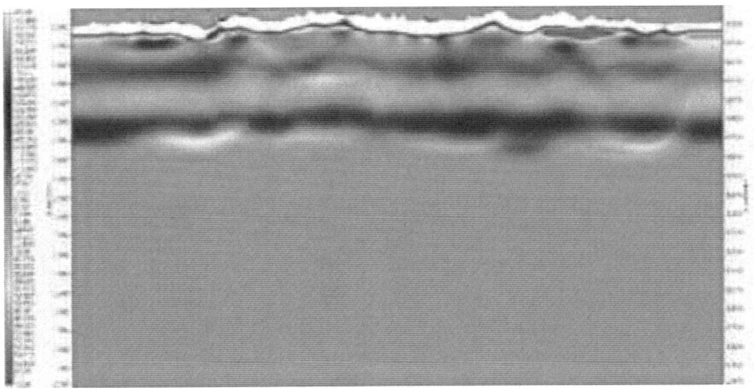

Fig. 5. Difference between two velocity model.

Velocity model obtained from both data seismic are used for static correction. Static correction from these velocity model shows better result than using elevation static method. Fig 6 shows gather data before static correction. Fig 7 shows gather after static correction using elevation static method. Fig 8 shows gather after static correction using travel time tomography. Rough topography in elevation static correction method can be corrected using velocity model from travel time tomography method.

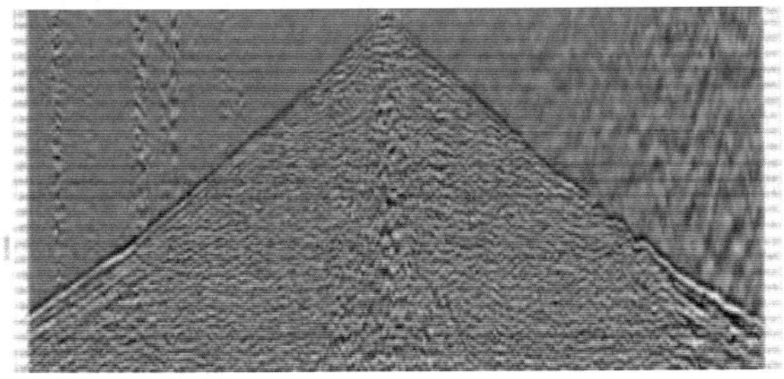

Fig. 6. Shot gather before static correction.

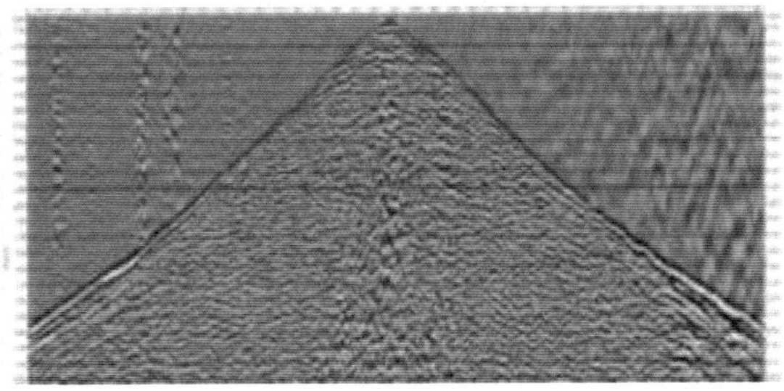

Fig. 7. Shot gather after static correction using elevation static method.

IOP Conf. Series: Earth and Environmental Science **132** (2018) 012011 doi:10.1088/1755-1315/132/1/012011

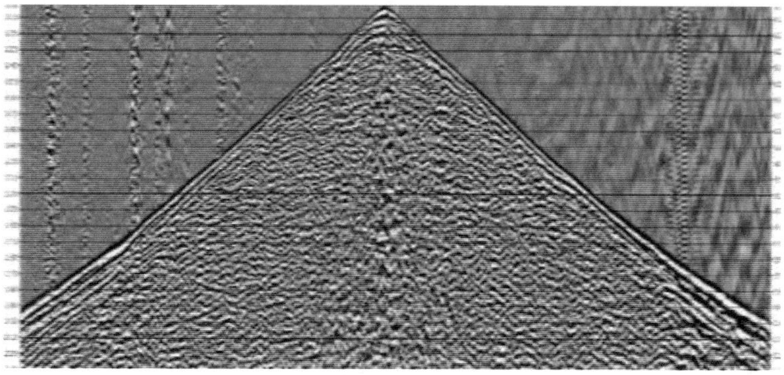

Fig. 8. Shot Gather after static correction using travel time tomography method.

Stacked data seismic with static correction using travel time tomography shows better result than using elevation static method.

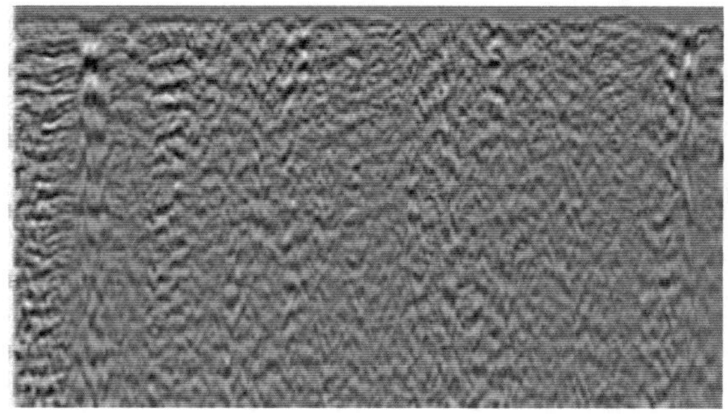

Fig. 9. Stacked seismis after correction static with conventional acquisition method

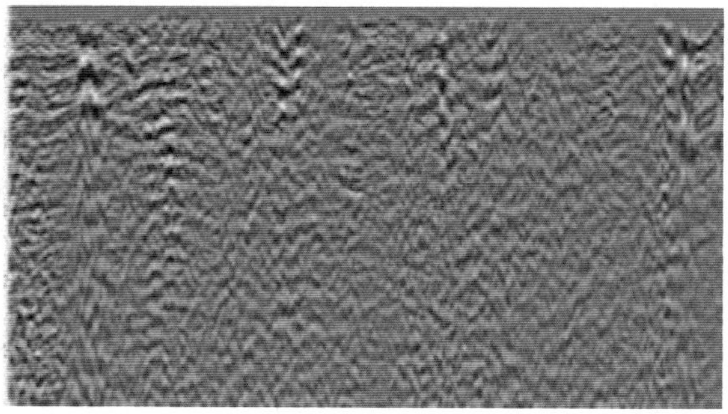

Fig. 10. Stacked seismic after elevation static correction with conventional acquisition method.

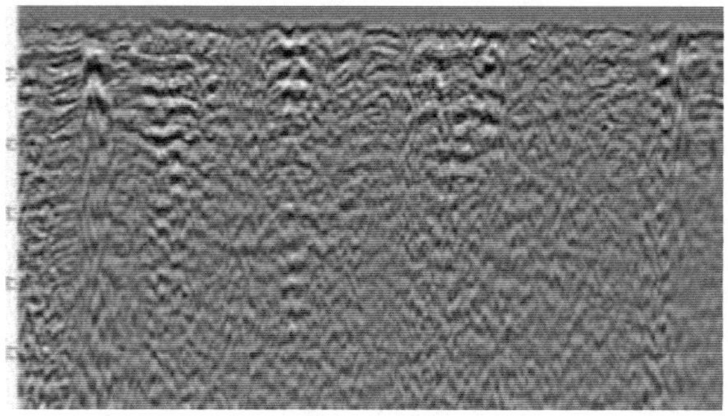

Fig. 11. Stacked seismic after static correction using travel time tomography and conventional acquisition method.

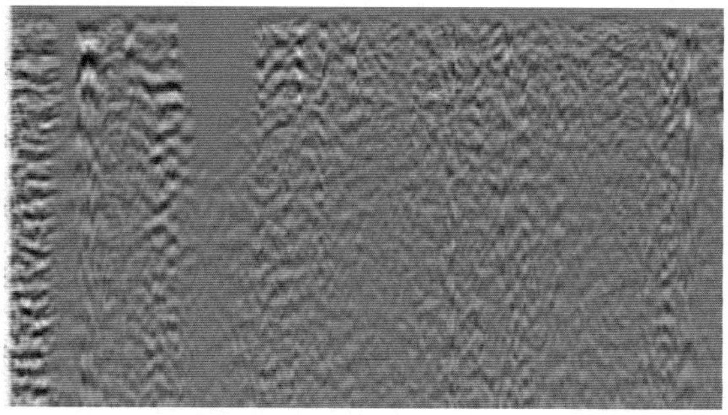

Fig. 12. Stacked seismic before static correction with buried receiver method.

IOP Conf. Series: Earth and Environmental Science **132** (2018) 012011 doi:10.1088/1755-1315/132/1/012011

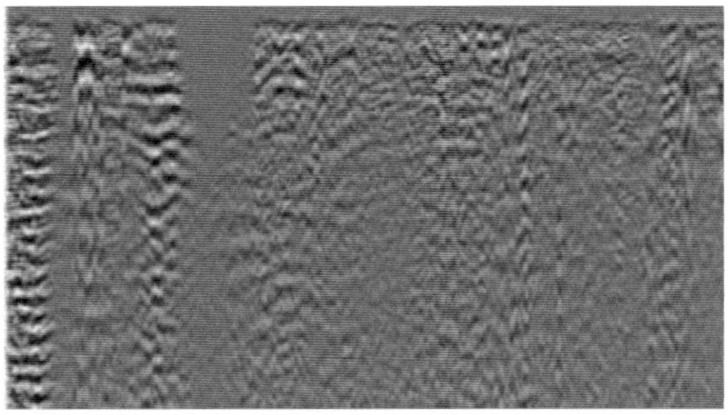

Fig. 13. Stacked seismic after static correction using travel time tomography and buried receiver method

5. Conclusion

Velocity modeling can be used for determine near surface problem. Velocity model that build form travel time tomography can represent near surface condition. The proper of velocity model can be used to static correction. Static correction method with velocity model that build from traveltime tomography batter than elevation static method.

Reference

[1] Delprat-Jannaud, F., and Lailly, P., 1993, Ill-posed and well-posed formulations of the reflection travel time tomography problem: Journal of Geophysical Research, Vol. 98.

[2] Fischer, R., and Less, J. M., 1993, Shortest Path Ray Tracing with Sparse Graphs: Geophysics, Vol. 58.

[3] Moser, T. J., 1991, Shortest Path Calculation of Seismic Rays.

[4] Rodi, W. L., 1989, Regularization and Backus-Gilbert Estimation in Nonlinear Inverse Problems: Application to Magnetotellurics and Surface Waves: Ph.D. thesis, Pennsylvania State Univ.

[5] White, D. J., 1989, Two-dimensional Seismic Refraction Tomography: Geophysical Journal International, Vol. 97.

[6] Zhang, J., 2015, Near-surface Traveltime Tomography with Long-wavelength Static Optimization.

[7] Zhang, J., Rodi, W., Mackie, R. L., and Shi, W., 1996, Regularization in 3-D DC Resistivity Tomography: Proceding SAGEEP 1996, Environment and Engineering Geophysical Society.

[8] Zhang, J. and Toksöz M.N., 1998, Nonlinear Refraction Traveltime Tomography.

[9] Zhu, X., and McMechan, G. A., 1989, Estimation of Two-dimensional Seismic Compressional-wave Velocity Distribution by Iterative Tomographic Imaging: International Journal of Imaging System and Technology, Vol. 1.

Correlation Equation of Fault Size, Moment Magnitude, and Height of Tsunami Case Study: Historical Tsunami Database in Sulawesi

Admiral Musa Julius[a] Sugeng Pribadi[b] Muzli Muzli[c]

Agency for Meteorology Climatology and Geophysics (BMKG)
Jl. Angkasa 1 No. 2, Kemayoran, Jakarta, Indonesia 10720

[a]*admiralmusajulius@gmail.com*
[b]*sugengpribadimsc@gmail.com*
[c]*muzli@bmkg.go.id*

Abstract. Sulawesi, one of the biggest island in Indonesia, located on the convergence of two macro plate that is Eurasia and Pacific. NOAA and Novosibirsk Tsunami Laboratory show more than 20 tsunami data recorded in Sulawesi since 1820. Based on this data, determination of correlation between tsunami and earthquake parameter need to be done to proved all event in the past. Complete data of magnitudes, fault sizes and tsunami heights on this study sourced from NOAA and Novosibirsk Tsunami database, completed with Pacific Tsunami Warning Center (PTWC) catalog. This study aims to find correlation between moment magnitude, fault size and tsunami height by simple regression. The step of this research are data collecting, processing, and regression analysis. Result shows moment magnitude, fault size and tsunami heights strongly correlated. This analysis is enough to proved the accuracy of historical tsunami database in Sulawesi on NOAA, Novosibirsk Tsunami Laboratory and PTWC.
Keywords: tsunami, magnitude, fault, height

1. Introduction

Sulawesi, one of the biggest island in Indonesia, located on the convergence of two macro plate that is Eurasia and Pacific. They are not only great plate, but also micro plate Philippine. Repercussions by this convergence, Sulawesi divided into two subduction zone that is Sulawesi sea and Molucca sea. They are not only subduction zone, Sulawesi also has more active tectonic fault such as Walanae, Palukoro, Gorontalo, Matano, Lawanopo, and Kolaka.

Tsunami happened because vertical deformation located on the sea floor suddenly. The deformation such as earthquake, landslide, volcano and fall of rock space. It make a movement of water column on the very long wavelength from the sea to the coastal point. Based on earthquake event, there are some parameter controlled tsunami problem such as magnitude. Magnitude strongly involved on the rupture size of fault that determine the tsunami size. Not only magnitude, tsunami strength also influenced by depth and slope between intra-plate. Study the correlation of moment magnitude, fault size and tsunami height has been done by Utsu (1954), Tocher (1958), Kanamori (1975), Bonilla (1984), Wesnousky (1986), Ambraseys (1988), Wells (1994), Wang (1998), Papazachos (2004) and Muzli (2015). This correlation use to predict the fault size and height of tsunami in Sulawesi by moment magnitude of earthquake. However, this study aims to determine the correlation between them.

2. Data and Method

This study using moment magnitude of earthquake and tsunami height data in Sulawesi region, sourced from Pacific Tsunami Warning Center catalogue. This catalogue found only 16 tsunami events that have height data. Based on this small number, we use empirical equation to determine tsunami height. We calculate tsunami Height using Supassri [1] equation:

$$\mathbf{H = 3E\text{-}5Mw^{6.2344}} \quad (1)$$

H = tsunami height
Mw = moment magnitude

Step we used are collecting, processing and assessing the correlation. Firstly, we converting surface magnitude (Ms) to moment magnitude (Mw) because there is only available Ms on the catalogue. We used Asrurifak [7] equation to convert this magnitude.

$$\mathbf{Mw = 0.143\ Ms^2 - 1.051\ Ms + 7.285} \quad (2)$$

Meanwhile, we used Papazachos [4] equation to determine length (L), wide (W), Area (S) and Slip (U). There are some categories for each magnitude:

$$\mathbf{LogS = 0.86M - 2.82,\ \sigma = 0.25,\ 6.7 \leq M \leq 9.2} \quad (3)$$

$$\mathbf{LogL = 0.55M - 2.19,\ \sigma = 0.18,\ 6.7 \leq M \leq 9.3} \quad (4)$$

$$\mathbf{LogW = 0.31M - 0.63,\ 6.7 \leq M \leq 9.2} \quad (5)$$

$$\mathbf{LogU = 0.64M - 2.78,\ 6.7 \leq M \leq 9.2} \quad (6)$$

L = fault length; W= fault width; U = fault slip; M= moment magnitude

After fault size has been done, we will found the correlation between moment magnitude, fault size and tsunami height using simple regression. We make correlation by linear and logarithm regression between moment magnitude – fault size, moment magnitude – tsunami height, and fault size-tsunami height.

3. Result and Discussion

The relationship between the magnitude of moments (Mw) with the parameters of fault extent such as fault length (L), fault area (S), fault width (W) and slip or displacement (U) can be expressed using equations from Papazachos et al. in the data and methods below.

3.1. Regression of Moment Magnitude – Fault Size

3.1.1. Moment Magnitude-Length

The graph of regression result below shows the linear relationship between the two parameters ie logarithmic fault length and moment magnitude (**FIGURE 1**).

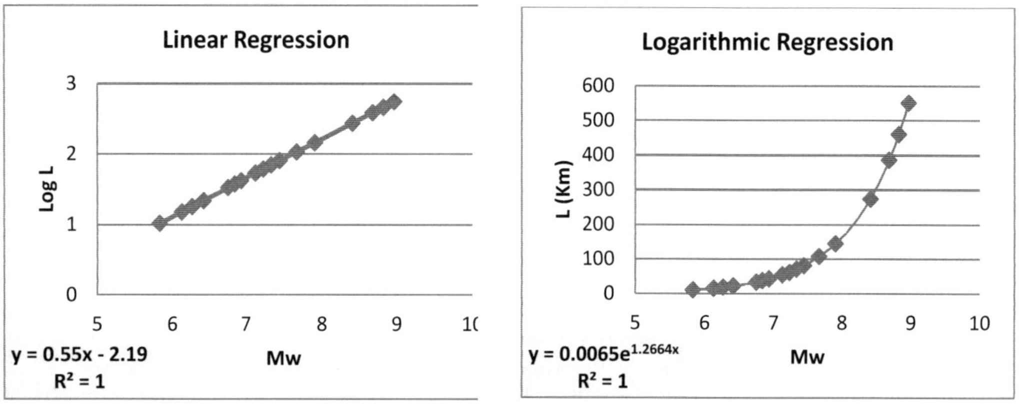

FIGURE 1. linear (left) and logarithmic (right) curve of moment magnitude – length regression

3.1.2. Moment Magnitude-Width

The graph of regression result below shows the linear relationship between the two parameters ie logarithmic fault width and moment magnitude (**FIGURE 2**).

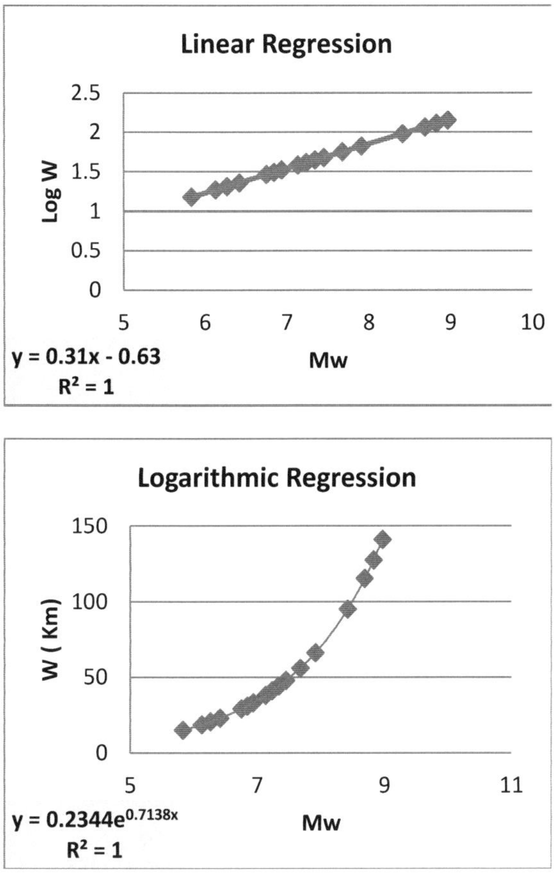

FIGURE 2. linear (left) and logarithmic (right) curve of moment magnitude – width regression

3.1.3. Moment Magnitude-Area

The graph of regression result below shows the linear relationship between the two parameters ie logarithmic fault area and moment magnitude (**FIGURE 3**).

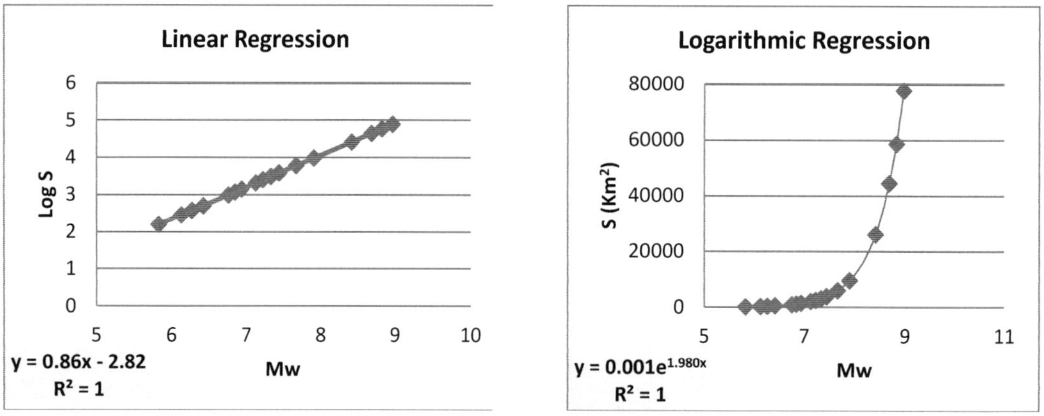

FIGURE 3. linear (left) and logarithmic (right) curve of moment magnitude – area regression

3.1.4. Moment Magnitude-Slip

The graph of regression result below shows the linear relationship between the two parameters ie logarithmic fault area and moment magnitude (**FIGURE 4**).

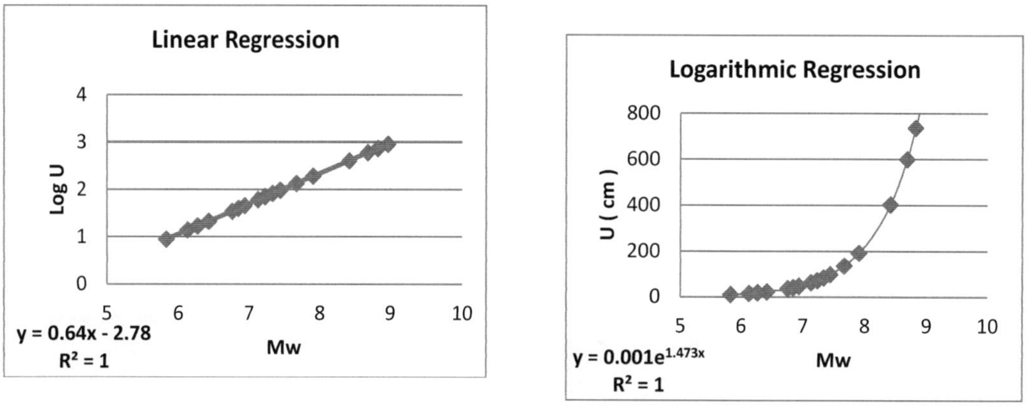

FIGURE 4. linear (left) and logarithmic (right) curve of moment magnitude – slip regression

3.1.5. Moment Magnitude-Fault Size

When combined, the relationship between moment magnitude and fault parameters is (**FIGURE 5**):

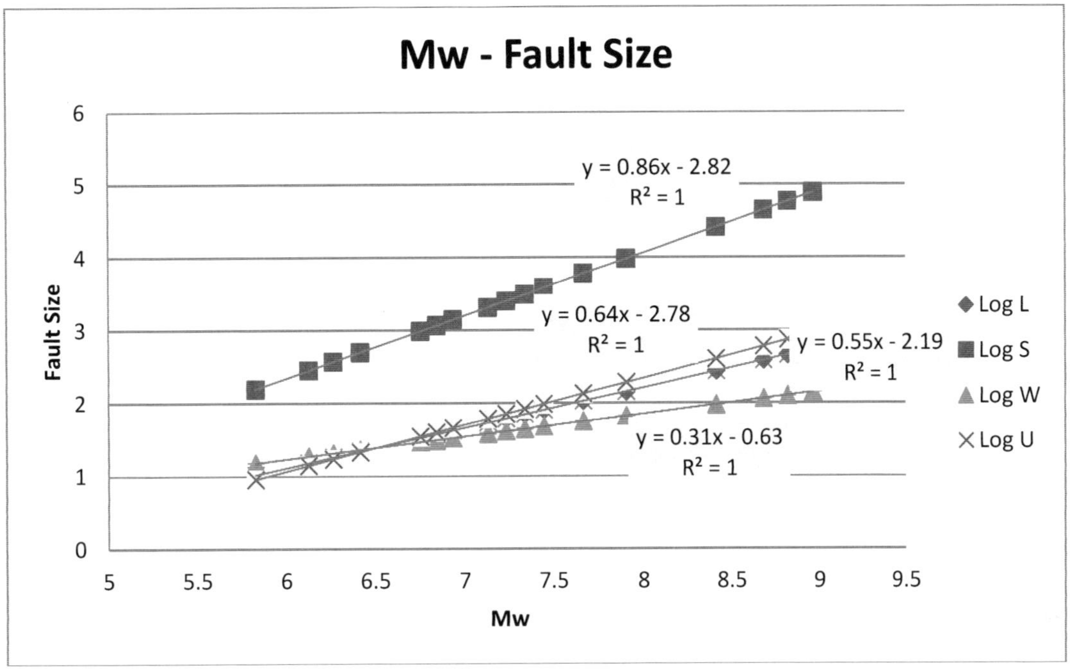

FIGURE 5. linear (left) and logarithmic (right) curve of moment magnitude – fault size regression

3.2. Regression of Moment Magnitude – Tsunami Height

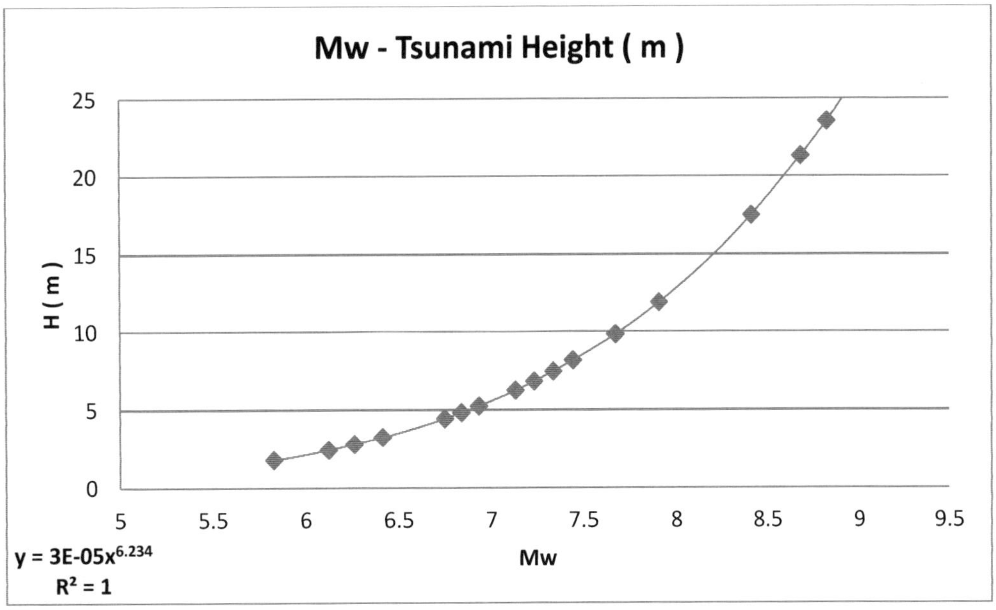

FIGURE 6. linear (left) and logarithmic (right) curve of moment magnitude – tsunami height regression

Based on the graphic images of the above regression results (**FIGURE 6**) it is known that the correlation value indicates a value of 1, which means there is a very strong relationship between the moment magnitude to the tsunami generated.

3.3. Regression of Fault Size – Tsunami Height

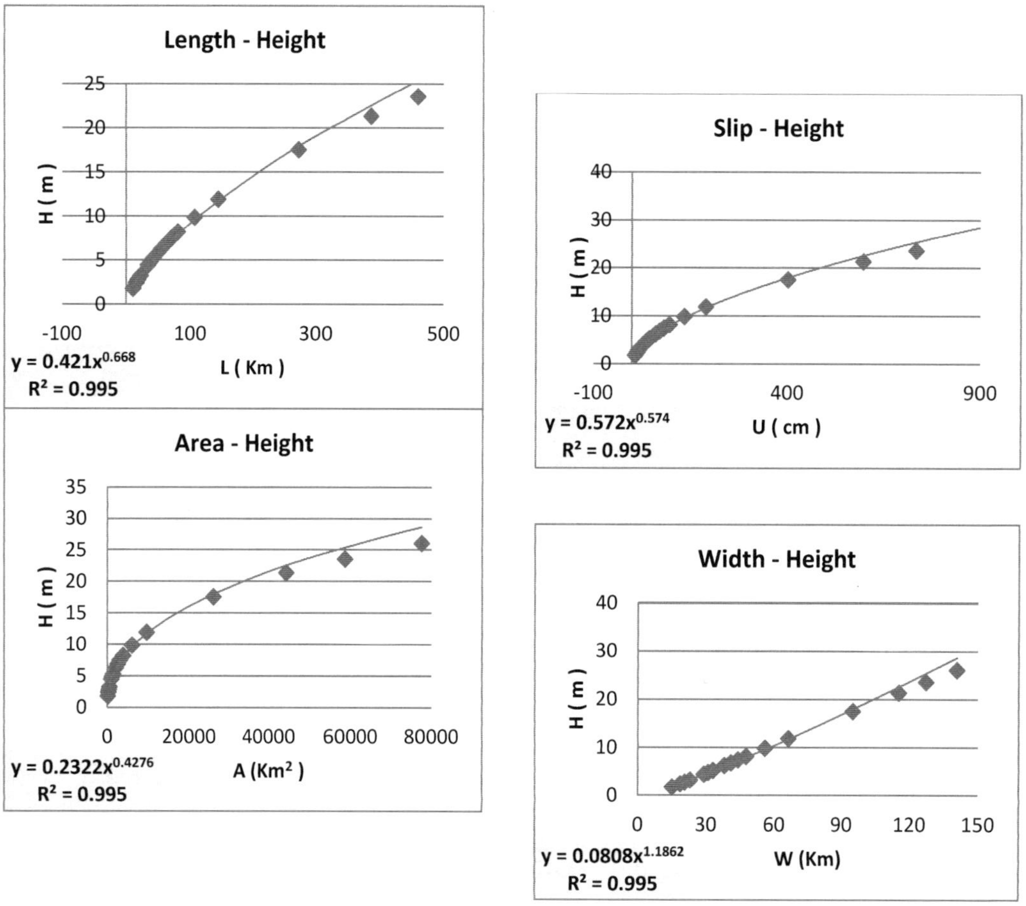

FIGURE 7. linear (left) and logarithmic (right) curve of fault size – tsunami height regression

Based on the graphic images of the above regression results (**FIGURE 7**), it is known that the correlation value indicates a value of 1, which means there is a very strong relationship between the fault sizes to the tsunami height.

Based on theory of statistics regression assessment, variable strongly correlated shown on the alignment model (r^2) near to 1. According to simple regression result, it known that moment magnitude, fault size (length, width, area, and slip) and tsunami height strongly correlated [6]. It showed on the value of r2 near to 1. Due to the fact that, with increasing the value of moment magnitude, value of fault size and tsunami height are also increase. This strong correlation means, all historical tsunami database in Sulawesi on NOAA, Novosibirsk Tsunami Laboratory and PTWC are correct.

This methods can be effectively used to predict tsunami height in order to tsunami risk reduction by land use planning such as plantation and constructing emergency building such as break water, emergency shelter, and water purification. In the other hand, tsunami risk reduction also strongly prevented by mitigation such as tsunami early warning system, community awareness, and education [5].

4. Conclusion

The result show that moment magnitude, fault size and tsunami height strongly correlated. Due to the fact that, with increasing the value of moment magnitude, value of fault size and tsunami height are also increase. This strong correlation means, all historical tsunami database in Sulawesi on NOAA, Novosibirsk Tsunami Laboratory and PTWC are correct.

References

[1] A. Supassri, et al., "Relationship Between Earthquake Magnitude and Tsunami Height Along the Tohoku Coast Based On Historical Tsunami Trace Database and the 2011 Great East Japan Tsunami", Report of Tsunami Engineering Vol. 30, 37-49 (2013).
[2] B. Priadi, "Geochimie du magmatisme de l'Oustet du Nord de Sulawesi, Indonesia: Tracage des sources et implications geodynamique," Ph.D. thesis, Universite Paul Sabatier, Toulouse, France 1993.
[3] B. Sunardi (private communication).
[4] C. Papazachos, et al., "Global Relations Between Seismic Fault Parameters and Moment Magnitude of Earthquakes", Bulletin of The Geological Society of Greece Vol. **36**, 1482–1489 (2004).
[5] K. Satake, "Earthquakes and Tsunamis," in *Tsunami Lecture*, edited by S. Pribadi (International Institute of Seismology and Earthquake Engineering, Tokyo, 2008), pp. 4.
[6] Muzli, et al., "Rapid Moment Magnitude Estimation using Strong Motion Derived Static Displacements" (Elsevier, Procedia Earth and Planetary Science, 2015), pp. 11–19.
[7] M. Asrurifak, et al., "Development of Spectral Hazard Map for Indonesia with a Return Period of 2500 Years using probabilistic Method", Civil Engineering Dimension Vol. **12 No. 1**, 52-62 (2010).
[8] NGDC/WDS Global Historical Tsunami Database available at https://www.ngdc.noaa.gov/hazard/tsu_db.shtml , accessed May 2015
[9] Novosivirsk Russia Tsunami Laboratory Catalogue available at http://tsun.sscc.ru/On_line_Cat.htm, accessed May 2015.

IOP Conf. Series: Earth and Environmental Science **132** (2018) 012014 doi:10.1088/1755-1315/132/1/012014

Ionospheric earthquake effects detection based on Total Electron Content (TEC) GPS Correlation

Bambang Sunardi[1], Buldan Muslim[2], Andi Eka Sakya[1], Supriyanto Rohadi[1], Sulastri[1], Jaya Murjaya[1]

[1] Indonesian Meteorological Climatological and Geophysical Agency, Kemayoran, Jakarta, Indonesia
[2] Space Science Center, Indonesian National Institute of Aeronautics and Space, Bandung, Indonesia 40173

Email : bambang.sunardi@bmkg.go.id

Abstract. Advances in science and technology showed that ground-based GPS receiver was able to detect ionospheric Total Electron Content (TEC) disturbances caused by various natural phenomena such as earthquakes. One study of Tohoku (Japan) earthquake, March 11, 2011, magnitude M 9.0 showed TEC fluctuations observed from GPS observation network spread around the disaster area. This paper discussed the ionospheric earthquake effects detection using TEC GPS data. The case studies taken were Kebumen earthquake, January 25, 2014, magnitude M 6.2, Sumba earthquake, February 12, 2016, M 6.2 and Halmahera earthquake, February 17, 2016, M 6.1. TEC-GIM (Global Ionosphere Map) correlation methods for 31 days were used to monitor TEC anomaly in ionosphere. To ensure the geomagnetic disturbances due to solar activity, we also compare with Dst index in the same time window. The results showed anomalous ratio of correlation coefficient deviation to its standard deviation upon occurrences of Kebumen and Sumba earthquake, but not detected a similar anomaly for the Halmahera earthquake. It was needed a continous monitoring of TEC GPS data to detect the earthquake effects in ionosphere. This study giving hope in strengthening the earthquake effect early warning system using TEC GPS data. The method development of continuous TEC GPS observation derived from GPS observation network that already exists in Indonesia is needed to support earthquake effects early warning systems.

1. Introduction

Advances in sciences and technology made the GPS technology as a study facility in Total Electron Content (TEC) anomaly detection related with earthquake [1, 2, 3]. Recent studies have shown that earthquake and tsunami effects in ionosphere through the atmospheric gravity and acoustic waves have been found in the case of major earthquakes such as Aceh Earthquake, December 26, 2004 and Tohoku Earthquake, March 11, 2011. A possible cause of TEC anomaly related with earthquake and tsunami is the atmospheric gravity wave and infrasonic wave which spread to ionosphere. So, the ionosphere fluctuates in order of atmospheric gravity and infrasound wave periods [4].

Earthquakes influence on ionosphere occurs through Lithosphere-Atmosphere-Ionosphere (LAI) coupling. Lognonne et al. [5] explained that the greatest amplitude of seismic waves in the earth surface were surface waves, one of which is the Rayleigh wave. The seismic waves propagate along the earth surface through the crust and upper mantle with a speed between 3 - 4 km/sec. These waves can produce atmospheric wave that propagates upward with a period greater than 10 seconds. The infrasonic waves spread in atmosphere, and when they reached ionosphere, their wave energy is transferred to ionosphere through collisions between ions and electrons so as to bring infrasonic waves in ionosphere. The infrasonic waves can be detected using some ionospheric observation equipment's including GPS.

In their study, Hao et al. [6] confirmed that ionospheric waves with a period of about 3 - 5 minutes caused by infrasonic waves due to Rayleigh waves, sourced from Tohoku earthquake, March 11, 2011, about 10 minutes earlier. The ionospheric disturbances phase velocity was about 3,6 km/sec which is equal to the speed of Rayleigh wave. Likewise, the results of Heki study [7] showed earthquake precursor as positive anomaly of ionospheric Total Electron Content (TEC) around the focal region detected by Japanese network of Global Positioning System (GPS). Huijun Le, et al. [8] investigated the ionospheric abnormal behaviors prior to the 2011 Tohoku-Oki earthquake by using TEC GPS data. The observation data revealed that on March 8, 2011, there was significant enhancement in TEC also in solar activity. They made conclusion that the significant enhancement in TEC on March 8, probably generated by the pre earthquake ionospheric disturbance. Woei Lin [9] used two dimensional principal component analysis (2DPCA) to detect Aegean-Sea TEC anomaly after magnitude M 6.4 earthquake on May 24, 2014. The results showed after earthquake, TEC anomaly became more intense, in at least 5 minutes time duration.

Study on ionospheric anomalies before and after earthquake in Indonesia was conducted by Muslim [10]. He conducted correlation techniques test to detect ionospheric earthquake effect. The results showed that most earthquakes can be known their precursors, and many earthquake effects cannot be detected. Sunardi, et al. [11] monitored Total Electron Content (TEC) anomalies related to 2015 around Java earthquake events. The monitoring results indicated the emergence of TEC anomalies in most of earthquake cases. By considering only the TEC parameter observation and Dst index, anomalies can be classified as detected earthquake precursors and detected earthquake precursors / geomagnetic storm effects. Sunardi, et al. [12] also conducted study on Total Electron Content (TEC) anomalies observed before strong earthquakes (Mw > 6) that occurred in Indonesia during 2014. The results showed that most of TEC anomalies appear before the earthquakes. In general, TEC anomaly occurs less than 10 days before the earthquake occurrence.

This paper discussed the ionospheric earthquake effects detection using TEC GPS data. Case studies used were Kebumen earthquake January 25, 2014, magnitude Mw 6.2, Sumba earthquake, February 12, 2016 with magnitude Mw 6.2, and Halmahera earthquake, February 17, 2016 with magnitude Mw 6.1. This initial research will be useful in efforts to strengthen earthquake effect early warning systems in Indonesia. Study and development of continuous TEC GPS observation data derived from GPS observation network that already exists in Indonesia is required to support these efforts.

2. Data and method

Total Electron Content (TEC) data were obtained from Center for Orbit Determination in Europe (CODE) in the Global Ionosphere Map (GIM) form. The TEC data generated every day using GPS data from more than 200 GPS stations around the world [13]. TEC modeled in the reference frame of geomagnetic sun using spherical harmonic expansion to degree and order of 15 [14]. GIM is based on a single layer model as shown in Figure 1 [15].

Assumption of GIM model is all free electrons were concentrated in a spherical shell of infinitesimal thickness. Conversion of vertical TEC *Ev* to slant TEC (*E*) can be written as [15] :

$$E = F(z)E_v = \frac{1}{\cos z'}E_v$$

with

$$\sin z' = \frac{R}{R_0+H}\sin z \qquad (1)$$

IOP Conf. Series: Earth and Environmental Science **132** (2018) 012014 doi:10.1088/1755-1315/132/1/012014

Where $F(z) = 1/\cos z'$ showed a single layer mapping function. The z and z' variables are zenith distance at the height of station and at single layer respectively (Figure 1). R is the radius of considered station. R_0 is the average radius of the Earth (~ 6,371 km) while H is the height of a single layer on the average surface of the Earth. Ideal altitude of this H layer is usually set with a high maximum electron density expected, for example, $H = 400$ km [15].

Electron density model E, surface layer density which represents TEC distribution on a global scale is expressed as follows:

$$E = (\beta, s) = \sum_{n=0}^{n_{max}} \sum_{m=0}^{n} \widetilde{P_{nm}} \; (\sin \beta)(a_{nm} \cos ms) + \; b_{nm} \sin ms \tag{2}$$

With E is vertical Total Electron Content (TEC), β is the latitude of geocentric from intersection point of satellite's receiver line and ionosphere, $s = \lambda - \lambda_0$ is sun fixed longitude from ionospheric penetrating point or sub ionospheric point, n_{mak} is maximum expansion level of the spherical harmonic expansion, $\widetilde{P_{nm}} = \Lambda_{nm} P_{nm}$ are normalized legendre function of degree n and order m by a normalization factor Λ_{nm} and classical Legendre function P_{nm}, and both a_{nm} and b_{nm} are TEC coefficients of spherical function [15].

The TEC E (β,s) is usually expressed in TECU (TEC Unit) where 1 TECU = 1 x 10^{16} free elektron/m^2. An equivalent set in the solar geomagnetic frame can be used for the coordinates (β,s). Making the mean sun as sun fixed reference frame and sun geographic latitude β_0 is set to 0, the geographic longitude of the sun may be written as [15] :

$$\lambda_0 = \pi - UT \tag{3}$$

With UT is the universal time. We can find the normalization factor Λ_{nm} which is defined as :

$$\Lambda_{nm} = \sqrt{\frac{(n-m)! \; (2n = 1) \; (2 - \delta_{0m})}{(n+m)!}} \tag{4}$$

where δ denoted the Kronecker delta.

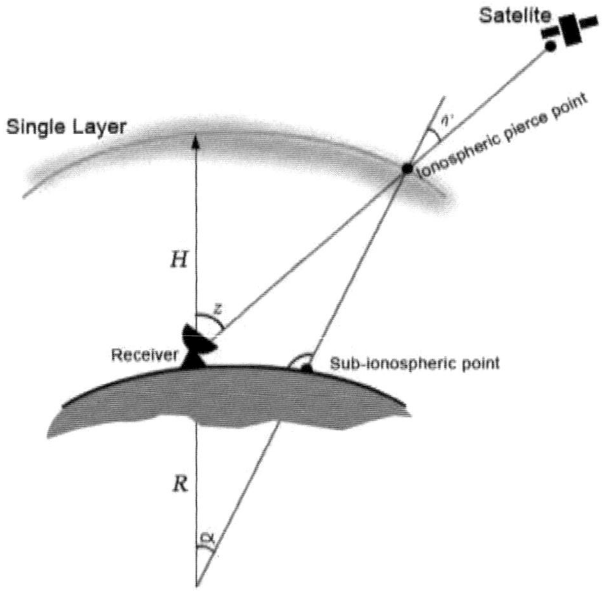

Figure 1. Single layer model of Global Ionosphere Map (GIM), modified from Schaer [15].

Ionospheric earthquake effect detection was done by using the closest to earthquake epicenter TEC data. TEC data from GIM are selected 31 days backward to represent normal conditions, so that there will always be 31 TEC data. At any observation time, it been calculated the average TEC at certain hours for 31 days in order to obtain the TEC monthly average diurnal variation. Correlation analyzes were conducted between TEC diurnal variation on a particular day of 31 days and TEC monthly average diurnal variation, which produces 31 correlation coefficients. Comparison between daily correlation coefficient deviation for 31 days and correlation coefficient average value (skk) divided by its correlation coefficient standard deviation (dskk) was used as an indicator of TEC anomalies in ionosphere. The anomaly threshold is when skk/dskk value was -1. TEC anomalies caused by earthquakes and or magnetic storms based on the value of skk/dskk less than -1. If TEC value met the requirement, then the day with skk/dskk less than -1 is set as ionospheric TEC anomalous condition. To determine the TEC anomalies occurred locally or globally, the TEC data closest to earthquake epicenter cross correlated with the vicinity TEC data. Geomagnetic disturbances like the magnetic storm will lead to global anomalies, otherwise the earthquake effect anomaly was more regionally.

Correlation method based on the assumption that major earthquake will affect the ionospheric layer which can make disturbances before and after the earthquake. However, ionospheric disturbances can also be caused by geomagnetic storms. Therefore, to determine whether ionospheric anomaly was caused by earthquake or magnetic storm, in addition to TEC data also required equatorial and low latitudes ionospheric disturbances index data. To clarify the TEC anomaly sources in low latitudes and the equator, it was needed Disturbance storm time (Dst) index data. Dst index can be used for characterize magnetic storms. Magnetic storm usually expressed by negative value indicates a decrease in earth magnetic field [16]. The magnetic storm intensity was classified into three classes as shown in Table 1 [17]. Dst index data can be real time obtained from the World Data Center for Geomagnetism, Kyoto [18].

Table 1. The intensity of geomagnetic storms based Dst index [17].

No	Dst Index	Category
1	-50 nT < Dst < -30 nT	Weak storm
2	-100 nT < Dst < -50 nT	Intermediate storm
3	Dst ≤ -100 nT	Strong storm

3. Result and analysis

In case of January 25, 2014 Kebumen earthquake event with magnitude Mw 6.2, the value of the correlation coefficient deviation comparison with standard deviation of the correlation coefficient (skk/dskk) and cross correlation are shown in Figure 2. The value of skk/dskk reached -4.2 occurred on Day of Year (DOY) 25 (January 25, 2014). This value is smaller than ionospheric TEC anomaly threshold that qualified as anomaly (skk/dskk =-1). Cross correlation results showed anomalies that occurred on January 25, 2014 around the earthquake epicenter, i.e. 8.23^0 S and 109.2^0 E. Dst index data on January 25, 2014 showed a value of about -9.6 nT, as shown in Figure 3. The value can be categorized as normal condition. Thus reinforces analysis that TEC anomaly on January 25, 2014 probably as Kebumen earthquake effect.

In case of February 12, 2016 Sumba earthquake with magnitude Mw 6.2, the value of the correlation coefficient deviation comparison with a standard deviation of the correlation coefficient (skk/dskk) and cross correlation were shown in Figure 4. The value of skk/dskk reached -4.5 occurred on DOY 43 (February 12, 2016). Cross correlation results showed anomalies that occurred in February 12, 2016 around the earthquake epicenter, i.e. 9.86^0 S and 119.32^0 E. Dst index data on February 12,

2016 indicated a value of about -11.3 nT, categorized as normal, as shown in Figure 5. This TEC anomaly could be the result of earthquake or the effect of Sumba earthquake which was detected in the ionosphere. TEC anomaly also appeared on DOY 34, but Dst index values showed the emergence of a geomagnetic storm that could be classified as intermediate storm. Thus the anomaly in the day could be geomagnetic storms effect.

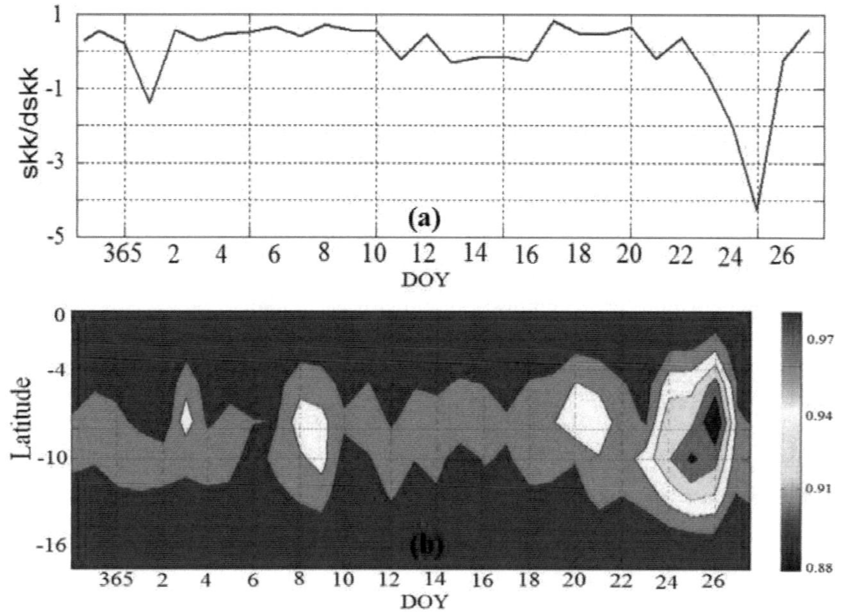

Figure 2. Skk/dskk value (a) and cross correlation of TEC data (b) from December 29, 2013 to January 28, 2014.

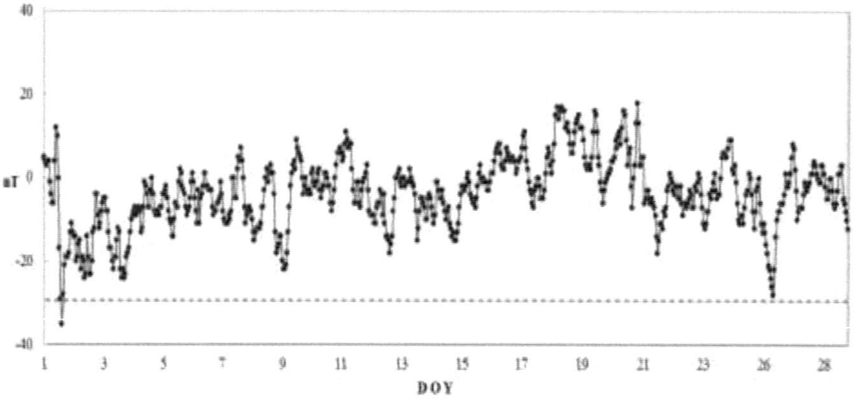

Figure 3. The Dst index values from January 1 to January 28, 2014.

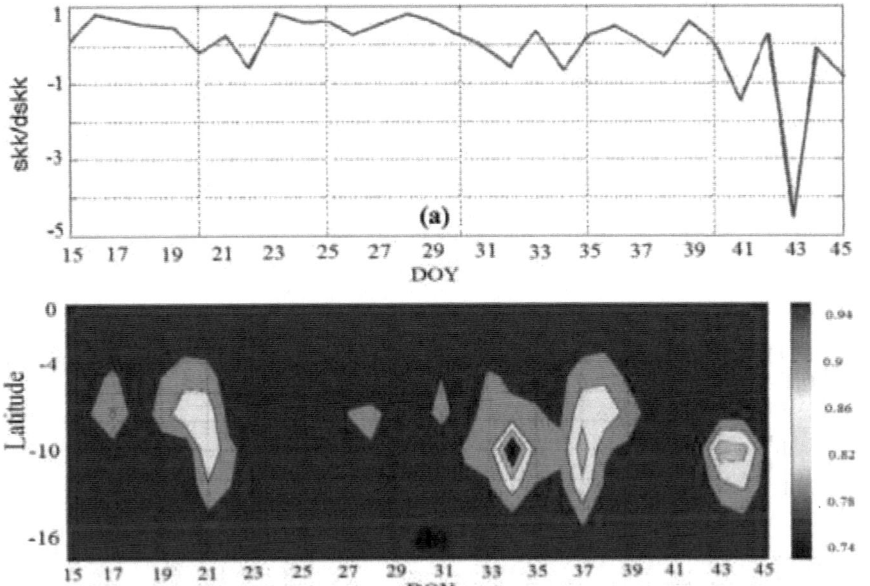

Figure 4. Skk/dskk values (a) and cross correlation of TEC data (b) from January 15 to February 14, 2016.

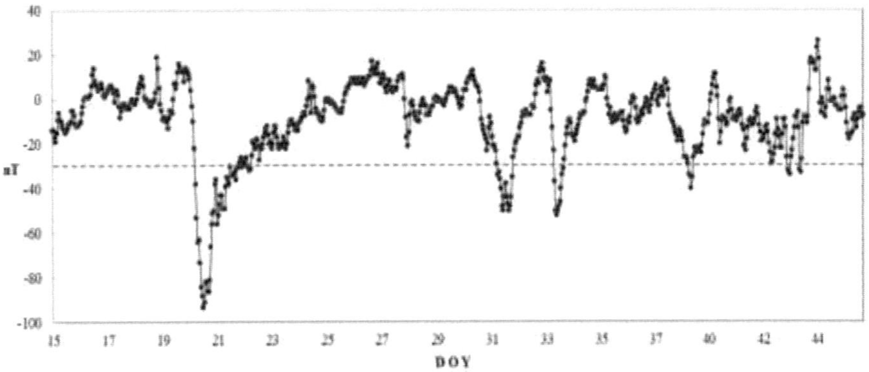

Figure 5. The Dst index values from January 15 to February 14, 2016.

Not all of earthquake effects can be detected using this method, for example in the case of February 17, 2016 Halmahera earthquake with magnitude Mw 6.1, as shown in Figure 6. In this case, it was not detected any TEC anomalies in the ionosphere as earthquake effect even though its magnitude almost similar to the previous earthquakes. TEC anomalies before an earthquake can be classified as earthquake precursor. The failure of correlation techniques to detect the earthquake effect was not caused by the incompatibility of this method for detecting the ionospheric effect, but due to the temporal resolution of the TEC data used i.e. 1 (one) hours, so the earthquake effects in gravitational waves acoustic form with period less than half an hour will not be appear in the one hour resolution TEC data. The results of this study provide hope to strengthen the earthquake and tsunami early warning system using TEC GPS data. The method development of continuous TEC GPS observation derived from GPS observation network that already exists in Indonesia is needed to support earthquake effects (or tsunami) early warning systems.

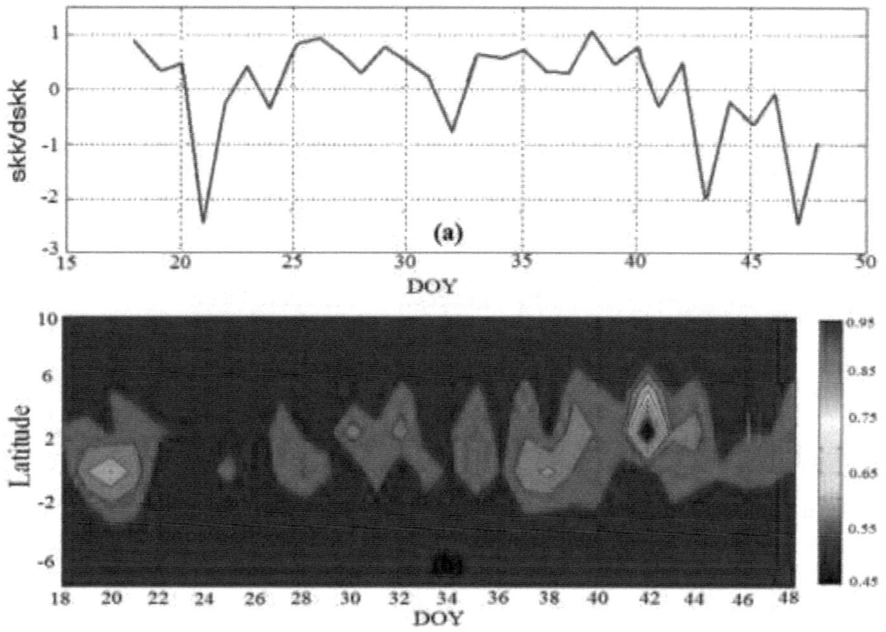

Figure 6. Skk/dskk values (a) and cross correlation of TEC data (b) from January 18 to February 17, 2016.

4. Conclusion

Based on Total Electron Content (TEC) data correlation, effects of The January 25, 2014 Kebumen earthquake and the February 12, 2016 Sumba earthquake could be detected in the ionosphere as TEC anomalous. However, not all earthquake effect could be detected in the ionosphere using this method. The failure of TEC data correlation method was due to poor temporal resolution of the TEC data. It was required the continuous monitoring of TEC GPS data to detect the earthquake effects in ionosphere.

References

[1] Lin J W 2015 Early warning of tsunami from seismo-ionospheric fluctuation after Japan's March 11, 2011, M=9.0 Tohoku earthquake using two-dimensional principal component analysis *Annals of Geophysics* **58**

[2] Khegai V V, Legen'ka A D and Kim V P 2013 Anomalous variations in the foF2 critical frequency Above Japan prior to the earthquake of March 9, 2011 *Geomagn. Aeronomy* **53** (4) 529-533.

[3] Jin S G, Jin R and Li J H 2014 Pattern and evolution of Seismo-ionospheric disturbances following the 2011 Tohoku earthquakes from GPS observations, *J. Geophys. Res. Space Physics* **119**, 7914-7927.

[4] Muslim B, Effendi J, Aldrian E, Fachrizal, Sunardi B, and Angga 2014 Pengembangan Sistem Monitoring Gelombang Ionosfer Terkait Gempa Bumi Menggunakan Data GPS (GPSIonoquake), *Prosiding Seminar Nasional Sains Atmosfer dan Antariksa* 12–29.

[5] Lognonné P, Artru J, Garcia R, Crespon F, Ducic V, Jeansou E, Occhipinti G, Helbert J, Moreaux G, Godet P E 2006 Groundbased GPS imaging of ionospheric postseismic signal *Planetary and Space Science* **54** 528–540.

[6] Hao, Y Q, Xiao Z, and Zhang D H 2012 Multi-Instrument Observation on Co-Seismic

Ionospheric Effects after Great Tohoku Earthquake *Journal of Geophysical Research* **117**.
[7] Heki K, Ionospheric electron enhancement preceding the 2011 Tohoku-Oki earthquake *Geophys. Res. Lett.* **38** L17312.
[8] Le H, Liu L, Liu J Y, Zhao B, Chen Y, Wan W 2013 The ionospheric anomalies prior to the M 9.0 Tohoku-Oki earthquake *Journal of Asian Earth Sciences* **62** 476–484.
[9] Lin J W 2015 Early warning from seismic ionospheric anomaly of the 24 May 2014, M = 6.4 Aegean-Sea earthquake: two-dimensional principal component analysis (2DPCA) *Geofísica internacional* **54**-2 191-198.
[10] Muslim B, Pengujian teknik korelasi untuk deteksi pengaruh aktivitas gempabumi pada ionosfer *Jurnal Sains Dirgantara* **12** 87-101.
[11] Sunardi B, and Sulastri 2016 Pemantauan anomali Total Electron Content (TEC) berkaitan dengan kejadian gempabumi di sekitar wilayah Jawa tahun 2015 *SPEKTRA: Jurnal Fisika dan Aplikasinya* **1** (2) 95-102.
[12] Sunardi B, Buldan M, Pakpahan S 2015 Anomali Total Electron Content (TEC) sebelum gempabumi kuat di Indonesia Tahun 2014, *Prosiding Seminar Nasional Jurusan Fisika FMIPA Unesa* Surabaya.
[13] Rothacher M, Beutler G, Brockmann E, Mervart L, Schaer S, Springer T A, Wild U, Wiget A, Boucher C, Seeger H 1996a Annual Report 1995 of the CODE Analysis Center of the IGS, Astronomical Institute, University of Berne.
[14] Schaer S 1999 Mapping and Prediction the Earth Ionosphere using the Global Positioning System Doctoral Thesis University of Bern.
[15] S. Schaer 1997 How to use CODE's Global Ionosphere Maps Astronomical Institute, University of Berne.
[16] Gonzalez W D 1994 What is a geomagnetic storm*? J. Geophys. Res.* **99** 5771.
[17] Gonzales W D, Tsurutani B T, Clua de Gonzales A L 1999 Interplanetary origin of magnetic storms *Space Sci. Rev.* **88**.
[18] Real-time (Quicklook) Dst index [cited 2016 July 30]; available from : http://wdc. kugi.kyoto-u.ac.jp/dst_real time/index.html.

Detecting overpressure using the Eaton and Equivalent Depth methods in Offshore Nova Scotia, Canada

Ernanda[1], A Q T Primasty[2] and K A Akbar[3]

Geophysics Sub-Department, Faculty of Mathematical and Science, Universitas Gadjah Mada

[1)] ernanda23@gmail.com, [2)] annequeentina@gmail.com, [3)] kevin.adelin.ab@gmail.com

Abstract. Overpressure is an abnormal high subsurface pressure of any fluids which exceeds the hydrostatic pressure of column of water or formation brine. In Offshore Nova Scotia Canada, the values and depth of overpressure zone are determined using the eaton and equivalent depth method, based on well data and the normal compaction trend analysis. Since equivalent depth method is using effective vertical stress principle and Eaton method considers physical property ratio (velocity). In this research, pressure evaluation only applicable on Penobscot L-30 well. An abnormal pressure is detected at depth 11804 feet as possibly overpressure zone, based on pressure gradient curve and calculation between the Eaton method (7241.3 psi) and Equivalent Depth method (6619.4 psi). Shales within Abenaki formation especially Baccaro Member is estimated as possible overpressure zone due to hydrocarbon generation mechanism.

1. Introduction

There are several aspects to do the geomechanics research such as overpressure, pore pressure and stress state relationship. Overpressure can be formed by several mechanisms including faulting, undercompaction, formation foreshortening, massive evaporite deposition, mineral phase change, repressuring, and hydrocarbon generation (Nguyen, 2013).

Pore pressure calculations is a function of distance. Equivalent depth, Eaton, and effective stress are several methods to determine pore pressure. In this research, we use equivalent depth and Eaton methods called as "vertical" and "horizontal" methods to identify the potential of abnormal pressure exist, according to many studies of overpressure in this location. The evaluation is limited by well data without seismic data as the method requires.

The location of research is in Penobscot area, Offshore Nova Scotia, Canada. The previous research indicates there are overpressured zone in this region. There are two wells that was drilled in Penobscot area, the first well is Penobscot L-30 with hydrocarbons were recovered by Repeat Formation Tester (RFT) and the second well is Penobscot B-41 (Clack, 1992). Different conditions with the Penobscot L-30 well, hydrocarbon in Penobscot B-41 were not encountered significantly. Both of well were drilled to 13950 feet for Penobscot L-30 and Penobscot B-41 to 11320 feet.

2. Geology Regional

The Scotian Basin is located on the northeastern flank of the Appalachian Orogen and covers an area of approximately 280,000 km^2 with a maximum sediment thickness of about 24 kilometers (Figure 1). The continental-size drainage system of the paleo-St. Lawrence River provided a continuous supply of sediments that accumulated in a number of complex, interconnected subbasins. Early synrift, carbonate margin, fluvial-deltaic-lacustrine and deep water depositional systems are all represented in the basin's stratigraphic succession (CNSOPB, 2013).

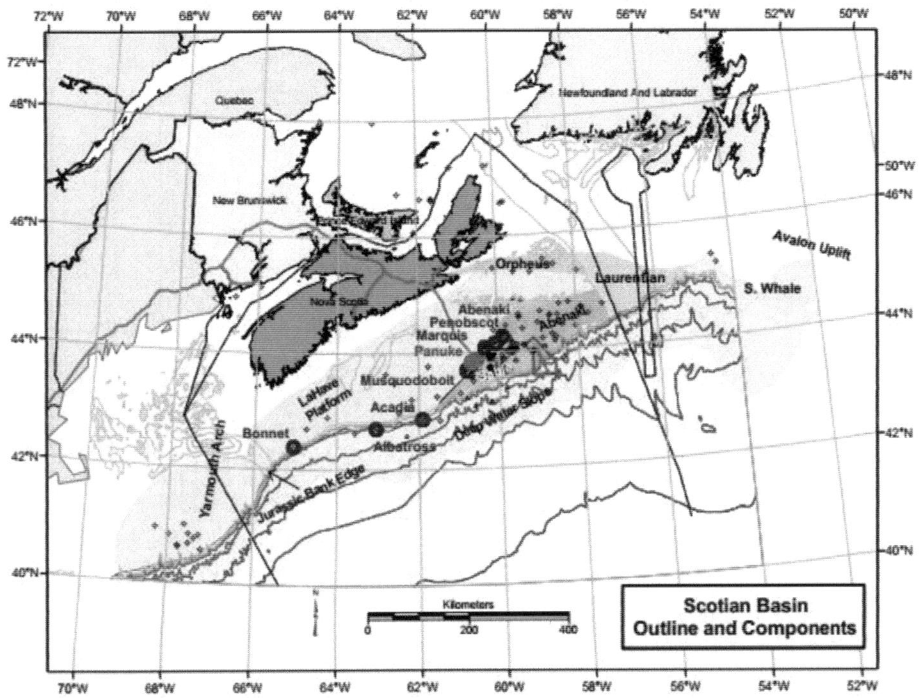

Figure 1. Research area located in Offshore Nova Scotia for Basin and Components (Penobscot L-30 and Penobscot B-41)

During the Middle Jurassic to Early Cretaceous, carbonate deposition dominated sedimentation around the entire North Atlantic margin, reflecting a period of continuous global sea level rise and gentle downwarping and thermal subsidence of rift-bounding continental crust (Kidston et al, 2005). Reef, platform and related facies were established along the margins of the Atlantic and Tethys Oceans under ideal conditions of broad, stable low relief and shallow shelves, warm oxygenated waters with limited oceanic circulation, limited background clastic sedimentation and low fluctuations of oxygen and nutrient levels.

2.1. Stratigraphy

Abenaki formation divided into 4 members (Figure 2) : Scaterie, Misain, Baccaro, and Artimon member (Wade et al, 1995). *Scaterie Member* is the most areally extensive sequence of the Abenaki Formation. It is a seaward-thickening wedge of predominantly platformal oolitic lime grainstones and packstones, with at least four deepening-upward transgressive sequences recognized. *Misaine Member* is the only clastic component of the Abenaki Formation and this neritic facies is representative of the Callovian global transgressive event in the Sable Basin. The Misaine is composed predominantly of dark grey, slightly calcareous shales with minor beds of pelleted and laminated lime mudstones. *Baccaro Member* is the thickest and best developed carbonate unit of the Abenaki Formation. The dominant lithology is limestone with minor shale and sand intervals representing the reworked remnants of lowstand events. *Artimon Member* is the youngest sequence of the Abenaki Formation which thinnest about 30-115 metres with a very modest and patchy areal distribution limited to the eastern part of Abenaki platform margin edge. Lithologically, the Artimon is composed of argillaceous, cherty limestones representing thrombolitic sponge and stromatoporoid mound deposition, with occasional interbedded calcareous shales (Wade et al, 1995).

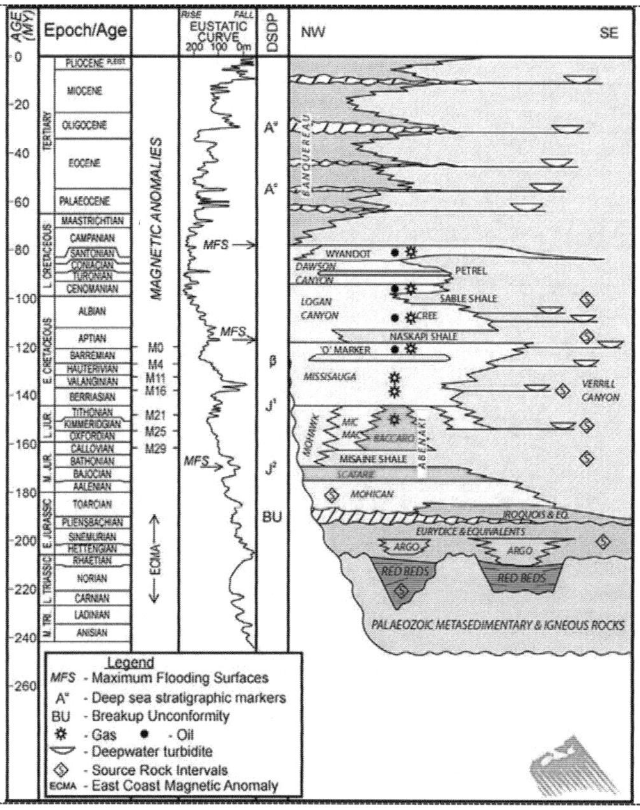

Figure 2. Scotian Basin Generelized Statigraphic Chart

3. Methodology

There are some pore pressure calculation methods are available, which differ in the way the effective stress is derived from data sets as logs and seismic. In this case, researcher estimating pore pressure in shales using the Eaton and Equivalent Depth methods based on well logs.

3.1. Equivalent Depth

Equivalent depth method is based on the reasonable assumption that formations having same interval velocities would have the same effective vertical stress irrespective of the depth (Ham, 1966). The initial method is defined by relationship between pore pressure, effective stress, and the overburden stress explained by the Terzaghi relationship.

When the data set deviates from the normal compaction trends, the approximation of the pore pressure at any depth can be calculated if the measured value isn't on normal compaction trend because measured values itself related to effective stress (Figure 3) and also represent depositional sequence. Pore pressure is calculated using the following equation 1:

$$Pza = Pzn + (Sza - Szn) \qquad (1)$$

Where,
Pza, Pzn = Pore pressure at za point (observed) and pore pressure at zn point (compacted normally)
Sza, Szn = Stress at za point (observed) and pore pressure at zn point (compacted normally)

IOP Conf. Series: Earth and Environmental Science **132** (2018) 012016 doi:10.1088/1755-1315/132/1/012016

The only unique assumption required by equivalent depth methods is that effective stress is a linear function of depth.

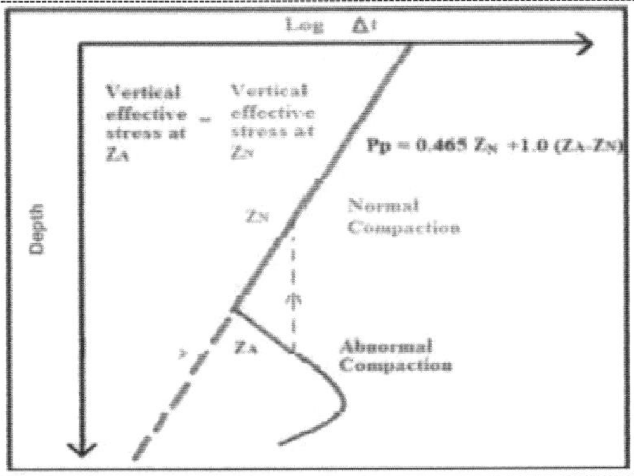

Figure 3. Graphical procedure of Equivalent Depth's method

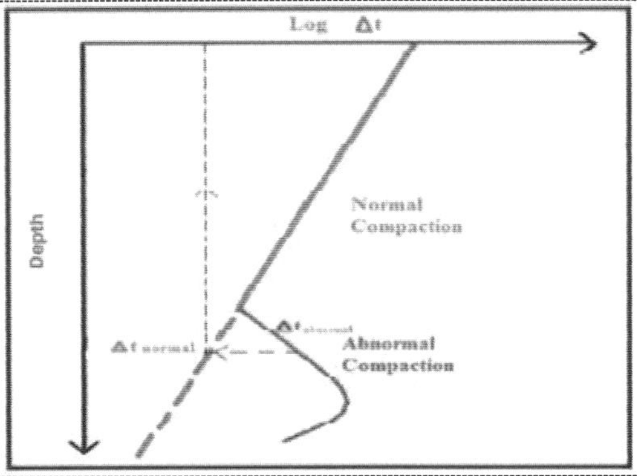

Figure 4. Graphical procedure of Eaton's method

3.2. Eaton

The Eaton method has been described as a "horizontal" pressure method because it compares an in-situ physical property to a "normally-compacted" equivalent physical property at the same depth (Figure 4). Eaton's method approximates the effective vertical stress with ratio of sonic log velocities or resistivity values. This method is based on the principle that the relationship between this ratio and pore pressure depends on changes in the vertical stress gradient (Eaton, 1972; Mouchet and Mitchell, 1989) in Ruth, 2002. The estimation value is calculated by following equation :

$$Pp = \sigma_{ob} - (\sigma_{ob} - Pp_{normal})(\frac{V\ observed}{V\ normal})^x \quad (2)$$

Where,

P_p = Pore pressure (psi)

σ_{ob} = Overburden pressure (psi)

$P_{p\ normal}$ = Normal Pressure (psi)

$V_{observed}$, V_{normal} = velocity value at the depth of observed zone and at normal pressure (feet/s)

The accuracy of equivalent depth method and the Eaton method is dependent on the extent to which the acoustic travel time anomalies are related to overpressure via changes in porosity.

4. Discussions

Eaton's and Equivalent depth methods require the use of a NCT (normal compaction trend), which means the trend of any physical property (example : travel time, resistivity, density, etc) is plotted as a function of depth through the normally pressured strata. According Figure 5 and 6, velocity data is plotted from both logs (Penobscot L-30 and Penobscot B-41) that has been filtered before, on gamma ray value using cut off 70 API. This analysis only evaluating pore pressure in shales due to high possibility become overpressured.

The NCT's analyze is used to identify abnormal pressure possibility by the existence of deflection data trend. Deflection exist at well Penobscot L-30, it's more represent possibility of overpressure and neither at well Penobscot B-41. Due to only Penobscot L-30 that has deflection data trend, so for next calculation only use Well Penobscot L-30. The NCT value in well Penobscot L-30 is y = 1.1707x – 5783.3. Overpressure possibility point is picked from software (Figure 7) to get accurate value yield top at depth 11804 feet as interest zone.

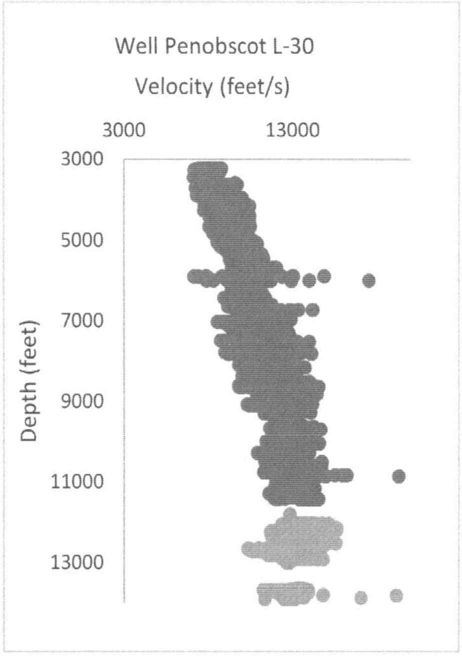

Figure 5. Depth versus Velocity curve on Penobscot L-30 (*X* = 734338,9 *Y* = 4893837,9) .

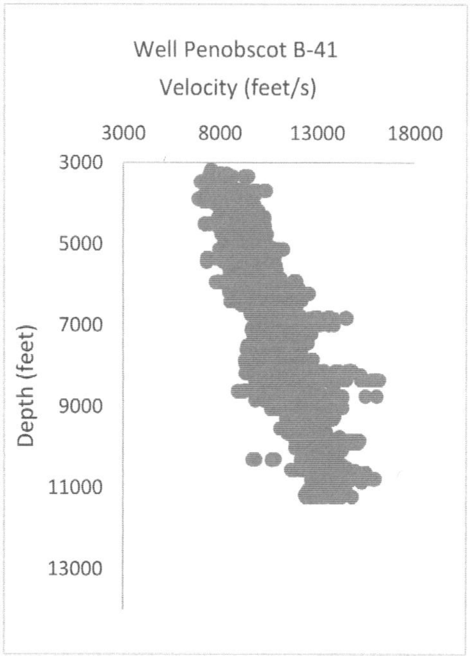

Figure 6. Depth versus Velocity curve on Penobscot B-41 (*X* = 731133,46 *Y* = 4894307,8)

In normal pressurised formation, porosity will decrease with depth as the fluid expelled out of the pore due to increasing overburden weight, it maintains the effective communication of pore fluids with surface. So at any depth pore pressure is simply the same as the hydrostatic pressure (1.03 g/cm^3 or 0.433 psi/ft) of the water column (Saran, 2011).

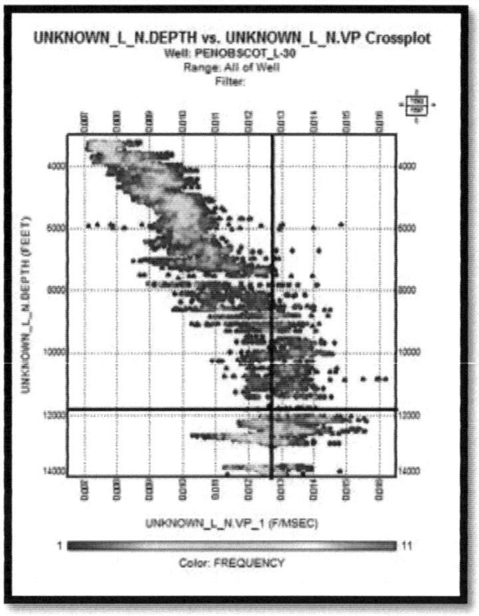

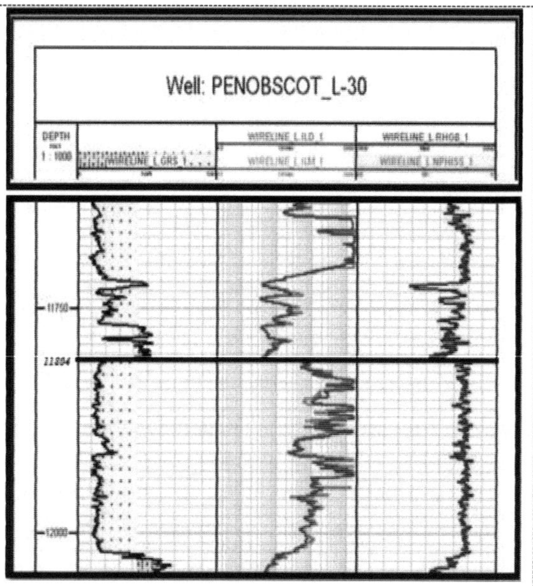

Figure 7. Picking depth point of overpressure on depth vs velocity well Penobscot L-30.

Figure 8. Predicted top overpressure zone in well log section at 11804 feet (Penobscot L-30).

4.1. Method Calculation

The main evaluation is method calculation, for Eaton method we have to calibrate the x value (exponent coefficient) using equation from graphic relation between travel time and effective stress shown Figure 9, it represents the value of Eaton exponent coefficient seems like $\Delta t = C\,\sigma^{-1/x}$. $y = 3082.8x^{-0.43}$ equation provide eaton exponent coefficient (x = 2.32). Although there are formation test data that measure pressure at some interval depth, but there is no fix relation between pressure test and calculation result. Eaton formula (Equation 2) assume that ratio (velocity or resistivity) influence the effective stress value, but in this case we only use velocity ratio, then the result of pore pressure is 7241.3 psi.

Equivalent depth is kind of pressure estimation method that simple enough for calculating using effective vertical stress principle, yield pore pressure value of zone interest 6619.4 psi. Generally both methods has each specific quality that related to another seems like Eaton method has been described as a "horizontal" method and the equivalent depth method has been described as "vertical" pressure method. The Eaton and Equivalent depth methods calculation must yield similar pore pressure estimation, but in this case result different value. It might be caused by the different approximation of both method.

The abnormal evaluation at 11804 ft describe pressure value 6619.4 psi (equivalent depth method) and 7241.3 psi (Eaton method) compare with normal pressure at the same depth is 5111.3 psi (Figure 10). It means there is possibility overpressure zone in Abenaki formation especially Baccaro Member. As we know Baccaro Member is part of carbonat platform from Abenaki and proven to produce hydrocarbon in this field.

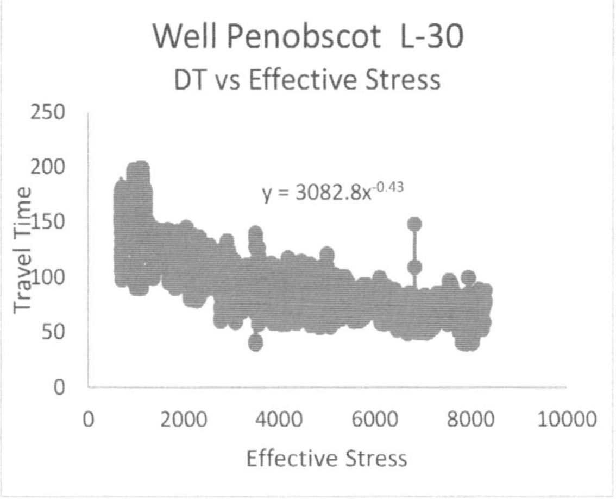

Figure 9. Equation from graphic relation between travel time and effective stress (Penobscot L-30) represent exponent coefficient value.

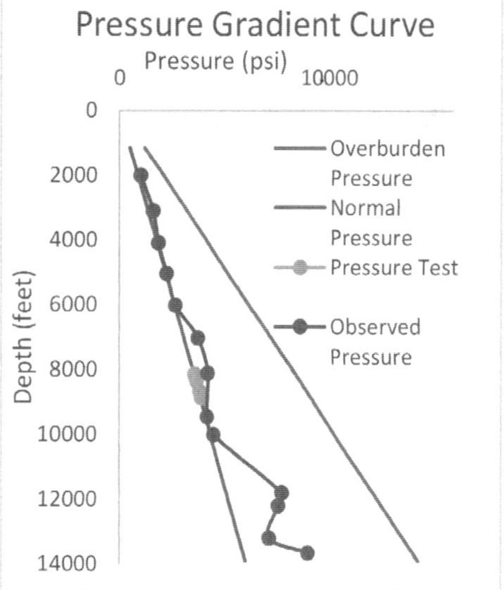

Figure 10. Pressure gradient curve.

Potential causes of overpressure in Nova Scotia are hydrocarbon generation and undercompaction (Dickson,2012). In this case, its tend to be caused by hydrocarbon generation occurs when pressure increase as a result of the addition of hydrocarbon fluids and gasses to a confined compartment and also types of hydrocarbon itself contribute to overpressure level. Due to lack of data sets and analysis makes detecting overpressure is too difficult to get the valid result, moreover this evaluation only use single well from two wells provided. And there is ambiguiity about detect abnormal pressure zone in Penobscot L-30 due to th deflection data trend caused by truly overpressure zone or it just caused by gas content within porous carbonate reservoir.

Overpressure analysis is important and needed for helping identify potential and produce resources. Moreover to increase safety for personnel during drilling and protect the environment from complication (i.e. well blowout due to uncontrolled pressure). It will more challanging in Offshore drilling include condition and cost.

5. Conclusions

Based on analysis before (Eaton and Equivalent Depth method), pressure evaluation only applicable on Penobsot L-30 well. Principally pressure estimation using the Eaton method is better than equivalent depth method caused considers physical property ratio (velocity) and has flexible exponent coefficient for specific formation character. An abnormal pressure is detected at depth 11804 feet as possibly overpressure zone, based on pressure gradient curve and calculation between the Eaton method (7241.3 psi) and Equivalent Depth method (6619.4 psi). Shales within Abenaki formation especially Baccaro Member is estimated as possibile overpressure zone due to hydrocarbon generation mechanism.

6. References

[1] Babu S and Sircar, A. 2011. A comparative study of predicted and actual pore pressure in Tripura, India. Journal of Petroleum Technology and Alternative Fuels **Vol. 2(9)** 150-160.

[2] Clack W J F and Crane J D T 1992 *Penobscot Prospect :Geological Evaluation and Oil Reserve Estimates Nova Scotia Resources (Ventures) Ltd*

[3] CNSOPB 2013 "*Regional Geology, Canada- Nova Scotia Petroleum Board"*

[3] Dickson C H, Silva R L, Grant W and Neil W 2014 *Overpressure, Seal Integrity, and Reservoir Connectivity in the Scotian Basin. GeoConvention Focus.*

[4] Ham Harold 1966 *A method of estimating formation pressures from gulf Coast Well Logs. Gulf Coast Association of Geological Societies Transactions* **Vol. 16** Pages 185-197

[5] Kidston A G, Brown D E, Smith B M and Altheim B 2005 *The Upper Jurassic Abenaki Formation Offshore Nova Scotia: A Seismic and Geologic Perspective Canada-Nova Scotia Offshore Petroleum Board* Halifax 165p

[6] Nguyen, T. 2013. Well Design: Formation Pressure. New Mexico Technical University

[7] Ruth P J V, Hillis R R and Swarbrick R E 2002 *Detecting Overpressure Using Porosity-Based Techniques in The Carnarvon Basin Australia APPEA Journal* 559-569.

[8] Wach, D G and Dickson C *Reservoir Connectivity and Overpressure - Sable and Porcupine Basins Petroleum Inffrastructure Programme*

[9] Wade J A, MacLean B C and Williams G L 1995 *Mesozoic and Cenozoic stratigraphy, eastern Scotian Shelf: new interpretations Canadian Journal of Earth Sciences* **Vol.32** no 9 pp 1462-1473

Acknowledgments

This team project as an undergraduate student of geoscience would like to acknowledge for providing data sets and informastion from the following : Nova Scotia Departement of Energy and Canada Nova Scotia Offshore Petroleum Broad ; dGB Earth Science B.V. c. The authors are thankful to Drs. Imam Suyanto, M.Si. for giving us permission to join PIT HAGI 2016.

IOP Conf. Series: Earth and Environmental Science **132** (2018) 012017 doi:10.1088/1755-1315/132/1/012017

Distribution Analysis of Sulphide Mineral (Pyrite) Using Induced Polarization Method in Libureng, Bone, South Sulawesi

Muhammad Altin Massinai[1]*, Lantu[1], Hidayat Latuconsina[2], Muhammad Fawzy Ismullah M[1]

[1]Geophysics Dept., Hasanuddin University, Indonesia
[2]Assistant Geophysics PT. Vale Indonesia, Tbk., Indonesia

*altin@science.unhas.ac.id

Abstract. Sulphide minerals are any member of a group of compounds of sulphur with one or more metals. Some of these sulphide minerals are economically important. This study used induced polarization method to identify distribution and mineralized zone of sulphide mineral (Pyrite), in Libureng, Bone Regency, South Sulawesi. The data processing yielded resistivity value, percent frequency effect (PPE) value, and metal factor (MF) value which were then used to produce 2-D and 3-D section model. Based on the data interpretation, an anomaly linked to pyrite deposits was seen in four trajectories with resistivity value <= 50, PFE => 3%, and MF >= 150, deposited in hydrothermal alteration zone, sericite zone.

1. Introduction

Sulphide minerals are a group of compounds of sulphur with certain substances such as metal, silver, copper, lead, zinc, and mercury. Some of these minerals are economically important in the form of ore, such as Pyrite, Chalcocite, Galena, and Sphalerite [1, 2]. The method able to detect the existence of subsurface sulphide mineral distribution is Geoelectric method. This method studies the nature of the flow of electricity in the earth and how to detect it in the earth's surface [3]. Geoelectric method consists of several methods such as method of geoelectric resistivity, IP (Polarization Index), self potential (Self Potential) and others. Induced method Polarization (IP) is a geoelectric method that is widely used in the exploration of base metals (base-metal) because of the phenomenon of polarization in a medium rock [4]. This phenomenon of polarization signifies metal content below the surface that are not detected properly when using the resistivity method [5].

Measurement and analysis of data in the form of a cross-induced polarization of 2D and 3D distribution of sulphide minerals. The data used is primary data obtained from research sites in Libureng, Bone, South Sulawesi. The main objective of this research is to create 2D and 3D cross sections based resistivity data, the data PFE and MF to identify the distribution of sulphide minerals (pyrite) [6].

Induced polarization measurements are usually made at or below a frequency of 10 Hz in order to remain in the area of non-induction. Two measurements are generally carried out namely Percent

Frequency Effect (PFE) and Metal Factor (MF). PFE is defined as the ratio between the difference in voltage at low frequencies with a voltage at high frequency is measured on the electrode voltage. FE value or PFE is a response from the existence of minerals contained in rock pores. The higher the concentration of minerals in rocks the greater the value of PFE and vice versa [7, 8, 9]. PFE value can be obtained using the following equation:

$$PFE = 100\frac{(\rho_2 - \rho_1)}{\rho_1} \quad (1)$$

Where
PFE : Percent Frequency Effect (%)
ρ_1: apparent resistivity measured at 10 Hz
ρ_2: apparent resistivity measured at 5 Hz

The following table shows PFE value as a reference in this study.

TABLE 1. PFE value for several rocks [5 & 10].

Material	PFE
Sulphide-containing rocks	>10
Cu ore porphyry (2-10) % Sulfida	5 – 10
Rocks with sulphide content 2 %	2 – 5
Volcanic Tuffs	2 – 4
Sandstone, Siltstone	1 – 3
Basalt	1 – 2
Granite	0.1 – 0.5

Metal Factor is defined as the amount of sulphide minerals contained in rocks, in which the amount depends on the value FE [8]. Metal factor can be formulated as follows:

$$MF = 2\pi \, X10^5 \frac{(\rho_2 - \rho_1)}{\rho_2 \rho_1} \quad (2)$$

Where
MF: Metal Factor
ρ_1: apparent resistivity measured at 10 Hz
ρ_2: apparent resistivity measured at 5 Hz

The following table shows MF value of some types of rocks

TABLE 2. MF value for several rocks [11].

Material	PFE
Massive Sulphides	10.000
Fracture – Filling Sulphides	1000 – 10.000
Massive Magnetite	3 – 3000
Porphyry Copper	30 – 1.500

Shale Sulphides	3 – 300
Clay	1 – 300
Sandstone 1 – 2% Sulphides	2 – 200
Fincly Dissem Sulphides	10 – 100
Tuffs	1 – 100
Graphite Sandstone and Limestone	4 – 60
Gravels	0 – 200
Alluvium	0 – 200
Precambrian Gneises	10 – 100
Granites, Monzonites, Diorites	0 – 60
Various Vulvanics	0 – 80

2. Methodology

2.1 Location

This research was conducted in Libureng, Bone Regency, South Sulawesi. The following picture shows a map of the location of the study [12].

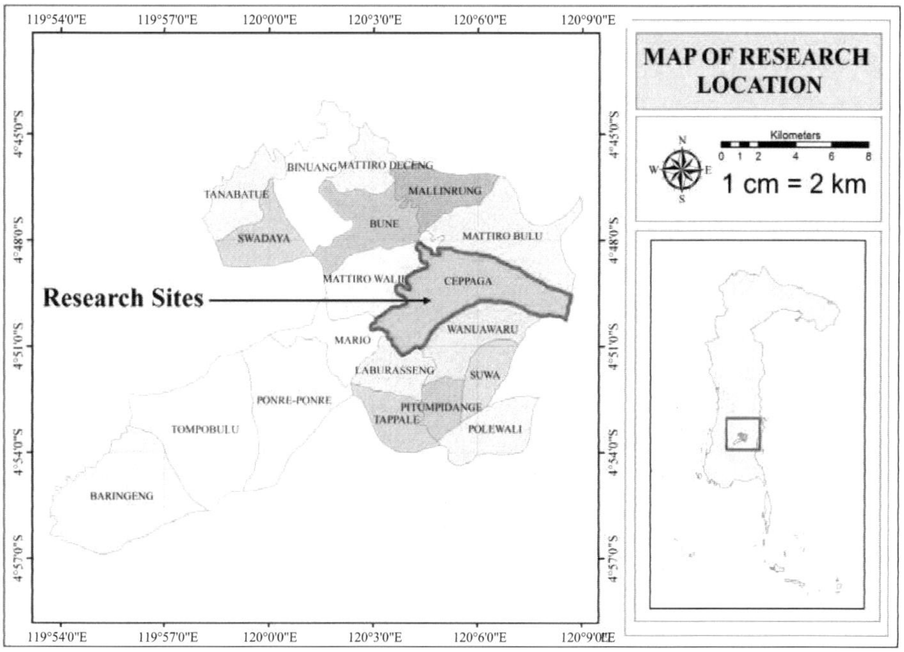

Figure 1. Location of The Study

2.2 Data Acquisition

- Data from field measurements processed using equation (1) for the PFE value and equation, (2) to obtain the value of M.
- Apparent resistivity, PFE and MF plotted using software for 2D section,
- 3D modelling done using inversion and pseudo

3. Analysis, Interpretation and Discussion

The process of analysis and interpretation of each track was based on the apparent resistivity values, PFE and MF supported by surface geological data. The difference of the apparent resistivity values, PFE and MF is proportional to the concentration of sulphide minerals contained in a rock [5, 13]. The

IOP Conf. Series: Earth and Environmental Science **132** (2018) 012017 doi:10.1088/1755-1315/132/1/012017

mineralized sulphide area was then determined by comparing the qualitative anomalies seen in the 2D cross section with the possibilities that exist in Table 3. Based on the table, it is generally assumed that sulphide minerals present in a rock if it has [14, 15]:

- Low resistivity value, for sulphide minerals have electrically conductive properties (conductive),
- PFE prices high, because the larger the presentation PFE value, the higher the concentration of sulphide minerals in a rock,
- High MF, because the higher the MF value, the greater the metal content in a rock.

Data processing produced resistivity values, PFE and MF which were then used to create 2D subsurface tomographic cross section in 4 lines. These 2D cross sections only show areas with resistivity values ≤50 Ωm, PFE values ≥3% and MF values ≥150 mhos / m which indicated the presence of pyrite. This area is characterized by the presence of an overlay among the three physical parameters. The four lines of each measurement have the same value for resistivity, PFE and MF. The results of the analysis of each line can be seen in the following presentation:

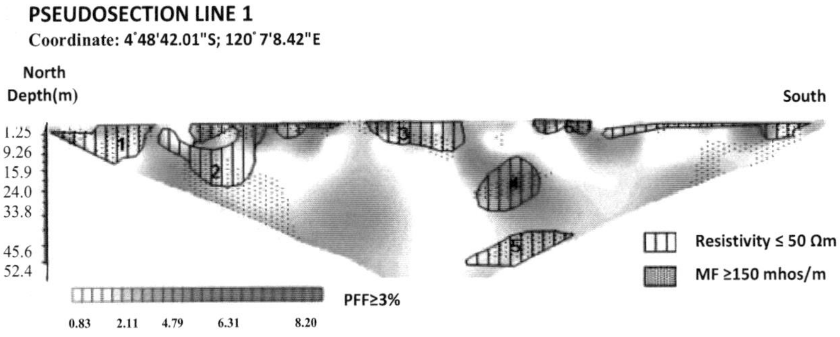

(a)

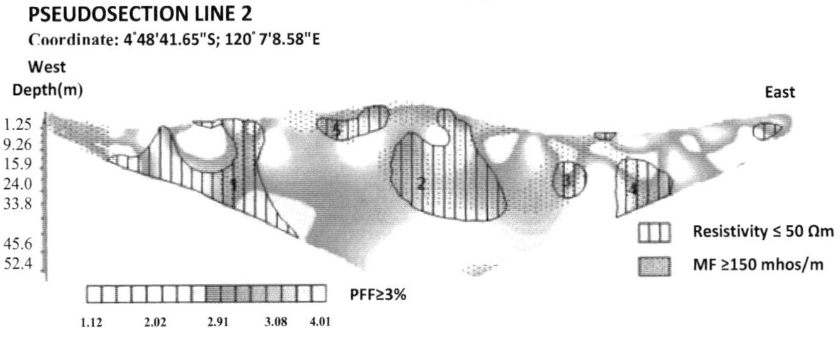

(b)

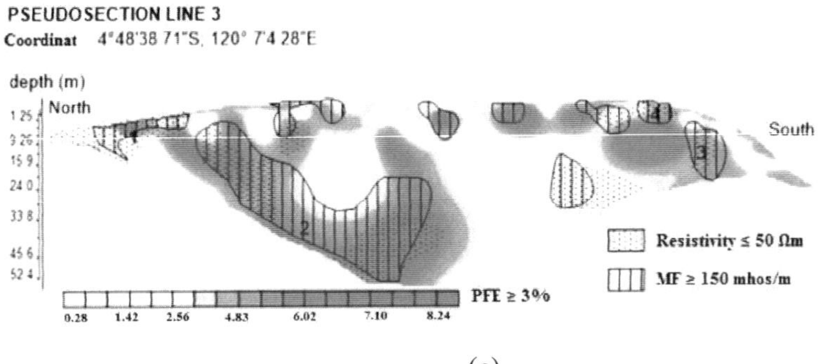

(c)

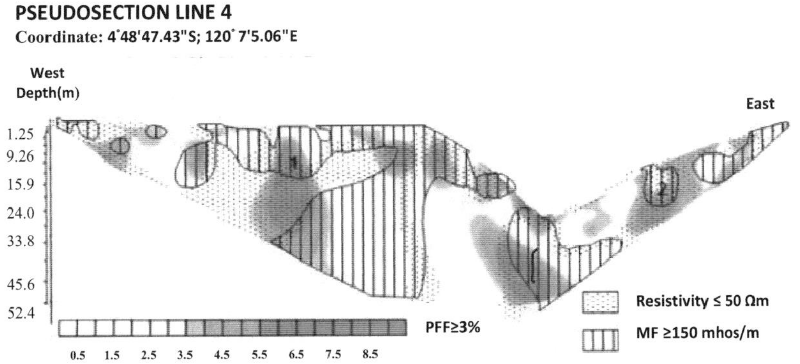

(d)

Figure 2. (a) Overlay Between Resistivity Section, PFE and MF For Track 1, (b) Overlay Between Resistivity Section, PFE and MF For Line 2, (c) Overlay Between Resistivity Section, PFE and MF For Line 3, (d) Overlay Between Resistivity Section, PFE and MF For Line 4

Figure 2 each has line has different direction, 310° NW, 220° SW, 310° NW, and 40° SW, respectively.

The figures are an overlay between apparent resistivity cross-section, PFE and MF on each trajectory measurements. The figures show that there are six locations that showed the prospect of sulphide minerals. Layers of sulphide minerals have 0.01-100 Ωm of resistivity values, 2-5% of PFE and 3-300 mhos/m of MF, bedrock layer has > 300 Ωm of resistivity value. In each image can be seen blue-stripped polygons which are the distribution of resistivity values with values ≤50 Ωm. Orange polygon is the distribution of sulphide minerals with PFE values ≥3% and polygons with black dots is the distribution of sulphide minerals with ≥150 mhos of MF / m.

The indicated area in the research location is a zone of sericite alteration. The main characterization of this zone is the emergence of intensive mineral sericite and quartz mineral alteration (secondary). This alteration zone is characterized by the emergence of minerals in the form of clay minerals, whereas secondary biotite and chlorite found in very small amounts. Sulphide minerals in this zone are dominated by pyrite whose content is growing towards the outside of this zone. The most prospective continuity of the mineralized zone is on line 1 and at the intersection of line 1 and line 2, also at the intersection of the line 2-3 and line 3-4.

Based on the analysis above, a 3D model was then made to map the prospect area of sulphide mineralization in the area, according to resistivity values anomaly, the value of PFE and the MF as shown in Figure 3. In this figure, only the area with resistivity values ≤50 Ωm, PFE values ≥3% and MF values ≥150 mhos / m is shown, which is interpreted as a zone of sulphide mineralization prospect in the study sites.

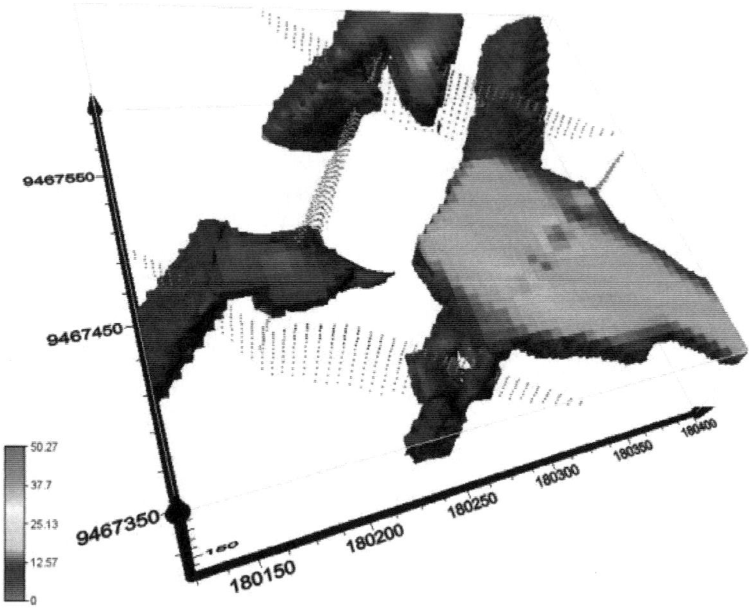

Figure 3. 3D Model Was Then Made To Map The Prospect Area of Sulphide Mineralization In The Area

Conclusion

The conclusions drawn from this study are as follow:

- The apparent resistivity value resulted is 0.28 – 14036, range of PFE value is 0.12 – 8.5 and MF is 20.3 – 1336.
- The 2D section of the apparent resistivity, PFE and MF indicates prospective sulphide mineralization i.e pyrite.
- Sulphide minerals in this zone are dominated by pyrite whose content is growing towards the outside of this zone. The most prospective continuity of the mineralized zone is on line 1 and at the intersection of line 1 and line 2, also at the intersection of the line 2-3 and line 3-4. This condition signed based on the 3D modelling of apparent resistivity, PFE and MF, with mineral distribution direction at 310° NW indicating hydrothermal alteration zone of sericite.
- From the test results of sulphide mineral sample containing pyrite using XRF analyser, the elements contained are SO3 72.13%, 27.16% Fe2O3, P2O5 0:51%, 0.043% MnO, ZnO 0.0014%.

Acknowledgement

The authors wish to express their gratitude to the Faculty of Mathematics and Natural Sciences, Hasanuddin University, for the permission to publish this article, which is part of the thesis carried out by the second author.

References

[1] Richard E C 1983 *American Mineralogist* **68** 245-254.
[2] Dutrizac J E 1976 *Canadian Mineralogist* **14** 172-181.
[3] Reynolds J M 1997 *An Introduction to Applied and Environmental Geophysics,* John Wiley and Sons Ltd 796 Chichester.
[4] Burtman V and Zhdanov M S 2015 *Resource Efficient Technologies* 34-48.
[5] Milsom J 2003 *Field Geophysics, The Geological Field Guide Series.* University College. 232p London.

[6] Park J O, You J Y and Kim H J 2009 *Geophysics and Engineering* **6** 73-78.
[7] Kearey P, Brooks P and Hill I 2002 *An Introduction to Geophysical Exploration* Blackwell Science Ltd. 281 London.
[8] Robb L 2005 *Introduction to Ore Forming Processes* Blackwell Science Ltd 386 Oxford.
[9] Yuval and Ordenburg D W 1996 *Applied Geophysics* **34** 187-198.
[10] White N C and Hedenquist J W 1995 *SEG Newsletter* **23** 1, 9-13.
[11] Telford W M, Geldart L P and Sheriff R E 1990 *Applied Geophysics* Cambridge Univ Press Cambridge.
[12] van Leeuwen T M 1981 *Geological Research and Development Centre Special Publicatian* **2** 277-304.
[13] Misra K C 2000 *Understanding Mineral Deposit*. Kluwer Academic Publishers. 845p Knoxville.
[14] Corbett G J and Leach T M 1998 *Economic Geology Special Publication* **6** 238.
[15] Wang Z 2001 *Geophysics* **66** 398-412.

IOP Conf. Series: Earth and Environmental Science **132** (2018) 012018 doi:10.1088/1755-1315/132/1/012018

Hydrocarbon Reservoir Identification in Volcanic Zone by using Magnetotelluric and Geochemistry Information

S I Firda*[1], A N Permadi[1], Supriyanto[1], B N Suwardi[2]

[1]Physics Department, University of Indonesia, Depok, West Java, 16424, Indonesia
[2]PT. Pertamina, Kwarnas Building, 13th Floor, Jakarta, 10110, Indonesia

*syaraphysics@gmail.com

Abstract. The resistivity of Magnetotelluric (MT) data show the resistivity mapping in the volcanic reservoir zone and the geochemistry information for confirm the reservoir and source rock formation. In this research, we used 132 data points divided with two line at exploration area. We used several steps to make the resistivity mapping. There are time series correction, crosspower correction, then inversion of Magnetotelluric (MT) data. Line-2 and line-3 show anomaly geological condition with Gabon fault. The geology structure from the resistivity mapping show the fault and the geological formation with the geological rock data mapping distribution. The geochemistry information show the maturity of source rock formation. According to core sample analysis information, we get the visual porosity for reservoir rock formation in several geological structure. Based on that, we make the geological modelling where the potential reservoir and the source rock around our interest area.

1. Introduction

Early efforts realized that MT was a potential tool for imaging conductive sedimentary rocks beneath thrust sheets of more resistive rocks such as volcanics overthrust (Anderson and Pelton, 1985). Several previous studies show how the effective using MT in clastic sedimentary rocks. Such as, the case study in Papua New Guinea by Christophersen (1991) that have a thick layer of clastic sedimentary rocks beneath 1-2 km of heavily karsted limestone. In volcanic zone, igneous rock that created by the solidification of magma have been ignored because of perceived lack of reservoir quality (Farooqui MY, et all, 2009). However, igneous rock can develop porosity and permeability (Srugoa P and Rubinstein P, 2007) by the activity that can influence every aspects of petroleum system, providing source rock , affecting fluid maturation and creating migration pathways, traps, reservoirs and seals (Schutter SR in Petford N and McCaffrey, 2003).

Source rocks results from a convergence of physical, biochemical and geologic processes that culminate in the formation of fine-grained sedimentary rocks containing carbon and hydrogen-rich organic matter (McCarthy K et all, 2011). Most hydrocarbons found in volcanic rocks come from sedimentary source rock, some volcanic rocks are also source rocks. Vegetation entrained in ash flows may contain enough water to protect it from the heat of emplacement (Farooqui MY, et all, 2009). The organic matter of it slowly cooks as pressure and temperature increase with the burial depth. Given sufficient heat, pressure and time and the organic matter contained within transform into kerogen. (McCarthy K. et all, 2011).

Some petroleum compounds within source rocks are released at temperature lower than those needed to break down the Kerogen (McCarthy K. et all, 2011). By monitoring the compounds of organic material, geochemical testing information of outcrop samples, formation cuttings, sidewall cores and conventional core can determine the amount of generated petroleum relative to a rock's total potential such as the type and thermal maturity of organic matter present in the rock. This study will show the source rock and reservoir potential from the geochemical information and apply it to magnetotelluric method by combine the 2-D MT modelling and inversion with geological structural data.

2. Materials and methods

In this research, we used 132 MT data points divided with two line at exploration area. There are Line-2 with 70 data points and line-3 with 62 data points. Each data have one kilometer interval. Line-2 is parallel with line-3 from North-East to South-West.

The geochemical information is from the previous study that Pertamina did. There are including kerogen types index, kerogen composition index and the maturity index. The kerogen types index is from TOC versus S_2. The kerogen composition index is from the fluorescene amorphous and lipthine and hydrogen index versus Oxygen Index that shows in modified of van krevelen diagram.

There are some steps for doing MT data processing. First, we used time series correction to reduce noise signal with SSMT2000. Second, we used crosspower correction to make each data points become smooth data by MT Editor and show the correct trends of the structure. After that, we used inversion of Magnetotelluric (MT) data that had been corrected to make a resistivity mapping and modelling by using Winglink.

We make an interpretation of geological structure by combining it with additional data, such as geological map data. The geological map can describe the structure and the formation information. Meanwhile, the source rock potential from the geochemical information and hydrocarbon reservoir potential from the core data sample information will describe the visual porosity in several formation.

3. Results and discussion

According to the previous study that Pertamina did, the geochemical data information show the TOC results of Ciletuh formation. Total organic carbon (TOC, wt. %) describes the quantity of organic carbon in the rock sample and includes both kerogen and bitumen. (Peters and Cassa, 1994). The kerogen types index show (figure 1) the Ciletuh formation is in the type III kerogen that derived primarily from terrigenous plant debris, which has been deposited in shallow to deep marine environtments.

IOP Conf. Series: Earth and Environmental Science **132** (2018) 012018 doi:10.1088/1755-1315/132/1/012018

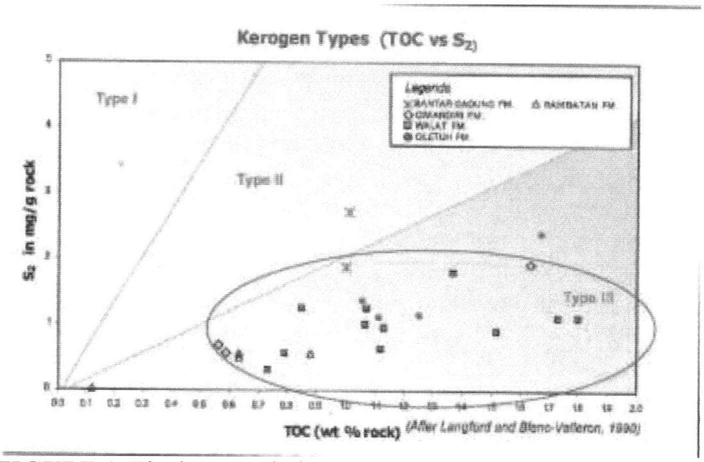

FIGURE 1. The kerogen index; TOC versus S_2 (Pertamina, 2007)

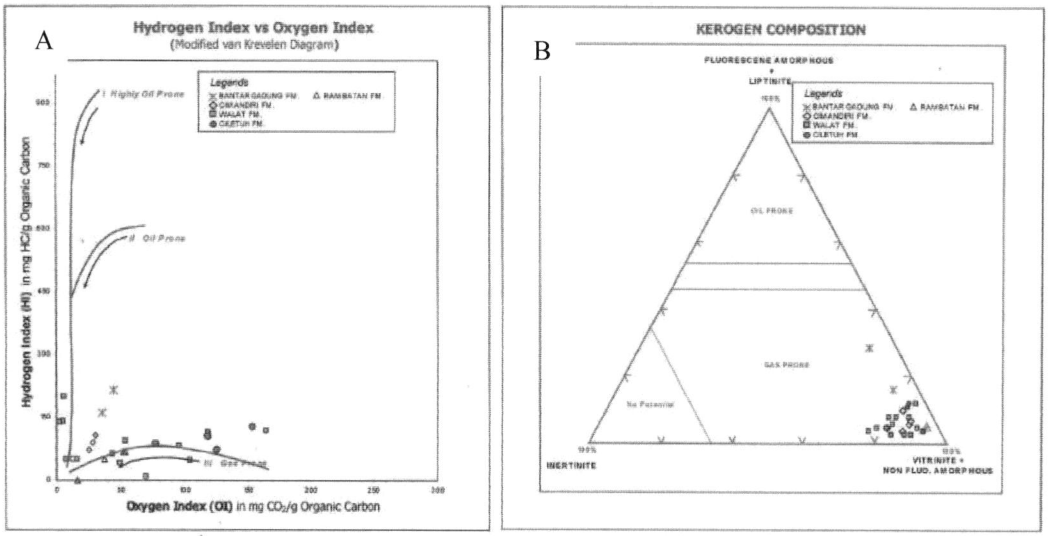

FIGURE 2. A. Hydrogen Index versus Oxygen Index (Pertamina, 2007), B. Kerogen composition (Pertamina, 2007)

TOC indicates the quantity, but not the quality, of the organic matter. Meanwhile, S2, corresponds to the hydrocarbons that evolve from the sample during the second programmed heating stage of pyrolysis. These hydrocarbons results from the cracking of heavy hydrocarbon and from the thermal breakdown of kerogen. S2 represent milligrams of residual hydrocarbon in one gram of rock, thus indicating the potential amount of hydrocarbon that the source rock might still produce if thermal maturation continue (McCarthy K. et all, 2011). Figure 2 show TOC versus S2 which have Type III kerogen that contain the residual hydrocarbon. The type III kerogen has lower hydrogen and higher oxygen content (figure 2) than types I or II; consequently, it tend to generate dry gas (figure 3).

In the other hand, the MT data inversion show the resistivity variation of the formation structure. Based on that, we combine it with the geological section map and show the model that represent several main formation. The inversion and interpretation of the Line-2 show (figure 4) the Gabon Faults that separated Ciletuh Formation and Penosogan Formation. The interpretation model of line-3 (Figure 5) show the Gabon Faults with the intrusion that separated Ciletuh Formation and Penosogan

Formation. This intrusion can make an effect for the maturity of the source rock in Ciletuh Formation. The maturity of the source rock will increase by the time and flowing to the potential reservoir area such as in the Jampang Formation (Tomj) that have sandstone, limestone and tuff in the syncline area rounded with anticline of the Ciletuh formation.

Based on the results from the inversion of MT data and the additional data for support the interpretation, we can see the indicate of the hydrocarbon reservoir in some enable geological area. S for Source rock and R is potential Reservoir. According to geochemical information and the geological structures, the source rock is around the Ciletuh Formation and the potential reservoirs divided with several formation of sedimentary rock. There are Jampang Formation (Tomj), Bentang Formation (Tmb) and Pamutuan Formation (Tmpa). According to the colors of those figures, the colors around green and blue representative of sedimentary rock that have potential reservoir with the resistivty value around 170-370 Rho (Ohm.m).

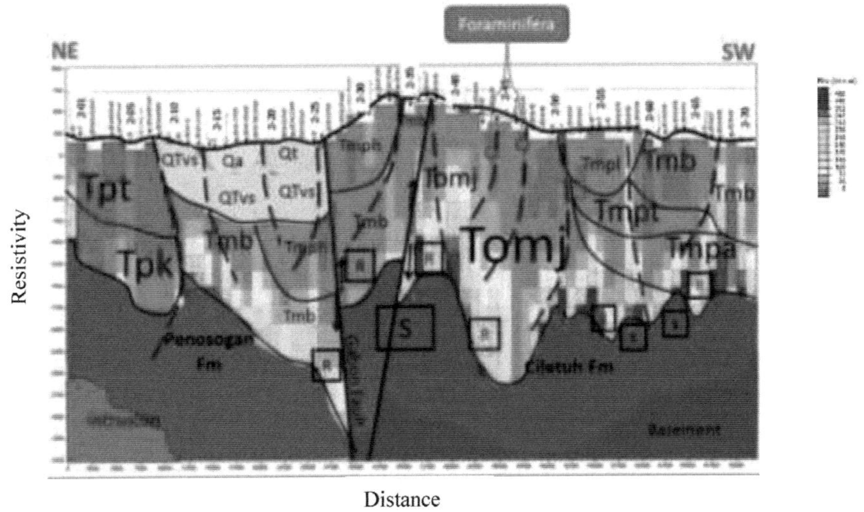

FIGURE 4. Interpretation model of line-2. The geological modelling section overlay with 2D inversion MT Data

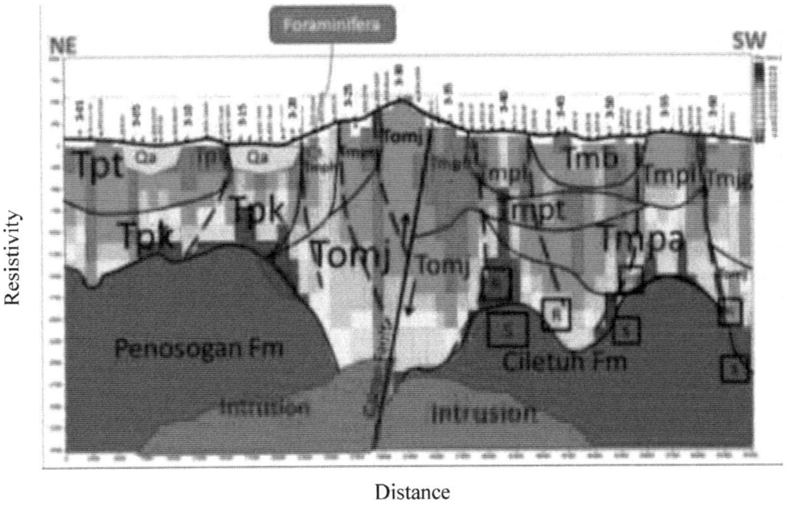

FIGURE 5. Interpretation model of line-3. The geological modelling section overlay with 2D inversion MT Data

4. Conclusions

The conclusion of this study show the resistivity mapping model that can be used for geological structure interpretation, especially in volcanic area zone with some additional data for support it, such as geochemical information. The interpretation of line-2 and line-3 show geological structure with the source rock and potential reservoir around the interest area. The identification of the hydrocarbon reservoir depend on source rock potential and the visual porosity from the core data sample. According to the limitations of MT method in the early days of its application to hydrocarbon exploration, the method will be more effective in certain niche environtments.

5. Acknowledgments

We would like to acknowledge PT. Pertamina as the data sponsor for let us learning and studying the data and information of this research. Furthermore, we would like to thank university of Indonesia as the accommodation sponsor and supervise the research that helped us with our research.

References

[1] Anderson, R. and Pelton, W.H., Magnetotelluric Exploration in Volcanic Cover, Over-thrust Belts, and Rift Zones (**55th** *Ann. Internat. Mtg.*, SEG, 1985), Session MT7.

[2] Christopherson, K. R., Application of Magnetotellurics to Petroleum Exploration in Papua New Guinea: A Model for Frontier Areas, (*Geophysics*, **56**, 21-27, April, 1991).

[3] Farooqui M.Y., Huijun H., Guoxin L., Nigel M., Tom N., etc. Evaluating Volcanic Reservoirs. (*Oilfield Review*, Shlumberger Spring 2009), pp. 36-47

[4] McCarthy, K., M. Niemann, D. Palmowski, K. Peters, and A. Stankiewicz, Basic Petroleum Geochemistry for Source Rock Evaluation, (*Oilfield Review Summer* **23**, no.2. 2011), pp. 32–43

[5] Passey, Q.R., S. Creaney, J.B. Kulla, F.J. Moretti, and J.D. Stroud, A Practical Model for Organic Richness from Porosity and Resistivity Logs, (*AAPG Bulletin,* **74**; no. 12, 1990) pp. 1777 – 1794

[6] Pertamina. Review Sistem Minyak dan Gas Bumi Berdasarkan Hasil Graviti & Magnetik dan Laporan Survai Geologi Permukaan Daerah Banjar, Jawa Barat. *Trisakti* 2007

[7] Schutter SR: "Hydrocarbon Occurrence and Exploration in and Around Igneous Rock," in Petford N and McCaffrey KJW, *Hydrocarbons in Crystalline Rocks*, (*Geological Society Special Publication* **214**, London:2013), pp. 7-33

[8] Srugoa P and Rubinstein P. Processes Controlling Porosity and Permeability in Volcanic Reservoirs from the Austral and Neuquen Basins, Argentina, (*AAPG Bulletin* **91**, no.1, January 2007), pp. 115-129

IOP Conf. Series: Earth and Environmental Science **132** (2018) 012019 doi:10.1088/1755-1315/132/1/012019

Identification of Detrital Carbonate in East Cepu High

R Sari[1,2] , I K Andika[1] , A Haris[2] and A Miftah[1]

[1]PT Pertamina EP Cepu, Indonesia

[2]Faculty of Mathematics and Natural Science, University of Indonesia, Indonesia

ramadan.sari@pertamina.com

Abstract. East Cepu High is a part of horst – graben series which formed by extensional tectonic processes during Paleogene in North East Java Basin. Due to excellent paleogeography position, the carbonate build-up was growth very well and as the main reservoir in East Cepu High. Sea level change have important factor to provide variation of facies in each carbonate buildup, one of emerging facies is detrital carbonate. Detrital carbonate indicated by onlap horizon featured with carbonate build up body. Based on paleogeography, fluctuation of sea level change and sediment source, detrital carbonate formed in leeward area in lowstand or highstand phases. Distinguish between detrital carbonate facies with other facies, advanced seismic processing performed by using continuous wavelet transform (CWT) and seismic inversion. CWT is one method of spectral decomposition used to find the frequency that represent a facies. The result from seismic inversion will support the interpretation for facies distribution. As the result, seismic data which have interval frequency 10 – 45 Hz and Acoustic Impedance (AI) value above 35000 (from cross plot between acoustic impedance and gamma ray) can be interpreted as detrital carbonate. Based on seismic interpretation, detrital carbonate facies distributed along leeward area with geometrical spreading. The lateral facies change from detrital carbonate to shale was identified which causing this facies become potential as hydrocarbon reservoir with stratigraphic trap. Based on the earlier studies, North East Java Basin have a strong hydrocarbon migration to fill the reservoir, therefore the detrital carbonate have high chance to be a new hydrocarbon prospect in this area.

1. Introduction

East Cepu High is a part of horst – graben series which formed by extensional tectonic process during Paleogene in North East Java Basin[1]. It is an excellent paleogeography position for reef growth and forming Isolated Carbonate Build-up (ICB) along platform during Late Oligocene until Early Miocene. It is surrounded by depocenter which are deep marine carbonate and re-worked material from carbonate build-up deposited.

ICB is the main reservoir in this area and it has various facies, one of them is re-worked carbonate build-up deposited well known as detrital carbonate facies. Environment and climate condition, sea level fluctuation, and basin configuration trigger re-worked and re-sedimentation in each carbonate build-up along platform and formed detrital carbonate on the slope in ICB. Based on regional geology detrital carbonate event occurred in Early Miocene[2,3,4,5].

IOP Conf. Series: Earth and Environmental Science **132** (2018) 012019 doi:10.1088/1755-1315/132/1/012019

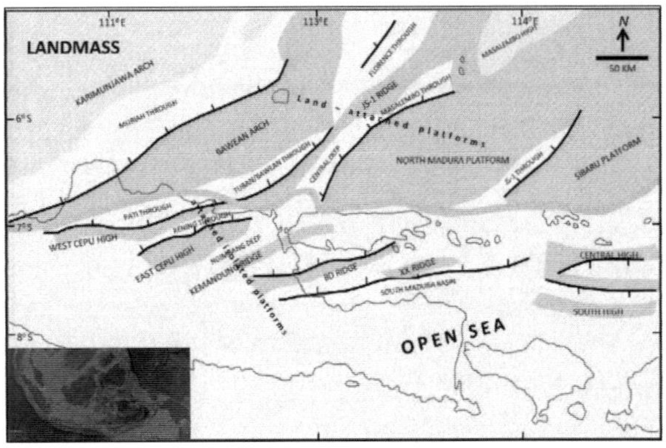

FIGURE 1. Horst and Graben configuration during Paleogene in North East Java Basin modified from Satyana[1]

Detrital carbonate can be deposited in three way, margin collapse, off-bank transport, and contourite. Each process have different characteristic in outcrop section, but indistinguishable in seismic section. In addition, each process can be happened in lowstand system track (LST) and highstand system track (HST). Margin collapse usually occurred when sediment not fully lithified and followed by change of pore pressure in sediment andwill cause scar evidence along pathway from source to depocenter. Second mechanism, off-bank transport is a common process in shelf area where sediment transport further away from shelf to open area. It has geometry sigmoid and oblique. The last one is contourite, which is a sedimentation process in below storm water base and it has characteristic thick and wider geometry.

Detrital carbonate in East Cepu High confirmed by penetration of J-2 well at flank of ICB J. Thin section from side wall core shown abundant of foraminifera mix with shallow marine bioclast from ICB. Most of the bioclast damaging and has orientation which is a characteristic of re-worked sediment. However, no conclusion about the dominant mechanism working in detrital carbonate ICB J and the thickness of detrital carbonate layer out of seismic resolution.

2. Method

Detrital carbonate penetrated by J-2 well at flank area from ICB J, the proven gas field in this area. The problem is the thickness of detrital carbonate out of seismic resolution. Advanced seismic processing performed by using continuous wavelet transform (CWT) and seismic inversion to distinguish between detrital facies with other facies in distribution mapping. Flow diagram of work can be seen in figure 2 below.

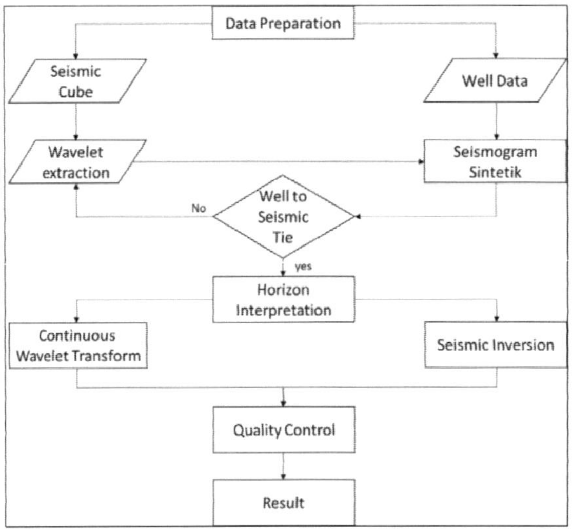

FIGURE 2. Workflow Diagram

IOP Conf. Series: Earth and Environmental Science **132** (2018) 012019 doi:10.1088/1755-1315/132/1/012019

3. Seismic Data Interpretation

Seismic data interpretation using analog from Ruby Field in Makassar Strait[6]. Seismic response of detrital carbonate is positive bright amplitude anomaly and indicated by onlap horizon featured with carbonate build-up body. Distribution mapping started from J-2 well tied and interpreted in 3D seismic cube area. Detrital carbonate existence very strong in flank of ICB K. There are two event with positive bright anomaly which onlap to ICB, upper detrital and lower detrital. In lateral succession, the positive bright anomaly has disappear, it is interpreted as lateral facies change from detrital carbonate to other facies with low impedance like shale. The connection between detrital carbonate and main body of build-up play an important role as well as lateral facies change to preserve hydrocarbon as hydrocarbon trap when migration occurred.

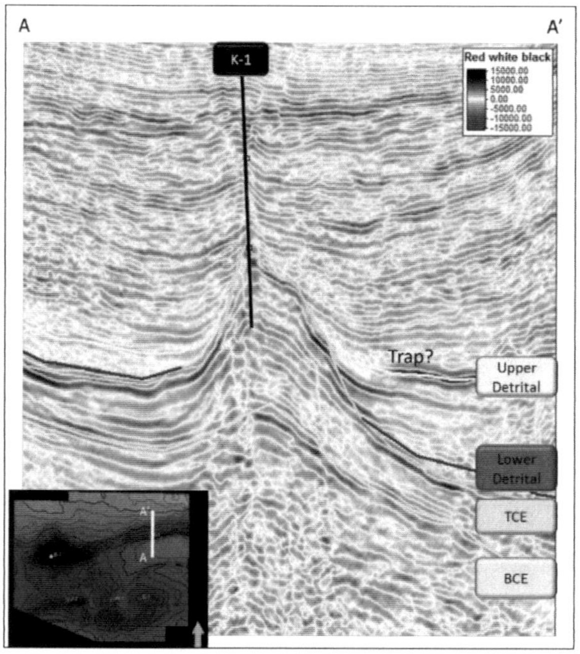

FIGURE 3. North – South section from ICB K there are two event detrital carbonate, upper detrital and lower detrital. In upper detrital event can be seen positive bright anomaly amplitude gradually disappear to the main body build-up and interpreted as lateral facies change. This lateral facies change play important role in stratigraphic trap reservoir like detrital carbonate.

4. Seismic Advance Processing

Lateral facies change will be identified with seismic advance processing which are continuous wavelet transform and seismic inversion.

4.1. Continuous Wavelet Transform

Spectral decomposition using to confirm the connectivity between detrital carbonate and main body build-up. There are many spectral decomposition method and this paper using Continuous Wavelet Transform (CWT). CWT analysis perform using TF Signal to analyzed frequency contain in trace seismic[7]. There are two area to be analyzed, detrital carbonate located in ICB J, where the detrital carbonate is penetrated and the best prospect of detrital carbonate in ICB K.

In trace seismic which penetrated main body build-up of ICB J has interval frequency 10 – 50 Hz with anomaly frequency in 10 Hz (Figure 4.a red line and red circle). In trace seismic which penetrated detrital carbonate has interval frequency 10 – 45 Hz with anomaly in 10 Hz and 25 Hz (Figure 4.a green line and green circle). In second area, trace seismic which penetrated main body build-up ICB K has interval frequency 10 – 25 Hz with anomaly frequency in 10 Hz (Figure 4.b white line and white circle). Next trace seismic to be analyzed coming from upper detrital carbonate event and it has interval frequency 10 – 45 Hz with anomaly frequency in 25 Hz (Figure 4.b green line and green circle). Both result showed consistency with ICB J. Last trace seismic to be analyzed in the position when upper

IOP Conf. Series: Earth and Environmental Science **132** (2018) 012019 doi:10.1088/1755-1315/132/1/012019

detrital carbonate event gradually disappear and interpreted as lateral facies change from detrital carbonate to shale (Figure 4.b red line and red circle).

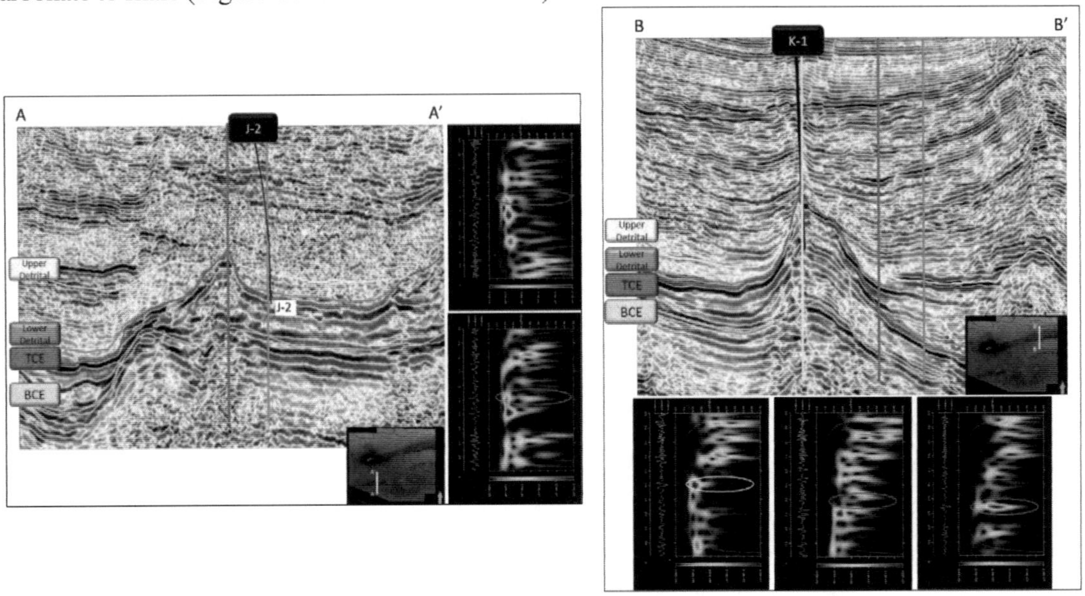

a b

FIGURE 4. Continuous Wavelet Transform using Kristekova[7]. In section ICB J, two trace seismic derived and in section ICB K, three trace seismic derived. The frequency from CWT may vary but have similarity in the same lithology.

Frequency anomaly on 10 Hz can be interpreted as Direct Hydrocarbon Indicator (DHI) in this area. Frequency anomaly on 25 Hz can be represented as detrital carbonate and not yet hydrocarbon proven. Lastly, shale doesn't presented by any anomaly frequency.

4.2. Seismic Inversion

Seismic inversion using to re-check and conform about detrital carbonate present in each anomaly shown in RMS Attribute and CWT. Cross plot AI vs density from J-2 shown detrital carbonate and shale can be separated. Detrital carbonate properties are AI more than 35000 and density more than 2.5 gr/cm^3.

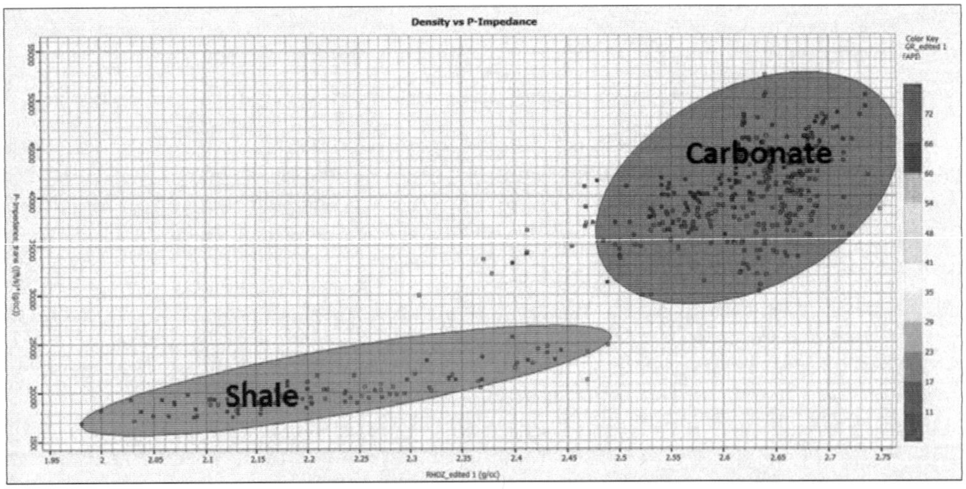

FIGURE 5. Cross plot AI vs density from well penetrated detrital carbonate

IOP Conf. Series: Earth and Environmental Science **132** (2018) 012019 doi:10.1088/1755-1315/132/1/012019

Seismic inversion running by model based method and the result show consistency with CWT result. P-impedance distribution shows the existences of lateral facies change from detrital carbonate and unconnected between detrital carbonate and main body build-up in detrital carbonate ICB K.

5. Discussion

Two advance seismic processing methods has been running and both shown similarity result. Detrital carbonate located in leeward area of each ICB. In theory, detrital carbonate can be deposited in three way, margin collapse, off-bank transport, and contourite. Each process have different characteristic in outcrop section, but indistinguishable in seismic section, in addition each process can be happened in lowstand system track and highstand system track.

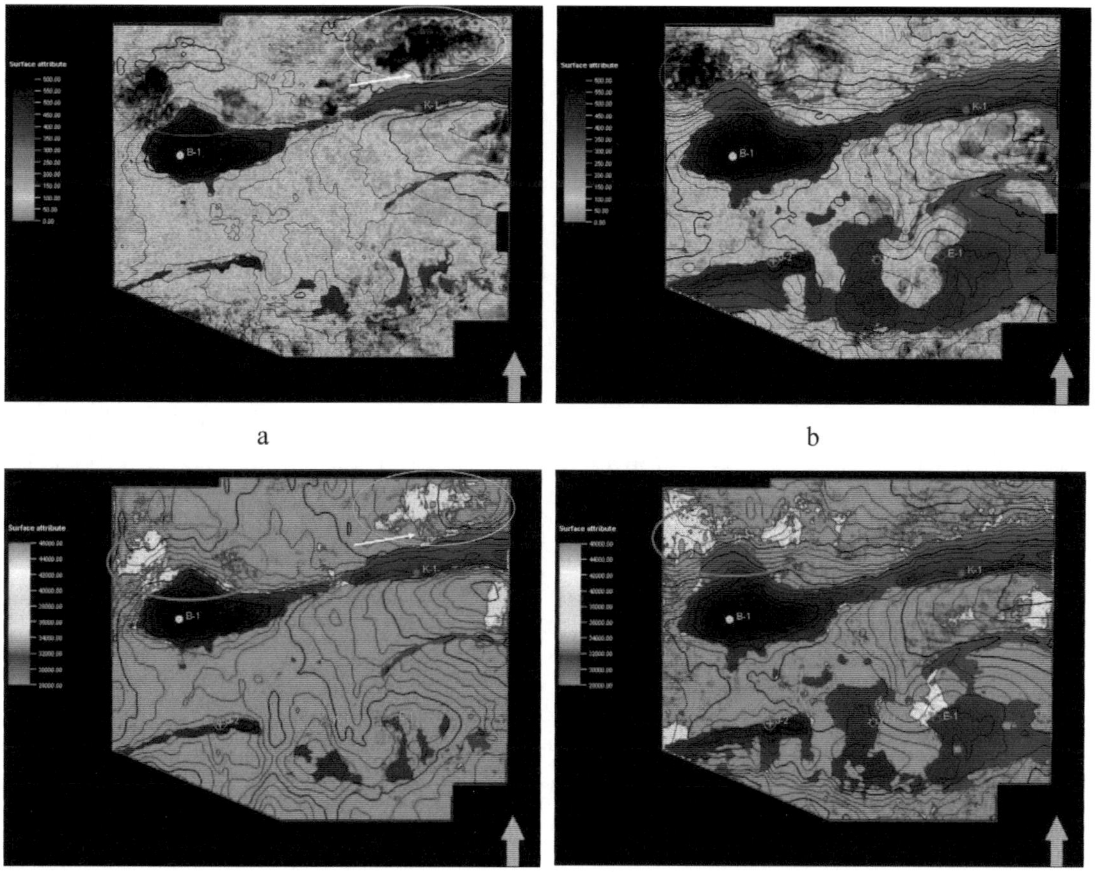

a b c d

FIGURE 6. Top Carbonate Envelope (TCE) Map overlay with Detrital Carbonate Map with CWT frequency 25 Hz. a) Upper detrital carbonate b) Lower detrital carbonate. Top Carbonate Envelope (TCE) Map overlay with Detrital Carbonate Map with P Impedance. c) Upper detrital carbonate d) Lower detrital carbonate.

Depositional process interpretation could be derived from geometrical of detrital carbonate deposit. In general geometrical spreading in detrital carbonate ICB B (red circle) or ICB K (white circle) can be seen. Detrital carbonate ICB B spreading in north – south direction and most likely attached to main body build-up. On other hand, detrital carbonate ICB K spreading in east – west direction and most likely not fully attached in main body build-up. It resemblance difference process between both of them. Detrital carbonate ICB B interpreted deposited by margin collapse process only, it shown by narrow spreading and attached with main body build-up. Detrital carbonate ICB K encounter various process to be deposited. It has wider area with east – west direction and a part of detrital carbonate attached with main body. It is like main body build-up has margin collapse in narrow space and the re-worked sediment spreading by contourite when it going to unconfined area.

Result from seismic advance processing strengthened the interpretation. In Figure 6.a and 6.c, there is a pathway feature in ICB K flank (white arrow). That pathway transport clastic sediment from margin collapse in body build-up to wider area. In ICB B detrital carbonate (red circle), there is no pathway seen. Seismic inversion also has similarity result and clearly showed closure of detrital carbonate in ICB K and attached detrital carbonate in ICB B. Detrital carbonate in ICB J which have been penetrated by well can't be discussed because out of seismic resolution. Existing field in surrounding area of detrital carbonate prospect make opportunity to be filled by hydrocarbon when migration occurred.

6. Conclusion

Oligocene – Miocene Carbonate in East Cepu High consist of various facies, one of them is detrital carbonate facies. Detrital carbonate in this area formed differentially and have different characteristic in each of ICB as main source. In general detrital carbonate in this area formed by margin collapse and sometime followed by contourite process. Detrital carbonate can be distinguished by other feature with anomaly frequency at 25 Hz and has property AI more than 35000 and density greater than 2.5 gr/cm3. The lateral facies change from detrital carbonate to shale was identified which causing this facies become potential as hydrocarbon reservoir with stratigraphic trap. Based on regional geology and other publication, North East Java Basin have a strong hydrocarbon migration to fill the reservoir, therefore the detrital carbonate have high chance to be a new hydrocarbon prospect in this area.

References

[1] Satyana, A H and Darwis A 2001 Recent Significant Discoveries within Oligo-Miocene Carbonate of the East Java Basin *IAGI 30th Conf. Proc.* (Yogyakarta: Ikatan Ahli Geologi Indonesia)

[2] Cahyono A B and Burgess C F 2007 Cepu 3D Seismic Variation in Oligo-Miocene Carbonate Buildip Morphology *IPA 31st Conf. Proc.* (Jakarta: Indonesia Petroleum Association)

[3] White J V, Derewetzky A N, Geary G C, Hohense V K, Jhonstone E M, Liu C, Pierce A C and Stevens J 2007 Temporal Control and Resulting Variation in Oligo-Miocene Carbonate from the East Java Basin Indonesia: Examples from the Cepu Area *IPA 31st Conf. Proc.* (Jakarta: Indonesia Petroleum Association)

[4] Apotria A, Weidmer M A, Walley D, Derewetzky A and Millman D 2009 Mass Wasting and Detrital Carbonate Deposition *IPA 33rd Conf. Proc.* (Jakarta: Indonesia Petroleum Association)

[5] Simo T, Weidmer M, van Simaeys S, Sekti R, van Gorsel H, Strohmenger C and Derewetzky A 2011 Sequence Stratigraphic Correlation and Sedimentological Implications, East Java Basin: Comparisons and Lesson Learned from Outcrop and Subsurface Studies *IPA 35th Conf. Proc.* (Jakarta: Indonesia Petroleum Association)

[6] Pireno G E, Cook C, Yuliong D and Lestari S 2009 Berai Carbonate Debris Flow as Reservoir in the Ruby Field, Sebuku Block, Makassar Straits: A New Exploration Play in Indonesia *IPA 33rd Conf. Proc.* (Jakarta: Indonesia Petroleum Association)

[7] Kristekova M 2006 *Time Frequency Analysis of Seismic Signal* (Brastilava: PhD Thesis Geophysical Institute, Slovak Academy of Science)

Acknowledgments

The authors wish to thank Directorate General Oil & Gas of Indonesia, SKKMIGAS for data and publication. The Pertamina EP Cepu management, EMCL and BKS for the support.

IOP Conf. Series: Earth and Environmental Science **132** (2018) 012020 doi:10.1088/1755-1315/132/1/012020

The Correlation between Radon Emission Concentration and Subsurface Geological Condition

Yudi Kuntoro[1], Herru L Setiawan[1], Teni Wijayanti[1] and Nandi Haerudin[2]

[1] Research and Development Center for Oil and Gas "LEMIGAS". Cipulir, Kebayoran Lama, South Jakarta, Indonesia

[2] Department of Geophysical Engineering, Faculty of Engineering, University of Lampung, Bandar Lampung, Indonesia

Corresponding author: yudik@lemigas.esdm.go.id

Abstract. Exploration activities with standard methods have already encountered many obstacles in the field. Geological survey is often difficult to find outcrop because they are covered by vegetation, alluvial layer or as a result of urban development and housing. Seismic method requires a large expense and licensing in the use of dynamite is complicated. Method of gravity requires the operator to go back (looping) to the starting point. Given some of these constraints, therefore it needs a solution in the form of new method that can work more efficiently with less cost. Several studies in various countries have shown a correlation between the presence of hydrocarbons and Radon gas concentration in the earth surface. By utilizing the properties of Radon that can migrate to the surface, the value of Radon concentration in the surface is suggested to provide information about the subsurface structure condition. Radon is the only radioactive substance that gas-phased at atmospheric temperature. It is very abundant in the earth mantle. The vast differences of temperatures and pressures between the mantle and the earth crust cause the convection flow toward earth surface. Radon in gas phase will be carried by convection flow to the surface. The quantity of convection currents depend on the porosity and permeability of rocks where Radon travels within, so that Radon concentration in the earth surface delineates the porosity and permeability of subsurface rock layers. Some measurements were carried out at several locations with various subsurface geological conditions, including proven oil fields, proven geothermal field, and frontier area as a comparison. These measurements show that the average and the background concentration threshold in the proven oil field (11,200 Bq/m^3) and proven geothermal field (7,820 Bq/m^3) is much higher than the quantity in frontier area (329 and 1,620 Bq/m^3). Radon concentration in the earth surface is correlated with the presence of geological faults. Peak concentrations of Radon takes place along the fault.

1. Passive Soil Radon Method

Radon is the only radioactive substance that gas-phased at atmospheric temperature. Radon has the atomic number 86 and is the result of the decay of Radium. Radium emits alpha rays along with the release of energy thus two protons and two neutrons will be lost. These events generate Radon particles and cause the change of phase from solid into gas.

IOP Conf. Series: Earth and Environmental Science **132** (2018) 012020 doi:10.1088/1755-1315/132/1/012020

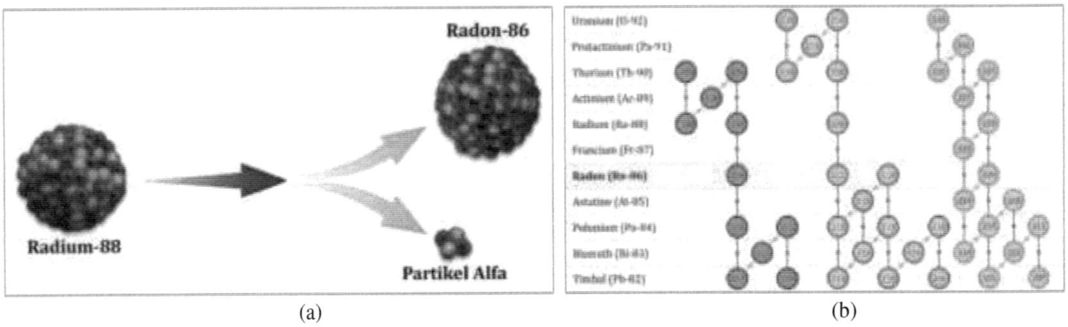

Figure 1. (a) Radium decays into Radon. (b) Radioactive decay chain

In the chain of decays, Radon can be formed by three main rows (series) of radioactive decay, namely Uranium series producing Radon-222 and Radon-218, Thorium series that produces Radon-220 (called Thoron) and Actinium series which produces Radon-219 (called Actinon). The concentration of the radioactive substance, in this case Radon, expressed in radioactivity per unit volume in units of Becquerel per meter cubic (Bq/m^3). Unit of Becquerel states the number of decays per unit time (count of decay per second). Thus the unit of Bq/m^3 has meaning the number of Radon gas atoms decay every second in a meter cubic of air volume. Other standard unit which is often used is Currie per liter (Ci/l). 1 Ci radioactivity is comparable to 37 Mega Bq.

Radon with mass number of 222 is the most stable isotope of Radon in nature. This isotope has a half-life of 3.8 days. In addition to Radon-222, there are at least 17 isotopes of radon gas found, which altogether have a very short half-life. Of the overall Radon isotope, Radon-222 is considered the most appropriate for the research with Passive Soil Radon method. On the periodic table of elements, Radon is classified in the category of noble gas, namely a gas which has a stable number of valence electrons (8 electrons in the outermost orbit) thus not combined with other particles, even with fellow Radon atoms.

1.1. Radon in Nature

Radioactive is the biggest contributor of earth heat. 50% of heat in the core of the earth comes from the energy produced in the decay of a radioactive substance activity. The existences of radioactive materials are abundant in the Earth mantle. In addition to surface through volcanic activity, Radon gas from the Earth mantle also moves through rock pores. Some Radon atoms are stuck together with oil and gas in the hydrocarbon trap, so it is believed that hydrocarbon reservoirs have high radioactive content.

Radon is the heaviest gas atom on the periodic table of elements, thus to move against the direction of gravitational force it needs external force. The main factor that helped Radon moves toward earth surface is the difference of Radon gas concentrations causing the radon gas transports (diffusion) and pressure from the center of the earth due to the high temperatures and pressures in the direction of the earth surface (convection flow). Radon migration process caused by convection flow is more dominant than the diffusion process.

Radon movement to the surface is strongly influenced by the media went through. Radon will be concentrated higher on media easily passed, namely a media with high porosity and permeability, and will be concentrated lower in media with low porosity and permeability. Through the mechanism of diffusion, Radon gasses in the rocks have a low transport speed, a maximum of only 155cm.

Table 1. Average distance of Radon and Thoron diffusion (Tanner, 1964).

Media	Average distance (Radon)	Average distance (Thoron)	Diffusion Coeff. (cm^2/s)
Air	220	2,8	10-1
Porous soil	155	2	5 x 10-2
Water	2	0,03	10-5
Saturated porous soil	1,55	0,02	5 x 10-6
Dry sand	160	2	5 x 10-2
Dry porous soil	155	2	5 x 10-2

Some carrier gas in the process of convection are CO_2, NO_X and SO_X. With the high temperature and pressure below the surface, CO_2 gas of combustion in the Earth mantle is pumped to the surface of the earth. Migration of CO_2 gas will be the convection currents for Radon atom to get to the surface.

2. Results Of The Previous Study

2.1. Application of Radon Detection Method

Radon gas detection method has been widely applied. Some of them are:

- Characterizing SGD (Submarine Groundwater Discharge).
- Preliminary exploration for Uranium prospecting.
- Exploration of geothermal.
- Early detection of earthquakes (Japan).

In the case of early detection of earthquakes, anomalous Radon concentration occurs a few days before an earthquake. This proves that changes in subsurface structure affect the value of Radon concentrations in the surface.

2.2. Phenomena of NORM in Oil and Gas

An interesting fact from the discovery of NORM (Naturally Occurring Radioactive Materials) in oil and gas has been found by researchers. The emergence of the value of radioactive concentration (Radon) in the oil well higher than the background concentration was first reported by Himstedt and Burton. In 1920 to 1930, the presence of NORM in the oil and gas also appeared in a number of scientific papers in Russia and Germany.

J.E Tilsley and P. R. J. Nichols (1991) stated that the tests of the 13 samples from field in South Wellcheburg, Ontario (3 gas fields and 10 oil fields) shows a strong correlation between the accumulation of radioactive gases with hydrocarbon deposit. The study was conducted by calculating the decay of Radium of soil samples taken.

A year later, the mapping of the concentration of Radon-222 gas conducted by Lu Zuhui using passive method in the area of Webei Hollow and Saihantal Hollow, Sangdong shows that there is the presence of suitability between Radon gas anomaly and hydrocarbon deposits. In 2001, a study from Al-Tamimi and Abu Murad succeeded in proving that the Radon concentrations near fault is much higher than background concentration.

The activities above lead to temporary conclusion that oil and gas field has a higher background concentration than frontier area. This is in line with the conclusions of the study conducted by Liu Zhi et. al. (1993) and David (2005). Based on the assessment of the theory, hypothesis developed later is: Radon concentration on the surface characterizing the structure and characteristics of the material underneath.

3. Methodology

The main instrument in the research entitled Implementation of Passive Soil Radon Method is the detectors of Radon, RAD7. RAD7 is a digital radioactive counter instrument based on "Geiger Muller counter". RAD7 instrument detects radon gas by capturing the entire alpha particles sucked out of the ground. Alpha particle is then detected by a Geiger Muller counter and sorted based on the energy carried by the alpha particles. At the time of Radon-222 atom decays to Polonium-218, it will be accompanied by the release of alpha particles that carry energy of 5.59 MeV. While on detecting Thoron, Radon-220 decays become Polonium-216 along with the release of alpha particles with an energy of about 6.4 MeV.

4. Data and Discussion

This activity is comprised of several data collection stages undertaken in different place under different geological conditions.

4.1. Frontier Area 1 (Kalimantan) and 2 (Sumatera)

The frontier area 1 has not been widely explored. There are limited data on both geological and geophysical subsurface. Radon measurement was conducted with a route along provincial highways from the west to the east and several tracks of smaller roads in the north and south of the main route. There is a total of 181 measurement points, in which the interval between the points varies between 1 and 4 km. The value and position of those measurement points if plotted will constitute a contour map as shown in figure 2(a). The average concentration of the entire observation points above is 3,320 Bq/m^3 and the highest concentration is 53,000 Bq/m^3.

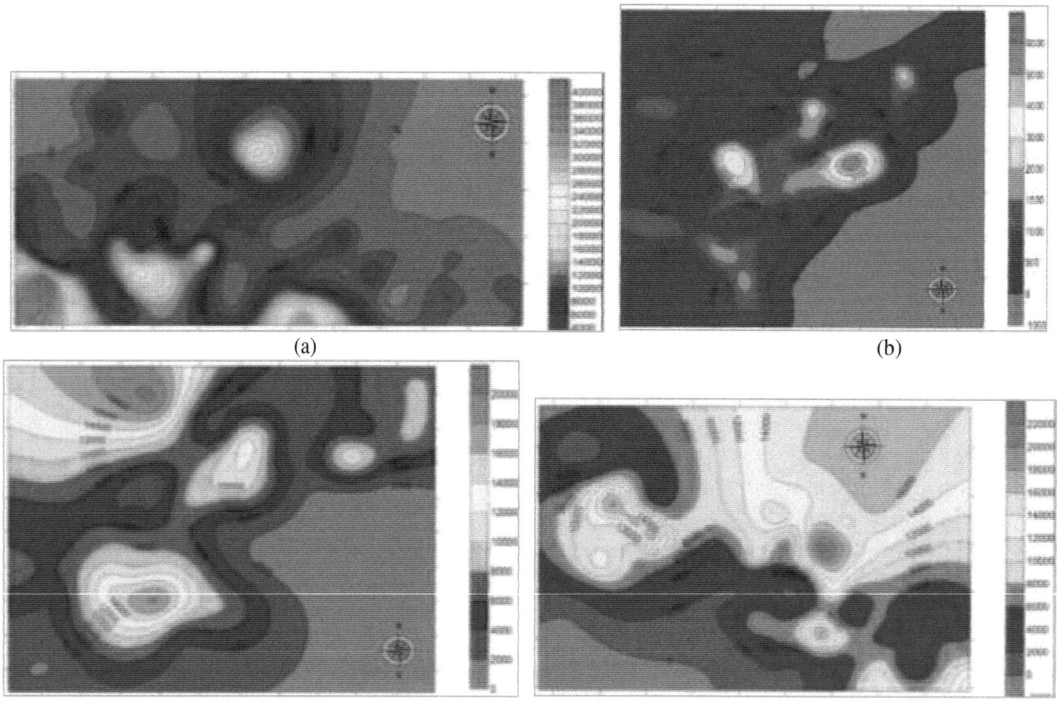

(a) (b) (c) (d)

Figure 2. The contour map illustrating Radon concentration in (a, b) frontier area, (c) geothermal and (d) oil field.

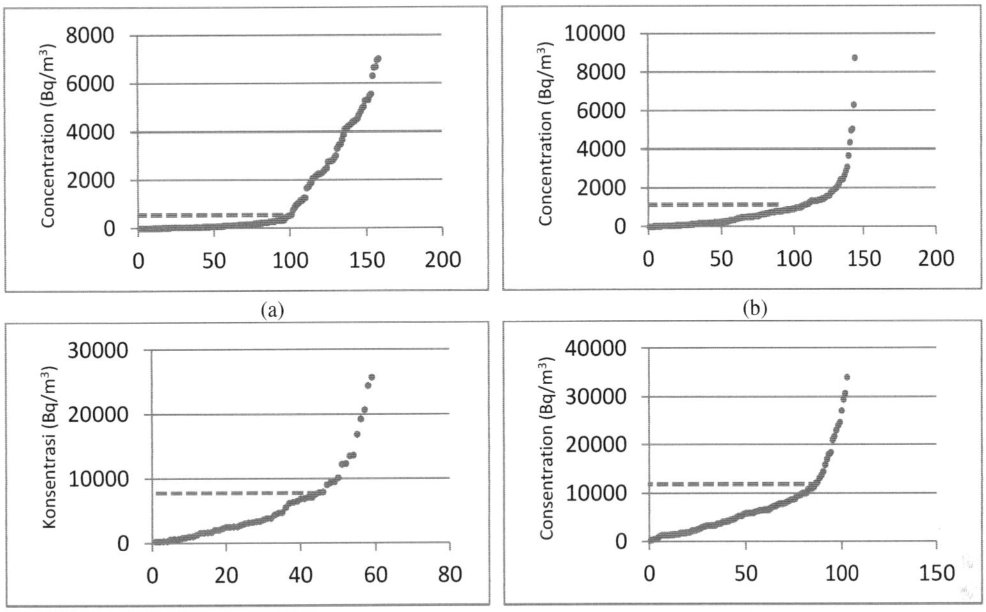

(a) (b) (c) (d)

Figure 3. Normal probability plot of (a, b) frontier area, (c) geothermal manifestation and (d) oil field.

This frontier area 1 is swampland covered with soils several decades ago. Both geological and geophysical data for the area remain unavailable. Soils at a depth of more than 50 cm have very high water content. Thus, measurement can be performed only at a maximum depth of 30 cm. It was done in 144 points. It took a grid form with a 2-km interval between the points. A contour map from plotting the Radon data of the area is presented in figure 2(b). Result of the field measurement on Radon concentration indicates that the highest value is 8,740 Bq/m^3 while the average value is 900 Bq/m^3.

4.2. Geothermal Manifestation Area, Sumatera

The province is rich in geothermal potential. Radon measurement carried out here aims to examine the existence of grabens. The Radon measurement in this area was undertaken in 59 points. Such measurement generates a contour map as presented in figure 2(c). Result of Radon measurement in this area generates data with a quite high average value by 5,880 Bq/m^3 and the highest value by 25,700 Bq/m^3.

4.3. Geothermal Well

The chosen area is a geothermal field that has been producing. The Radon measurement was conducted around the operating geothermal production and injection wells. The purpose of selecting the location is to confirm and to test the hypothesis that the geothermal reservoir has a great energy to produce the convection flow that bring the Radon to the surface. In order to test the hypothesis, the measurement was conducted in 6 location points, they were 2 production wells, 2 injection wells, and 2 abandoned geothermal wells. The results of Radon Measurement on the location obtained the highest value of 89.000 Bq/m^3 in production well 2. Those 6 location points generated the data as shown in figure 4.

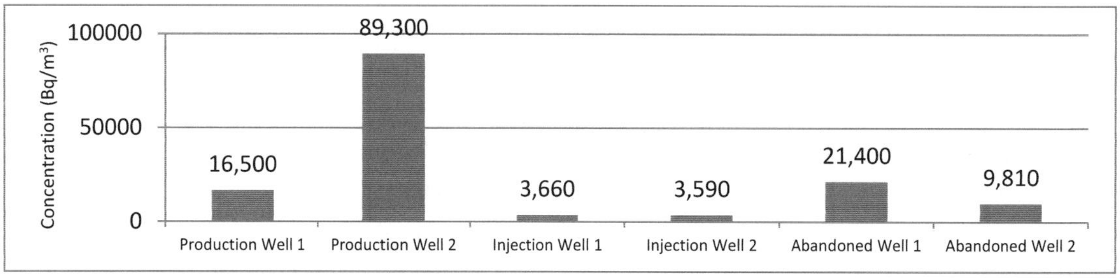

Figure 4. Radon concentration on geothermal wells

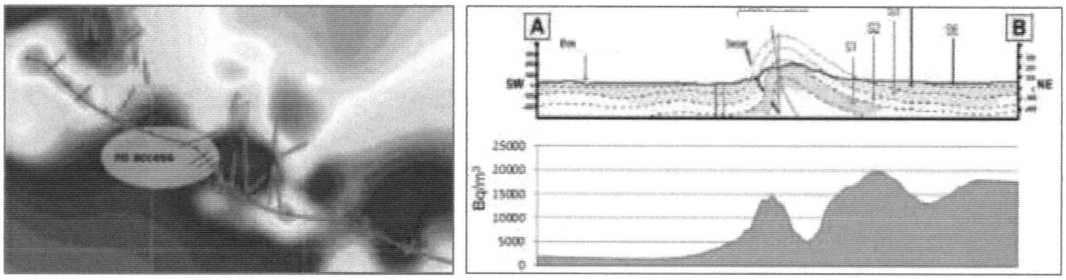

(a) (b)

Figure 5. (a) Radon and the distribution of fault based on IKONOS. The red line shows the fault. (b) Comparison of the geological cross-section and Radon one

Based on the map above, the fault is distributed from the southeast to the northwest coincide with of Radon concentration in some locations. Continuity of Radon concentration is cut in areas that are not accessible so it can not be measured. We can make a conclusion that the radon concentrations is correlating positively to the fault structure. From figure 5 it can be seen that the Radon concentration rises drastically in the fault zone. High Radon concentration around point B is the background concentration which is influenced by the content factor (subsurface materials). The cross section also provides a positive correlation between the Radon concentration towards the river. The value measured near point A (river) is very low because Radon cannot be measured in soil with high humidity.

Statistical test with a normal probability plot method is also applied to the data of Radon to separate the anomaly and background concentration. Threshold zone on normal probability plot is shown in the area with the data to form a linear trendline. Data that is away from the linear trend line has entered an anomalous zone. The boundary between the two zones is named threshold. Normal probability plot of oil field data is shown in the graph in figure 3(d). Based on the graph, the value of threshold in the oil field is 11,200 Bq/m^3.

4.4. Producing Oil Field

This is the only field with the subsurface information that has been known before this activity. The field has been producing oil for a long time. There are thousand wells that has been drilled and still produce oil until now. The measurement was conducted with the intervals between the measurement points are 500m. The data of Radon measurement results generated the contour map as shown in figure 2(d). The data has an average value of 7.663 Bq/m^3 with the highest value of 34.000 Bq/m^3.

Based on the comparison chart of threshold concentration above, it can be concluded that the background concentration on the oil field is the highest. It also make an evidence that the field with geothermal manifestations have a high background concentrations. Thus, Radon method has the potential to be used as an initial exploration of geothermal.

5. Conclusion

Radon measurement results showed a strong correlation between Radon concentrations in the surface and the faults. In the oil field, the value of background concentration (threshold) of Radon is higher than the frontier one and the results of measurements in geothermal field shows a very strong correlation between Radon concentrations and geothermal reservoir. In its application for oil and gas exploration, the role that Passive Soil Radon Method plays is just supporting data.

References

[1] Al-Tamimi, M.H. dan Abumurad. K.M., 2001, *"Radon Anomalies along Faults in North of Jordan"*, Radiation Measurements, Vol.34, Issues.1–6, p.397–400.

[2] Alif, S. M., Meilano, I., Gunawan, E. and Efendi, J., 2016. "*Evidence of Postseismic Deformation Signal of the 2007 M8.5 Bengkulu Earthquake and the 2012 M8.6 Indian Ocean Earthquake in Southern Sumatra, Indonesia, Based on GPS Data*", Journal of Applied Geodesy, 10(2), 103-108, DOI: 10.1515/jag-2015-0019.

[3] Anugrah, B., Meilano, I., Gunawan, E. and Efendi, J., 2015. "*Estimation of postseismic deformation parameters from continuous GPS data in northern Sumatra after the 2004 Sumatra-Andaman earthquake*", Earthquake Science, 28(5-6), 347-352, DOI: 10.1007/s11589-015-0136-x.

[4] Ardika, M., Meilano, I. and Gunawan, E., 2015. "*Postseismic Deformation Parameters of the 2010 M7.8 Mentawai, Indonesia, Earthquake Inferred from Continuous GPS Observations*", Asian Journal of Earth Sciences, 8, 127–133, DOI: 10.3923/ajes.2015.127.133.

[5] Baradacs, E., Csige, I., Dezsö, Z., Hunyadi, I. dan Szerbin P., 2001, "226Ra in Geothermal and Bottled Mineral Waters of Hungary", Radiation Measurements, No.34, p.385-390.

[6] Chon, H.T., Je, H.K. dan Kang, C., 1999, *"A Preliminary Study on Soil–Gas Radon Geochemistry According to Different Bedrock Geology in Korea"*, Environmental Geochemistry and Health, No.21, p.117-131.

[7] Dung, L. T., Thang, N. X., dan Vuong, L. V., 1995, *"Radon Transport Equation in Earth's Crust"*, Radiation Measurements, Vol.25, No.1, pp.661-663

[8] Gunawan, E., Kholil, M. and Meilano, I., 2016. "*Splay-fault rupture during the 2014 Mw7. 1 Molucca Sea, Indonesia, earthquake determined from GPS measurements*", Physics of the Earth and Planetary Interiors, 259, 29-33, DOI: 10.1016/j.pepi.2016.08.009.

[9] Gunawan, E., Maulida, P., Meilano, I., Irsyam, M. and Efendi, J., 2016. "*Analysis of coseismic fault slip models of the 2012 Indian Ocean earthquake: Importance of GPS data for crustal deformation studies*", Acta Geophysica, 64(6), 2136-2150, DOI: 10.1515/acgeo-2016-0106.

[10] Gunawan, E., Meilano, I., Abidin, H. Z., Hanifa, N. R. and Susilo, 2016. "*Investigation of the best coseismic fault model of the 2006 Java tsunami earthquake based on mechanisms of postseismic deformation*", Journal of Asian Earth Sciences, 117, 64-72, DOI: 10.1016/j.jseaes.2015.12.003.

[11] Gunawan, E., Ghozalba, F., Syauqi, Widiastomo, Y., Meilano, I., Hanifa, N. R., Daryono and Hidayati, S., 2017. "*Field Investigation of the November to December 2015 Earthquake Swarm in West Halmahera, Indonesia*", Geotechnical and Geological Engineering, 35(1), 425-432, DOI: 10.1007/s10706-016-0117-4.

[12] Gunawan, E., Meilano, I., Hanifa, N. R. and Widiyantoro, S., 2017. "*Effect of coseismic and postseismic deformation on homogeneous and layered half-space and spherical analysis: Model simulation of the 2006 Java, Indonesia, tsunami earthquake*", Journal of Applied Geodesy, DOI: 10.1515/jag-2017-0009.

[13] Ito, T., Gunawan, E., Kimata, F., Tabei, T., Meilano, I., Agustan, Ohta, Y., Ismail, N., Nurdin, I. and Sugiyanto, D., 2016. "*Co-seismic offsets due to two earthquakes (Mw 6.1) along the Sumatran fault system derived from GNSS measurements*", Earth, Planets and Space, 68(1), 1, DOI: 10.1186/s40623-016-0427-z.

[14] Kristiansson, K. dan Malmqvist, L., 1982, *"Evidence for Nondiffusive Transport of 222Rn in*

the Ground and A New Physical Model for the Transport", Society of Exploration Geophysics, Vol.47, No.10, p.1444-1452.

[15] Monnin, M., Morin, J.P. dan Siedel, J.L., 1993, *"A Tri-Dimensional Model for Radon Transport in A Porous Medium"*, Radiation Measurements, Vol.22, No.1-4, pp.415-418.

[16] Ohkura, T., Tabei, T., Kimata, F., Bacolcol, T. C., Nakamura, Y., Luis, Jr., A. C., Pelicano, A., Jorgio, R., Tabigue, M., Abrahan, M., Jorgio, E. and Gunawan, E., 2015. *"Plate Convergence and Block Motions in Mindanao Island, Philippine as Derived from Campaign GPS Observations (Special Issue on Enhancement of Earthquake and Volcano Monitoring and Effective Utilization of Disaster Mitigation Information in the Philippines)",* Journal of Disaster Research, 10(1), 59-66.

[17] Olofsson, B. dan Skeppström, K., 2006, *"A Prediction Method for Radon in Groundwater Using GIS and Multivariate Statistics"*, Science of the Total Environment, No.367, p.666–680.

[18] Pratama, C., Ito, T., Sasajima, R., Tabei, T., Kimata, F., Gunawan, E., Ohta, Y., Yamashina, T., Ismail, N., Nurdin, I., Sugiyanto, D., Muksin, U. and Meilano, I., 2017. "*Transient rheology of the oceanic asthenosphere following the 2012 Indian Ocean Earthquake inferred from geodetic data*", Journal of Asian Earth Sciences, 147, 50-59, DOI: 10.1016/j.jseaes.2017.07.049.

[19] Puchala, R., Solecki A.T. dan Tchorz, D., 2007, *"Radon and its Decay Product Activities in the Magmatic Area of the Karkonosze Granite and the Adjacent Volcano-sedimentary Intrasudetic Basin"*, Annals of Geophysics, Vol.50, No.4.

[20] Raharja, R., Gunawan, E., Meilano, I., Abidin, H. Z. and Efendi, J., 2016. "*Long aseismic slip duration of the 2006 Java tsunami earthquake based on GPS data*", Earthquake Science, 29(5), 291-298, DOI: 10.1007/s11589-016-0167-y.

[21] Sikka, D.B. dan Shives. R.B.K., 2001. *"Mechanisms to Explain the Formation of Geochemical Anomalies Over Oilfields"*, AAPG Hedberg Conference "Near-Surface Hydrocarbon Migration: Mechanisms and Seepage Rates", Vancouver.

[22] Soonawala, N.M. dan Telford, W.M., 1980, *"Movement of Radon in Overburden"*, Society of Exploration Geophysics, Vol.45, No.8, p.1297-1315.

[23] Tanner, Allan B., 1964, *"Radon migration in the ground: A review"*, The Natural Radiation Environment: Chicago Univ. Press, Chap.9, p.161-190.

[24] Yakovleva, V.S., 2005, *"A Theoretical Method for Estimating the Characteristics of Radon Transport in Homogeneous Soil"*, Annals of Geophysics, Vol.48, No.1.

IOP Conf. Series: Earth and Environmental Science **132** (2018) 012021 doi:10.1088/1755-1315/132/1/012021

Analysis of Self-Potential Response beyond the Fixed Geometry Technique

Harry Mahardika

Kelompok Keahlian Fisika Bumi dan Sistem Komplekss, FMIPA, Institut Teknologi Bandung, Jl. Ganesha No. 10 Bandung 40132, Indonesia

Corresponding author: hmahardi@fi.itb.ac.id

Abstract. The self-potential (SP) method is one of the oldest geophysical methods that are still available for today's application. Since its early days SP data interpretation has been done qualitatively until the emerging of the fixed geometry analysis that was used to characterize the orientation and the electric-dipole properties of a mineral ore structure. Through the expansion of fundamental theories, computational methods, field-and-lab experiments in the last fifteen years, SP method has emerge from its low-class reputation to become more respectable. It became a complementary package alongside electric-resistivity tomography (ERT) for detecting groundwater flow in the subsurface, and extends to the hydrothermal flow in geothermal areas. As the analysis of SP data becomes more quantitative, its potential applications become more diverse. In this paper, we will show examples of our current SP studies such as the groundwater flow characterization inside a fault area. Lastly we will introduce the application of the "active" SP method - that is the seismoelectric method - which can be used for 4D real-time monitoring systems.

1. Introduction

Self-potential (SP) method corresponds to the passive measurements of electric signals associated with variety of source current mechanisms in the conductive subsurface of the earth. It is a geophysical method that everyone knows about but nobody seems to appreciate [1], hence it's named the ugly duckling of environmental geophysics. In the local scope, such as in the past HAGI annual meetings there are only few papers on SP method: one is applied in hydrothermal system study [2] and two are focusing in groundwater flow in porous media [3, 4]. In SP data analysis, there are two known methods of quantitative analysis (inversion) of SP data: 1. fixed geometry technique, and 2. continuous model inversion. The fixed geometry inversion is based on the solution of Poisson equation for electrical problem that is created from electrically-polarized structures (electrical dipole) along a principal profile over the body [5]. Here, the physical mechanism that is creating the electrical dipole however is not taken into account in the Poisson equation. On the other hand, the continuous model inversion is based on the solution of Poisson equation that is generated from either redox-based contribution [6, 7] and streaming current contribution related to groundwater flow [7]. The two inversion methods were applied for various near surface applications where the former are relatively more widely used for SP studies in Indonesia due to its simplicity and effectiveness in terms of computation time [8, 9]. Therefore, we present our current research on continuous model inversion for SP data by doing a quick review on the fixed geometry technique and its limitation, and then describing the formulation of continuous model inversion with application example, and introducing the time-lapse expansion of SP inversion with prospect application in water encroachment monitoring.

2. SP and Fixed Geometry Inversion

The fixed boundary inversion for electric potential formulation is based on a polarized structure along a principal profile over an ore body [10]. Considering a polarized body located at the 2D homogeneous medium and its polarization axis pointing towards geographical North direction we will have an electric potential described at an observation point:

$$\varphi(x,z,K,\theta)=K\frac{(x-x_s)\cos\theta+(z-z_s)\sin\theta}{\left[(x-x_s)^2+(z-z_s)^2\right]^q} \quad (1)$$

where K is an electric dipole moment (in mV), θ is the dipole orientation from positive vertical axis, and (x_s, z_s) are the horizontal and vertical dipole source location (in m). In this formulation, geometrical factor q is a quantity related to the shape of the anomaly such as sphere (q = 1.5), horizontal cylinder (q = 1), and vertical cylinder (q = 0.5). Given a specific geometrical factor that we want to recover, we estimate four quantities that is shaping the model vector $\mathbf{m}=[x_s, z_s, K, \theta]$.

The application of the fixed geometry technique is presented in the following example where real SP data was collected from the Intra-Mural Field of Colorado School of Mines, shown in figure 1a. In this case, a preliminary SP surface map is obtained and we chose SP profile line A-B for inverse calculation. Applying initial guess of (x_s, z_s, K, θ) = (0 m, -10 m, 10 mV, 45°) and fix the geometrical factor q = 0.5 representing a vertical cylinder structure, we get a good fit of SP field response with the inverse calculation after 333 iterations as it shown in figure 1b. The inverse calculation, done using Gauss-Newton technique, is also recover the structure location in x_s = 15 m and z_s = -24 m and the electric dipole parameters K = -14.5 mV, and θ = 73°. The inversion result shows differences with the previous interpretation of the data, which suspect that the location of a buried pipe is 12 m further from the current estimation (grey boxes at figure 1a).

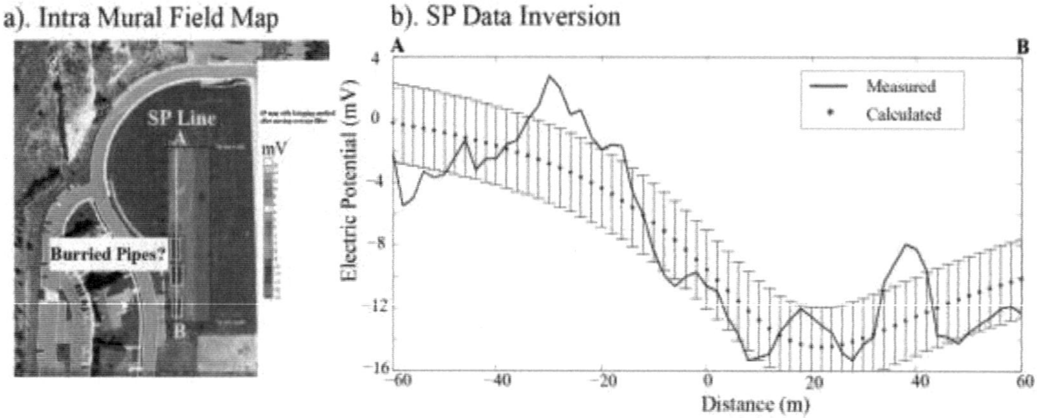

Figure 1. Intra Mural SP example. As it shows here, a). is the location of Intra Mural Field, CSM, where surface SP measurements conducted and profile line A-B was selected. And in b). is showing SP data along profile A-B and the inverse calculation result showing a fairly good agreement between them

IOP Conf. Series: Earth and Environmental Science **132** (2018) 012021 doi:10.1088/1755-1315/132/1/012021

There are several variations that can be done for fixed geometry technique application. Candra et.al [9] shows the usage of singular value decomposition technique for finding the inverse solution of several synthetic and measured SP test cases that is similar to the Intra Mural example. Other example [11] expands the formulation of equation 1 such that it representing inclined sheet geometry of infinite horizontal extent to mimics the surface of a fault. In one example they reconstruct the center location and the dimension of the fault as well as the dipole density and its inclination angle. From these example however, it is clear that the fixed-geometry inversion of SP data with its entire variant is relatively limited in characterizing of the electric dipole and its position and orientation in the subsurface, and therefore could not be used for interpreting physical processes that really happen there. And this is where the continuous model approach plays a bigger role in the game.

3. Continuous Model Inversion for SP Data Analysis

The concept of continuous model inversion of SP response is based on the solution of electric Poisson equation that is generated from either redox-based contribution and streaming current contribution related to groundwater flow [7]

$$\nabla \cdot \sigma \nabla \varphi = \nabla \cdot \left[{}^{(1)} -\sigma \nabla E_H -{}^{(2)} \frac{1}{e}\left(\sigma_{(+)} - \sigma_{(-)}\right) \nabla \mu_f +{}^{(3)} Q_v \mathbf{u} \right] \tag{2}$$

where σ is the conductivity of porous medium whereas $\sigma_{(+)}$ and $\sigma_{(-)}$ are the conductivities of anions and cations respectively (all in S/m), μ_f is the electrochemical potential of pore fluid (in V), E_H is the redox potential (in V), and $\mathbf{u}$ is the fluid flow or Darcy velocity [m/s]. The first term in the right hand side of the eq (2) is related to the physics of redox potential created in an ore body due to the presence of water table. The second one is related to the gradient of the chemical potential of the pore water solution (e.g., associated with a gradient of salinity or ionic strength) and the third is related to the flow of the pore water. These contribution are also called the diffusion current density and streaming current density respectively. The excess of charge Q_v (in C/m^3) is controlling the electrokinetic process and related to the saturated permeability k_0 of the porous medium via an empirical relation [12] given

$$\log\left(Q_v\right) = -9.23494 - 0.8219 \log\left(k_0\right) \tag{3}$$

For simplicity, we can focus on the electrical signals of electrokinetic nature that are due to the streaming current density. We consider an isotropic but possibly heterogeneous medium such that the governing equation for the electrical problem [13] is given as

$$\nabla \cdot \sigma \nabla \varphi = \Im \tag{4}$$

where equation (4) is basically a Poisson's equation for the self-potential problem φ (expressed in V), and $\Im = \nabla \cdot \left(Q_v \mathbf{u}\right)$ denotes the (scalar) volumetric current density (in A/m^3). Therefore, the electric potential distribution at an observation points along the surface is given by

$$\varphi(x, y, z) = \int_V \mathbf{K}\left(x, y, z, x_s, y_s, z_s\right) \Im\left(x_s, y_s, z_s\right) dV \tag{5}$$

where $\mathbf{K}\left(x, y, z, x_s, y_s, z_s\right)$ is the kernel and dV is a small volume around the source point $\Im\left(x_s, y_s, z_s\right)$.

The continuous model technique formulated from equations (4 and 5) implies the importance of resistivity data in the inverse-model calculation. In most applications, resistivity information of the subsurface is given by the electrical dc-resistivity measurement or electromagnetic methods done in active or massive mode. From equation (5), the elements of the kernel are the Green functions connecting the self-potential data at a set of measurement stations located at the surface and the sources of current density located in the conducting volume. The kernel computation accounts for the electric resistivity distribution and the boundary conditions applied to the system. Therefore, the data are defined using 3D coordinates (x, y, z), hence equation (5) can be written as

$$\mathbf{d} = \mathbf{Km} \tag{6}$$

where $\mathbf{m}$ is a model vector containing the *M*-source current density terms $\Im$, and $\boldsymbol{d}$ denote the data vector of electric potentials φ observed in *N*-receiver locations. Subsurface model $\mathbf{m}$ is expressed as $\mathbf{m}=[\mathbf{m}_1, \ldots, \mathbf{m}_M]^T$ where $\mathbf{m}_i$, $i = 1, 2, \ldots, M$; while the data vector $\boldsymbol{d}$ is also written in the form of $\mathbf{d} = [\mathbf{d}_1, \ldots, \mathbf{d}_N]^T$. The data misfit vector is defined in the spatial domain by

$$\mathbf{e} = \mathbf{Km} - \mathbf{d} \tag{7}$$

In equation (7), forward modeling response $\mathbf{Km}$ is specified for the electrical problem where kernel matrix $\mathbf{K}$ is related to the analytical Green's function of the electric Poisson equation shown in equation (5). Using the least square solution of the misfit we construct an objective function ψ [13] that can be expressed as

$$\psi = \mathbf{W}_d \|\mathbf{Gm} - \mathbf{d}\|^2 + \alpha\|\mathbf{m}\|^2 \tag{8}$$

where $\mathbf{W}_d$ correspond to the data weighting matrix, $\| \ \|^2$ denotes the L-2 norm, α is the (spatial) regularization parameter and $\mathbf{G}$ is the Green's matrix computed from $\mathbf{G} = \mathbf{PK}^{-1}$ as a product of the inverse kernel matrix times the sparse selector matrix $\tilde{\mathbf{P}}$ that contains value of 1 on each row in the column that correspond to the location of the receivers.

The least square solution of equation (8) in now applied for analysis of an SP data collected at the Pinggirsari county, Banjaran, of the Bandung Regency. The goals of the survey are to identify the subsurface fault location and orientation, as well as to characterize the geologic strata over the area. In this study we conduct not only SP but also dc resistivity (or electric resistivity tomography / ERT) that we use for SP inversion input. In that sense, the SP data and the resistivity data are collected in the same profile line along the Pinggirsari - Majalaya county road. The result of the resistivity measurements is shown in the tomogram of figure 2a where possible existence of a fault on the subsurface is observed. In figure 2b we show the good fit of the inverse calculated electrical response (presented in black dots with error bars) and measured SP data (presented in blue line). Whereas in figure 2c we show the reconstructed volumetric current density that shows a dipole-like anomaly on the edge of the possible fault, with stronger positive magnitude. This could show the existence of possible groundwater flow along the fault plane from top to bottom direction. Another dipolar anomaly on the top middle-right of current density section will be analyzed for later study, but possibly not related to groundwater flow but rather electrochemical one.

4. The Future is Time-Lapsed Inversion?

The expansion of SP analysis from fixed geometry into continuous model inversion is necessary due to the SP inverse problem is similar in essence to electroencephalography (EEG) in medical imaging.

In this case, the subsurface is defined as a space-time model, which encompasses all space models during the entire monitoring periods. So now, the entire monitoring data are defined using 3D coordinates (x, z, t). In this case, subsurface model $\tilde{\mathbf{m}}$ is sparsely sampled at some pre-selected times and is expressed as $\tilde{\mathbf{m}}=\left[\mathbf{m}_1,\ldots,\mathbf{m}_t\right]^T$ where $\mathbf{m}_i$, $i = 1, 2, \ldots, \mathbf{t}$ is the model for the i-th time step and $\boldsymbol{t}$ is the number of monitoring times. The data misfit vector is defined in the space-time domain by

$$\mathbf{e}=\tilde{\mathbf{K}}\tilde{\mathbf{m}}-\tilde{\mathbf{d}} \quad (9)$$

In eq (10), the vector $\tilde{\mathbf{d}}$ corresponds to the electrical data vector defined in the spatial coordinate system by $\tilde{\mathbf{d}}=[\mathbf{d}_1,\ldots,\mathbf{d}_t]^T$. Forward modeling response $\tilde{\mathbf{K}}\tilde{\mathbf{m}}$ is specified for the electrical problem where kernel matrix $\tilde{\mathbf{K}}$ is related to the analytical Green's function of the electric Poisson equation.

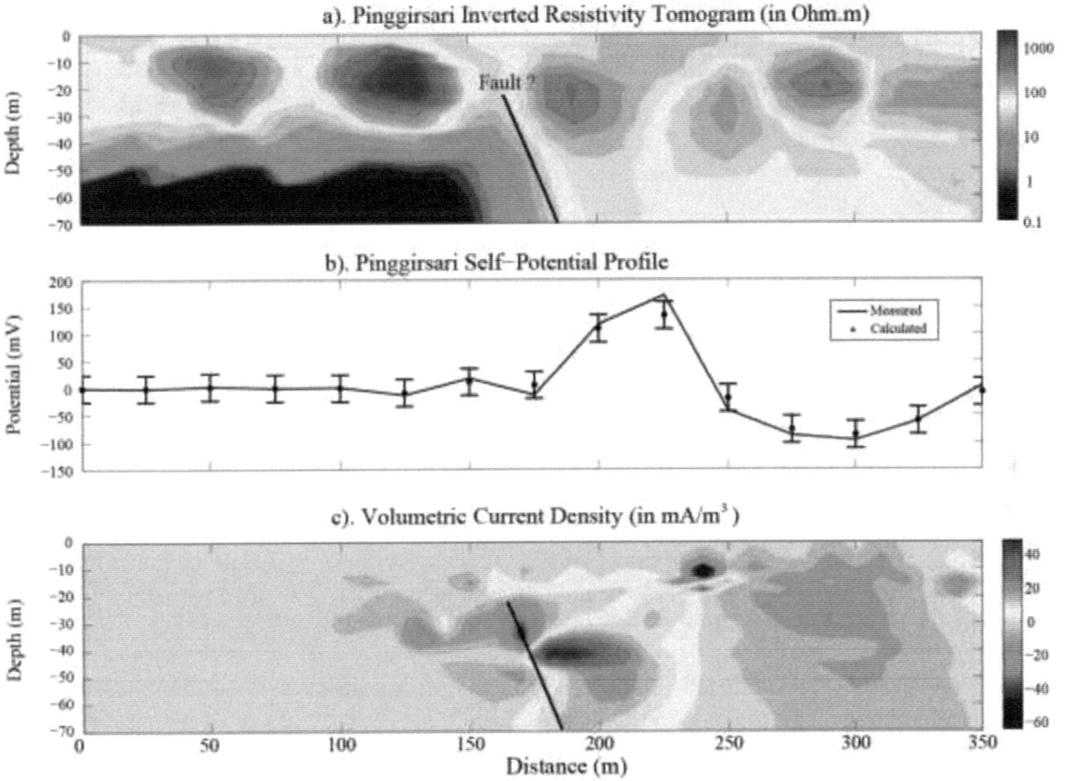

Figure 2. Results from Pinggirsari Survey. a). The resistivity tomogram shows possible existence of a fault on the subsurface. b). The inverse calculation of the electrical response (black dots) is in a good fit with the measured SP data (blue line). c). The reconstructed volumetric current density show dipole like anomaly on the edge of the possible fault.

The solution of eq (10) is computed with some modification of the 4D-ATC algorithm such that it will adopt two regularizations in space and time domains to stabilize the inversion calculation. The objective function $\Psi_{\tilde{m}}$ [14, 15] can be expressed as

$$\Psi_{\tilde{\mathrm{m}}}=\mathbf{W}_{\mathrm{d}}\left\|\tilde{\mathbf{G}}\tilde{\mathbf{m}}-\tilde{\mathbf{d}}\right\|^2+\alpha\left\|\tilde{\mathbf{m}}\right\|^2+\gamma\left\|\tilde{\mathbf{P}}\right\|^2 \quad (10)$$

where $\mathbf{W}_d$ correspond to the data weighting matrix, $\| \|^2$ denotes the L-2 norm, α is the (spatial) regularization parameter, Σ is a smoothness function for the time variation, γ is the time regularization parameters, and $\tilde{\mathbf{G}}$ is the Green's matrix computed from $\tilde{\mathbf{G}}=\mathbf{P}\tilde{\mathbf{K}}^{-1}$as a product of the inverse kernel matrix times the sparse selector matrix $\tilde{\mathbf{P}}$ that contains value of 1 on each row in the column that correspond to the location of the receivers.

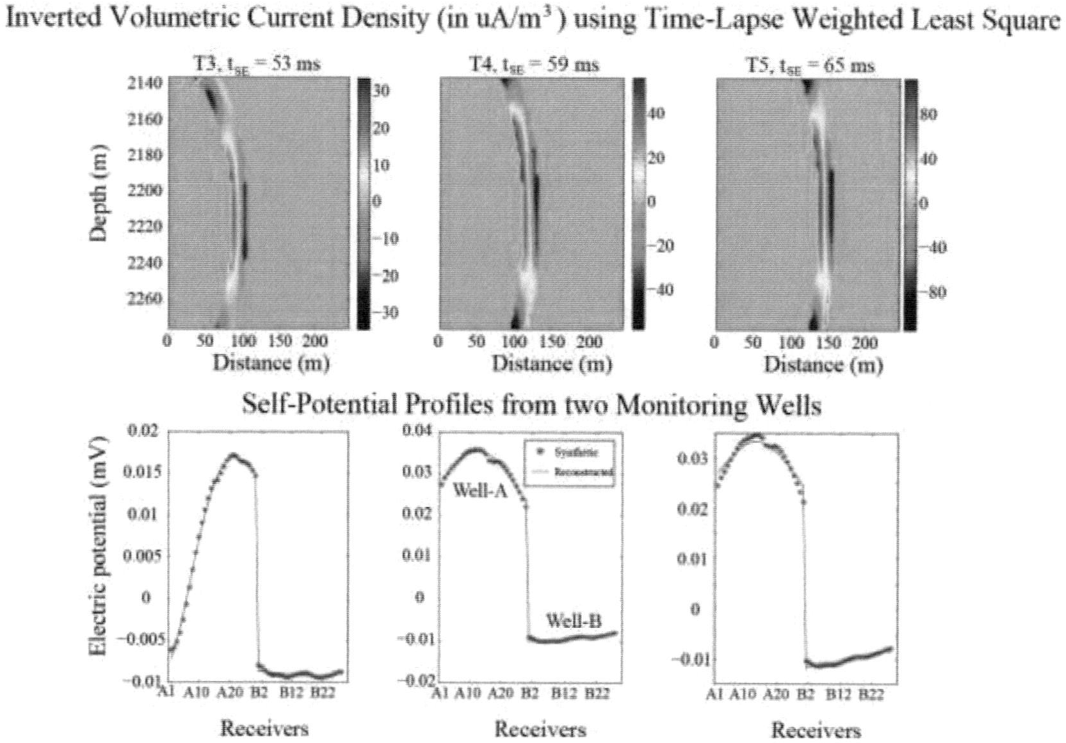

Figure 3. The inverted volumetric current densities of seismoelectric signals at any given arrival times for each saturation profiles (t_{SE} = 53 ms from saturation profile T3, t_{SE} = 59 ms from saturation profile T4, and t_{SE} = 65 ms from profile T5).

To illustrate the usefulness of time lapse inversion for SP response we use one example from water encroachment problem given in Revil and Mahardika (2013) [16]. Here, we use seismoelectric method which describes the generation of electrical and electromagnetic disturbances associated with the occurrence of seismic sources and seismic wave propagation in partially or totally water-saturated porous media. In other word, it is the active version of the typical SP method that we know. The author wish the reader to look at the given reference [16] for further details on the geometrical setting of the study, the process of the flooding monitoring, and the picking of the saturation profiles. In this problem, the volumetric current density (taken from the seismoelectric response due to water saturation contrast) calculated using the time independent least square solution did not give a satisfactory inversion result. So, we apply time constraining factor computed from the average radiuses of the wavefronts from saturation profiles T3 to T5. Using this adaptive function we get a new inversion result shown in figure 3. In the top half of figure 3 we show the inverted volumetric current densities concentration which all of the current densities are located at the weighted region or close to the waterfront boundary. Here, the dipolar shape-like current density solution is physically realistic due to seismoelectric effect is created from imbalance of charge distribution across the saturation contrast that induced electrical current via electrokinetic effect [17, 18]. While in the bottom half of figure 3 we show the computed electrical response is fairly in a good fit with the synthetic data taken from Well-A and Well-B.

5. Conclusion

In this paper we have shown examples of our current SP studies, one is the characterization of groundwater flow along a possible fault surface and another is the synthetic real time monitoring of fluid injection experiment. Through the expansion of fundamental theory, computational methods, and field-and-lab experiments in the last fifteen years, SP method has become a complementary package alongside electric-resistivity tomography (ERT) for detecting fluid flow for hydrological application, which also extends to the hydrothermal flow in geothermal areas [19]. As the analysis of SP data becomes more quantitative its potential applications becomes more diverse. Analysis of SP data should not done only with fixed geometry technique, but it have to expand to continuous model inversion due to its similarity to electroencephalography (EEG) in medical imaging. In the last decade, the recording and inversion of EEG signals has been instrumental in our understanding of how the brain works and in the mapping of its various functions. Therefore we can apply this technique in hydrology and geoscience in general in order to get similar purposes as EEG in neuroscience.

Acknowledgements

The author thanks the head of KK Fisika Bumi dan Sistem Kompleks Prof. Doddy Sutarno for his support on the project, Head of the Lab Permodelan dan Inversi EM Dr. Wahyu Srigutomo for his permission on using the data and the members of the Banjaran field-team for their collaboration on the project.

References

[1] J.E. Nyquist and C. E. Corry, The Leading Edge 21, pp. 446–451(2002).
[2] W. Srigutomo, A. S. Sunarya, C. M. Sufyana, A. Singarimbun, P. M. Pratomo, and E. C. Novana, "Pemodelan Self-Potensial untuk Investigasi Sistem Hidrotermal Dangkal Kawah Domas Gunung Tangkuban Parahu Bandung Jawa Barat," in Prosiding PIT HAGI 2009.
[3] M. H. Syahrudin, "Detecting Groundwater Flow In Porous Media Feeding Into A Nearby Spring," in Prosiding PIT HAGI 2010.
[4] M. Hamzah, D. Santoso, W. Parnadi, "Self-Potential Study For Laboratory Measurements Of Electrokinetic Potentetial From Fluid Flow In Porous Media," in Prosiding PIT HAGI 2007.
[5] S. Yungul, Geophys. 15, 237-246 (1950).
[6] M. Sato and H. M. Mooney, Geophys. 25, 226–249 (1960).
[7] A. Revil and A. Jardani, The Self-Potential Method Theory And Applications In Environmental Geophysics (CUP, Cambridge, 2013), pp. 23-77
[8] W. Srigutomo, E. Agustine, and M. H. Zen, Indones. J. Phys. 17, 49–55 (2006).
[9] A. D. Candra, W. Srigutomo, and B. J. Santosa, "A Complete Quantitative Analysis of Self-Potential Anomaly Using Singular Value Decomposition Algorithm," in Proc. of the IEEE Int. Conf. on Smart Instr., Meas. and Appl. (ICSIMA), DOI: 10.1109/ICSIMA.2014.7047419
[10] B. B. Bhattacharya, and N. Roy, Geophys. Prosp. 29, 102-107 (1981).
[11] S.P. Sharma and A. Biswas, Geophys. 78, WB3–WB15 (2013).
[12] A. Jardani, A. Revil, A. Boleve, Geophys. Res. Letters 34, L24403 (2007).
[13] B. J. Minsley, J. Sogade, and F. D. Morgan, J. Geophys. Res. 112, B02202 (2007).
[14] Y. Zhang, A. Ghodrati, and D.H. Brooks, Inv. Prob. 21, 357–382 (2005).
[15] J. H. Kim, M.J. Yi, S.G. Park, and J.G. Kim, J. Appl. Geophys. 68, 522-532 (2009).
[16] A. Revil and H. Mahardika, Water Resour. Res. 49, 744–766 (2013).
[17] Haartsen, M.W. and S. Pride, J. Geophys. Res. 102, 24745–24769 (1997).
[18] Haines, S., "Seismoelectric imaging of shallow targets", PhD. dissertation, Stanford Univ, 2004.
[19] A. Revil and A. Jardani, The Self-Potential Method Theory And Applications In Environmental Geophysics (CUP, Cambridge, 2013), pp. 192-282.

IOP Conf. Series: Earth and Environmental Science **132** (2018) 012022 doi:10.1088/1755-1315/132/1/012022

Gabor Deconvolution as Preliminary Method to Reduce Pitfall in Deeper Target Seismic Data

M Oktariena[1] and W Triyoso[2]

[1] Geophysical Engineering Department, Institut Teknologi Bandung
[2] Global Geophysics Group, Geophysical Engineering Department, Institut Teknologi Bandung

Corresponding author: madaniyaoktariena@outlook.com

Abstract. Anelastic attenuation process during seismic wave propagation is the trigger of seismic non-stationary characteristic. An absorption and a scattering of energy are causing the seismic energy loss as the depth increasing. A series of thin reservoir layers found in the study area is located within Talang Akar Fm. Level, showing an indication of interpretation pitfall due to attenuation effect commonly occurred in deeper level seismic data. Attenuation effect greatly influences the seismic images of deeper target level, creating pitfalls in several aspect. Seismic amplitude in deeper target level often could not represent its real subsurface character due to a low amplitude value or a chaotic event nearing the Basement. Frequency wise, the decaying could be seen as the frequency content diminishing in deeper target. Meanwhile, seismic amplitude is the simple tool to point out Direct Hydrocarbon Indicator (DHI) in preliminary Geophysical study before a further advanced interpretation method applied. A quick-look of Post-Stack Seismic Data shows the reservoir associated with a bright spot DHI while another bigger bright spot body detected in the North East area near the field edge. A horizon slice confirms a possibility that the other bright spot zone has smaller delineation; an interpretation pitfall commonly occurs in deeper level of seismic. We evaluates this pitfall by applying Gabor Deconvolution to address the attenuation problem. Gabor Deconvolution forms a Partition of Unity to factorize the trace into smaller convolution window that could be processed as stationary packets. Gabor Deconvolution estimates both the magnitudes of source signature alongside its attenuation function. The enhanced seismic shows a better imaging in the pitfall area that previously detected as a vast bright spot zone. When the enhanced seismic is used for further advanced reprocessing process, the Seismic Impedance and Vp/Vs Ratio slices show a better reservoir delineation, in which the pitfall area is reduced and some morphed as background lithology. Gabor Deconvolution removes the attenuation by performing Gabor Domain spectral division, which in extension also reduces interpretation pitfall in deeper target seismic.

1. Introduction

In its ideal form, seismic trace should be a product of convolution between subsurface reflectivity and source waveform. However, seismic data recorded is a filtered result of subsurface reflectivity during the wave propagation process infiltrating layers of lithology below the earth surface. The filter is a combination between the source wavelet used during seismic acquisition along with the attenuation effect during seismic wave propagation through various lithology condition. Anelastic attenuation process during seismic wave propagation is the trigger of seismic non-stationary characteristic. An absorption and a scattering of energy are causing the seismic energy loss as the depth increasing. In extension, attenuation leads to a loss in amplitude as well as frequency content of seismic data, or known as characteristics to the non-stationary condition of seismic data. As a function of energy loss, a deeper target of seismic data would suffer the most of attenuation effect. It is a common knowledge that a reservoir located deep below earth surface such as Talang Akar Fm. is difficult to delineate due to the poor seismic resolution and its typical thin reservoir grouped as a sand/shale crossover. The exploration challenge is how to optimize and quite possible, to enhance the seismic resolution in order to show a better imaging of hydrocarbon reservoirs through a conventional seismic data.

Responding to this challenge, preliminary conditioning process is the next sensible step to do in order to enhance seismic data resolution before it is being used as an input for interpretation or advanced reprocessing workflow. Deconvolution is a common method to boost seismic resolution during seismic data processing. Several assumption takes place in deconvolution process. One of them is assuming a stationary condition in seismic data. This assumption allows the method to simply calculate one deconvolution operator value applied to the whole seismic data. For many years, this assumption serves a great approach to the real condition, albeit it is still quite not right to apply stationary assumption for a seismic data which almost always identified as having a variability in time, commonly known as non-stationary characteristic.

Gabor Deconvolution is a non-stationary deconvolution method done in frequency domain. It is a modification of Wiener method for a non-stationary condition such in seismic data. Using Lamoureux window, Gabor Deconvolution divides the non-stationary seismic trace into a smaller localized signal called Gabor Wave Packets. By doing Fourier analysis in each localized packages, Gabor Deconvolution allows a non-stationary filtering in seismic trace represented by the various deconvolution operator generates for one seismic trace [1]. On its wake, Gabor Deconvolution also estimates Spectrum of Propagating Wavelet, which is a combined function between source wavelet and the attenuation occurred during propagation. The purpose of applying Gabor Deconvolution as preliminary conditioning method is to restore the amplitude and frequency content diminished by attenuation effect in order to get the seismic data recorded resembling its' supposed to be subsurface condition. In extention to the enhancement of seismic data resolution, Gabor Deconvolution could also reduce the interpretation pitfall of deeper target seismic data caused by amplitude failure in identifying Direct Hydrocarbon Indicator (DHI).

2. Methodology

Deconvolution is a main step in seismic data processing which aim to enhance the data resolution and produce an adequate imaging of subsurface condition through a conventional seismic data. The main purpose of deconvolution is to compress source waveform effect in the data recorded so the output seismic would increase its vertical resolution and resemble the actual reflectivity series. The most common method of deconvolution is Wiener Deconvolution and Predictive Deconvolution that categorized into statistical deconvolution. Both assumes seismic data as a stationary signal, in which against the fact that seismic data is a non-stationary signal that suffers energy loss as the depth progressed. Gabor Deconvolution, which works in frequency domain, is a deconvolution method that approach this non-stationary condition of seismic data using a modification of Wiener method.

Dennis Gabor in 1946 brought up the idea of applying Fourier analysis in partition sense [2]. A series of non-stationary seismic trace is divided into smaller signal packet which was localized by a certain window. The smaller signal packet is called Gaussian Wave Packet, localized using Gaussian window at first attempt by the inventor. Later on, Gabor Deconvolution has developed further by Margrave et. al. [3] and now using Lamoureux window to decompose trace into Gabor Wave Packet. The Lamoureux window provides a more localized signal due to its zero condition on each end point whereas the Gaussian window only decays to each end point; never touches zero condition.

Gabor Deconvolution has a similar workflow to the conventional seismic deconvolution process since it is basically only a modification in which the process is applied to a smaller and more localized signal. Fourier transform is applied to each Gabor Wave Packet making it possible to do a non-stationary filtering to seismic trace. Hyperbolic time-frequency smoothing is applied to seismic in

IOP Conf. Series: Earth and Environmental Science **132** (2018) 012022 doi:10.1088/1755-1315/132/1/012022

Gabor Spectrum in order to estimate Spectrum of Propagating Wavelet, or in the conventional deconvolution model known as the deconvolution operator. The source waveform as well as attenuation effect are removed by doing a spectral division between seismic in Gabor Spectrum and its deconvolution operator. This leaves only the estimated reflectivity in its output making it suitable to apply in a case where seismic energy loss clouding the real amplitude response commonly found in deeper level of seismic data.

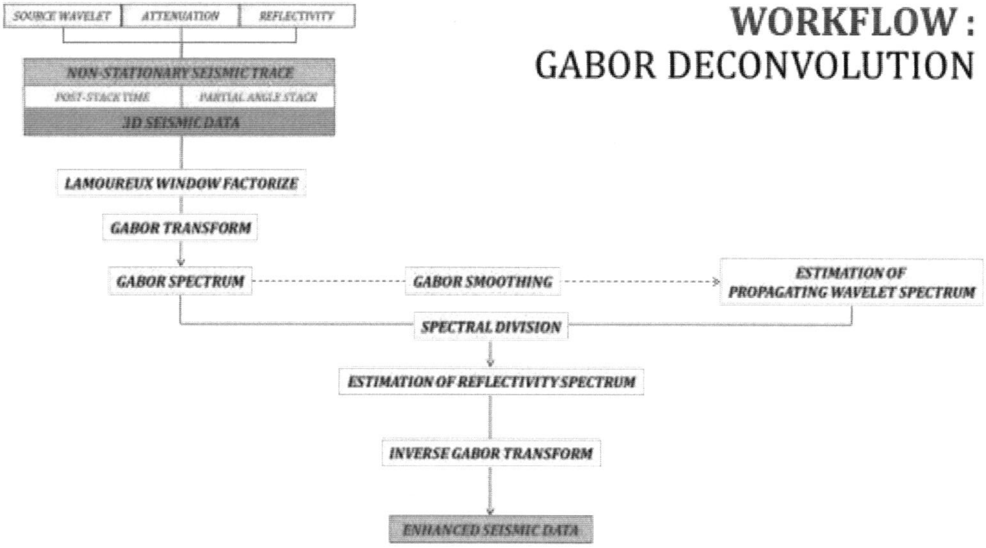

Figure 1. Gabor Deconvolution workflow (after Margrave et. al., 2002)[3] in general sense.

Gabor Deconvolution consists of several step that illustrated in Figure 1. The basic steps are choosing width of stationary packets from non-stationary seismic signal, converting seismic data into Gabor domain, generating a Gabor slice using a moving Lamoureux window, smoothing the seismic data in Gabor domain into the Spectrum of Propagating Wavelet, deconvolving initial seismic signal by doing a spectral division with the propagating wavelet, and performing time-variant filter if necessary. The algorithm is quite complex, but has a benefit of approaching a solution for non-stationary seismic data.

The final goal of this study is to compare the benefit in applying Gabor Deconvolution as a preliminary conditioning method before embarking a further interpretation and advanced reprocessing workflow. The enhanced seismic data would be used as an input of workflow for detecting a possible reservoir delineation in deeper target seismic. A production report and log sensitivity analysis confirm that the reservoir in this study area associated with a bright DHI area and small value of Vp/Vs Ratio. Vp/Vs Ratio volume would be obtained from Aki-Richard Partial Stack Inversion algorithm formulated by Supriyono [4]. P&S-Reflectivity volume would be created from Partial-Angle Stack Seismic Data (Near & Far) and inverted to P&S-Impedance using Model-Based Hard Constraint method. The output Vp/Vs slice from the enhanced seismic data and the initial seismic data would then be compared to showcase the reduction of reservoir area caused by amplitude pitfall commonly found in deeper target seismic data.

3. Results & Discussions

Gabor Deconvolution is evaluated as seismic enhancement method focusing on its reducing pitfall benefit in deeper target seismic data. The data used in this study is a marine seismic data with proven hydrocarbon reservoir status in Talang Akar Fm. The study will focus on PRAIA sequence which is part of Talang Akar Fm showing a sand/shale crossover with the producing level located exactly in the middle of its thickness. LJB-01 is an oil-producing well associated with a bright-spot DHI and a small Vp/Vs Ratio value. Gabor Deconvolution has proven could resolute a local fluid trap problem of the upper level reservoir and minimize the attenuation effect when it is applied to Post-Stack Seismic Data as shown in Figure 2 [5]. Referring to previous study, Gabor Deconvolution in Ensemble Mode produces a more stable output and preserves the DHI well. Gabor Deconvolution Ensemble Mode would be used as preliminary conditioning method for Partial Angle Stack Seismic data in this study.

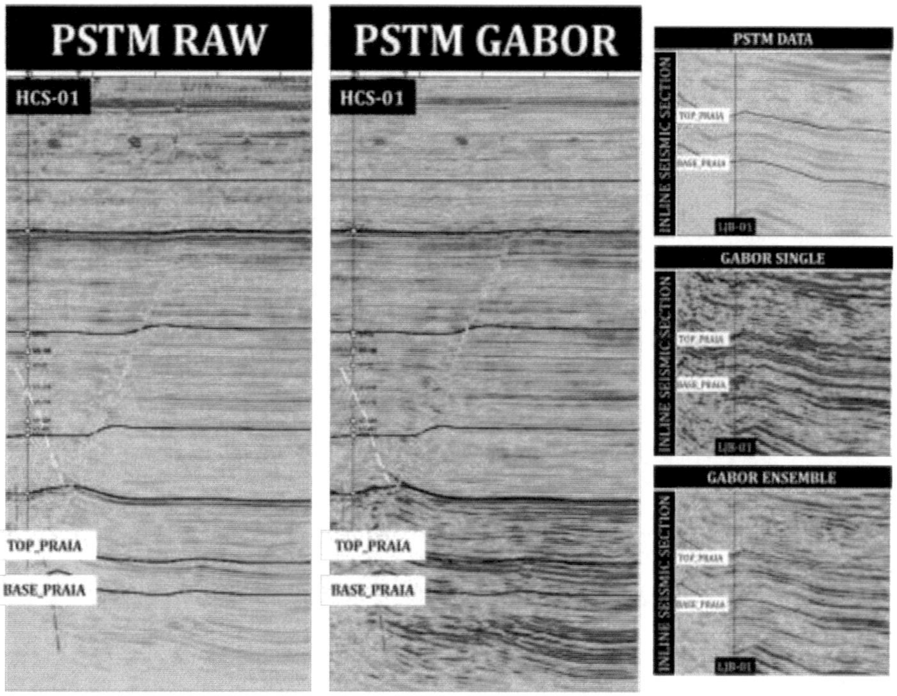

Figure 2. Previous study had researched the benefit of Gabor Deconvolution as preliminary conditioning technique applied to Post-Stack Seismic Data. Gabor Deconvolution in Ensemble Mode has proven to be a suitable technique for seismic enhancement purpose in this study area [6].

There are several main parameter that influence the quality of output result during Gabor Deconvolution process. To find a suitable value, a parameter optimization is a must step to undergo in which we could determine the most effective value for each parameter and also find the best value combination between all parameters. The optimization is analyzed for half-width analysis window, window overlap factor, slope exponent for window, and corridor width for hyperbolic smoothing. The optimization focuses on the Lamoureux window parameter and a smoother parameter to determine the attenuation function. As the input data in this study consists of two different datasets, Near Angle Stack and Far Angle Stack, the Gabor Deconvolution parameter would be slightly different from one another. Figure 3 shows a comparison between initial Partial Angle Stack datasets with its enhanced result. Notice that after applying Gabor Deconvolution, the number of event between TOP_PRAIA and BASE_PRAIA is increasing. Several thin layers could be resolute. It is a given because the tuning thickness has been lowered by Gabor Deconvolution. The initial seismic data has 19,899 m tuning thickness and the enhanced seismic data has a 14,225 m tuning thickness while the thinnest reservoir layer in PRAIA sequence is around 15 m [6]. The most noticeable thing, after Gabor Deconvolution, the amplitude level around PRAIA level has been restored to be as strong as a shallow level of seismic data. It is one of indication that the attenuation effect has been reduced from the initial seismic data. If we take a look at the frequency content and amplitude content of the enhanced seismic datasets as shown in Figure 4, it has proven that Gabor Deconvolution method could restore the amplitude and frequency content of seismic data back resembling the supposed to be stationary condition. The initial Partial Angle Stack datasets has a decaying frequency content as well as an amplitude loss both in lower and higher side. After applying Gabor Deconvolution, the amplitude spectrums are back resembling the perfect boxcar and the frequency content appears in all time-frequency plot. The diminishing in frequency and loss in amplitude content are often the cause of DHI pitfall in deeper level seismic data.

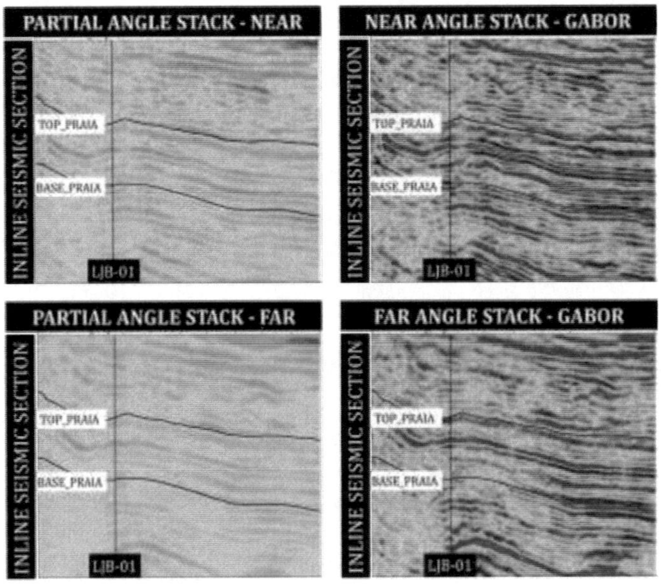

Figure 3. Comparison of seismic section in inline 211 around CDP well LJB-01. PRAIA sequence is located within Talang Akar Fm. Left side shows the initial Partial Angle Stack Seismic. Right side shows the enhanced datasets after applying Gabor Deconvolution as preliminary conditioning method.

Reservoir delineation would be analyzed through Vp/Vs Ratio maps with assistance from reflectivity volume. Vp/Vs Ratio is generated from Partial-Angle Stack Seismic datasets using Aki-Richard Partial Stack Inversion [4]. P&S Reflectivity volumes are produced from Partial-Angle Stack datasets by a least-square inversion of Zoeppritz Aki-Richard simplification for two fold datasets case as the Partial-Angle Stack data we use has only the near and far part. The formulation of reflectivity volume could be studied further in the reference research. Because the input for calculating the reflectivity volume is in fact two different datasets, we must ensure that the two has a similar amplitude and bandwidth level before joining the two together. This purpose is achieved through a spectral matching step. Then P&S Reflectivity are calculated as an input for Model-Based Hard Constraint seismic inversion to produce P&S Impedance. Gabor Deconvolution shows its benefit in increasing the correlation between estimated impedance from inversion result and hard-data impedance from well-log as shown in Figure 5. The crossplot shows that after applying Gabor Deconvolution, the data is more focusing around the red line and the scattered point is decreasing. The estimation error also reduces from 2588,04 to 2296,82 point.

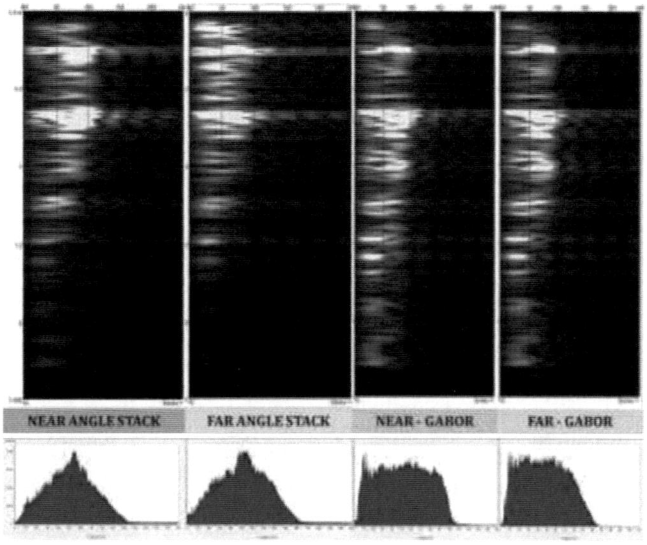

Figure 4. Frequency and amplitude content comparison between initial Partial-Angle Stack Seismic and the Gabor Deconvolution results. Above part shows time-frequency plots using Spectral Decomposition technique for a -10 to 10 Hz window. Below part shows amplitude spectrum with amplitude in Y-Axis and frequency in X-Axis.

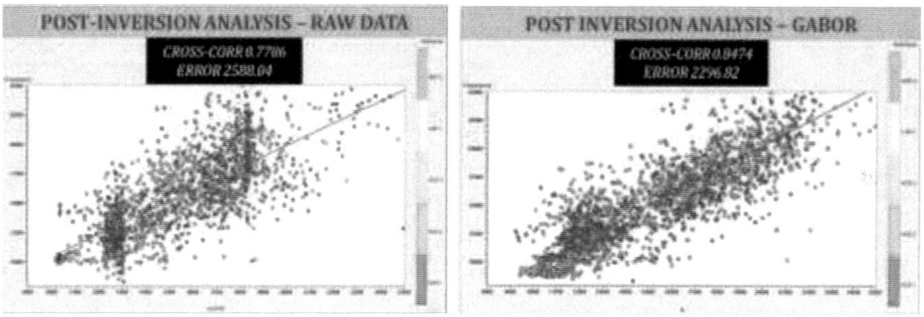

FIGURE 5. Comparison of post-inversion analysis before and after applying Gabor Deconvolution.

The inversion result shows that after applying Gabor Deconvolution, P&S Impedance along with Vp/Vs Ratio section could resolute more event compared to the initial seismic. The enhancement result is strongly appointed around the black arrow area where several thin layer appears above the arrow (Figure 6). The arrow itself points at a single thick layer in initial seismic data but the layer turns out to be a doublet layer that could be separated into two different layers after applying Gabor Deconvolution. The pitfall found in this exact layer. This layer is the producing layer in PRAIA sequence with LJB-01 proven as oil-producing well. Post-Stack Seismic Data (PSTM) associates this oil saturated layer with a bright spot area [6]. To the north-east area, the PSTM data shows a similar bright spot that potentially saturated with oil. The PSTM slice maps a vast bright spot area while the Partial-Angle Stack Data shows that the bright spot is actually smaller. The horizon slice of enhanced seismic result after Gabor Deconvolution shows a more detailed reservoirs delineation (Figure 7). Area above the arrow in initial seismic data detects as a bright zone and its produced Vp/Vs Ratio also shows a small value indicated the area as a hydrocarbon saturated zone. If the development decision is based from these information, that above the arrow area would be choose as the next well location. But when we analyze the map after Gabor Deconvolution enhancement, that area actually contains a blotchy saturated area surrounded by non-saturated zone with the saturated intensity is yellow to green. The most possible hydrocarbon location is indicated by a red area in Vp/Vs Ratio slice. This is the amplitude pitfall that could mislead the development plan if the decision solely relies on DHI. Even if we don't produce Vp/Vs Ratio slice as a justification for development plan, the enhanced seismic after Gabor Deconvolution itself could already help to delineate the reservoir better. So by using Gabor Deconvolution as a preliminary method for further interpretation or advanced reprocessing workflow, the pitfall and interpretation uncertainties could be reduced further.

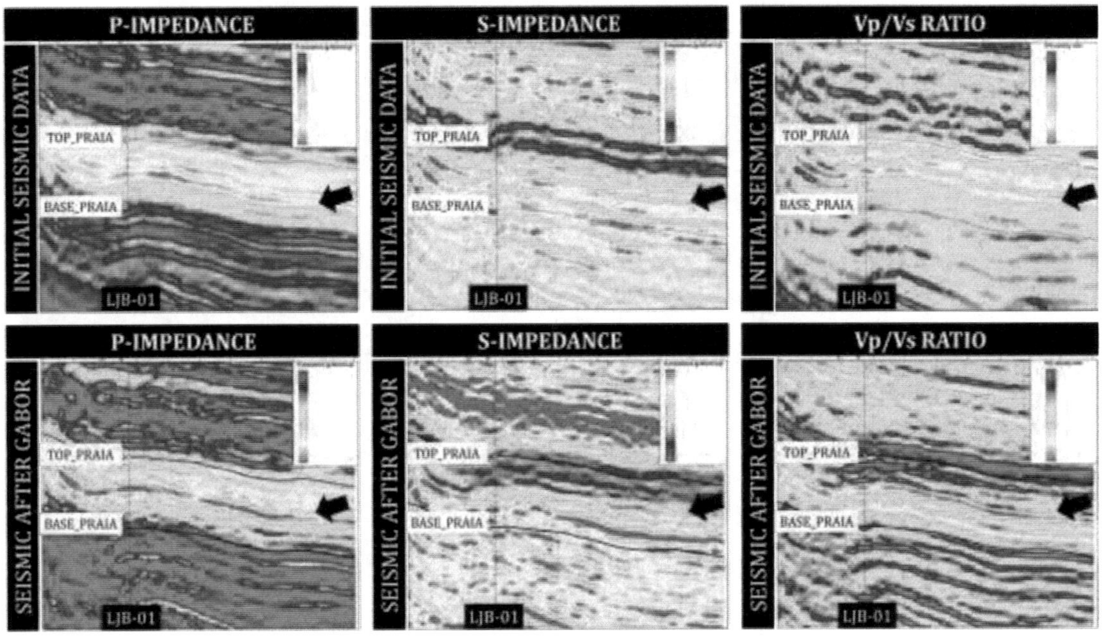

Figure 6. Output comparison of Aki-Richard Partial Stack Inversion. Upper side shows P-Impedance, S-Impedance, and Vp/Vs Ratio of initial seismic data. Lower side shows the results from enhanced seismic after applying Gabor Deconvolution. Black arrow shows oil-producing layer in PRAIA sequence.

4. Conclusions

A loss in amplitude and frequency content caused by attenuation effect are proved to increase interpretation uncertainty in deeper level of seismic data. Amplitude failure in detecting hydrocarbon indicator is a common pitfall occurred in area with strong attenuation influenced such as Talang Akar Fm. Without any delicate treatment, the amplitude could lead to a bizarre development plan assisted by a false accuracy of reservoir delineation. Gabor Deconvolution as a preliminary conditioning method has benefit in enhancing seismic imaging and vertical resolution in deeper target seismic data by removing attenuation effect that cause a non-stationary condition in seismic data. The enhanced seismic from Gabor Deconvolution could recover a thin layer and separate the doublet within PRAIA sequence in Talang Akar Fm for this study area. Gabor Deconvolution has proven an aid to reduce a possible development risk by delineating the reservoir in more detailed manner even before the advanced reprocessing method is applied. By using Gabor Deconvolution as preliminary conditioning method, the result of further reprocessing workflow, such as Aki-Richard Partial Angle Stack Inversion, could achieved even more satisfying justification to determine hydrocarbon delineation in deeper target seismic data.

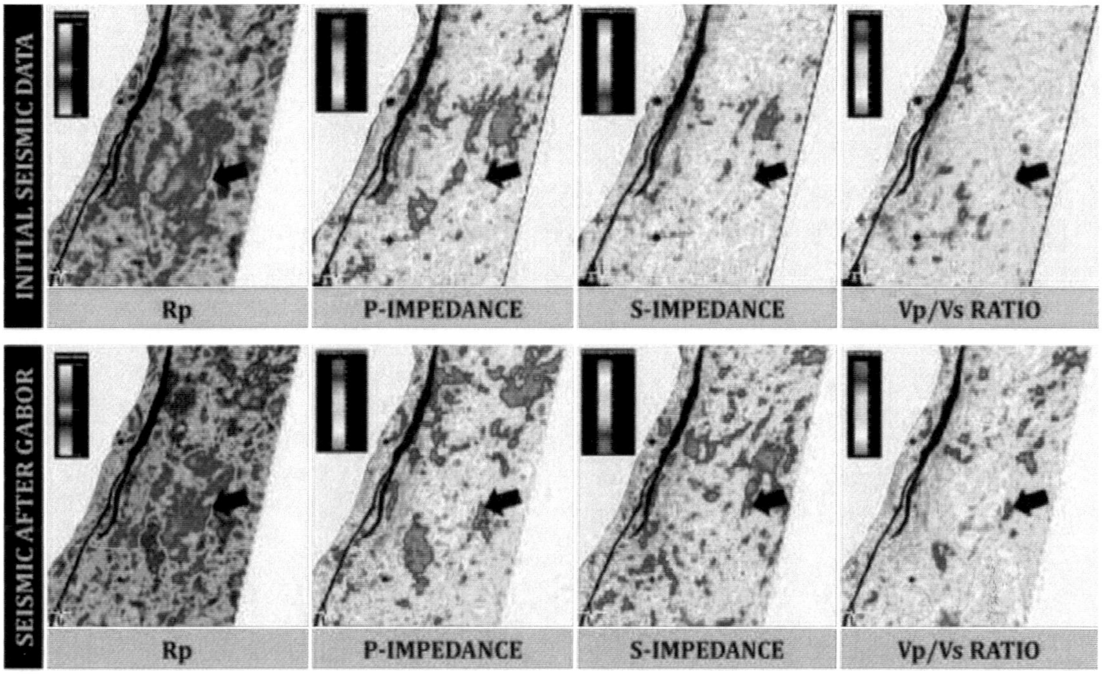

Figure 7. Horizon slice of oil-producing layer in PRAIA sequence. Upper side shows Reflectivity P, P-Impedance, S-Impedance, and Vp/Vs Ratio of initial seismic data. Lower side shows Reflectivity P, P-Impedance, S-Impedance, and Vp/Vs Ratio of enhanced seismic result after applying Gabor Deconvolution as preliminary conditioning method. Black arrow shows the possible hydrocarbon delineation from well LJB-01 area.

References

[1] Margrave G F, Dong L, Gibson P, Grossman J P, Henvel D, and Lamoureux M P 2003 Gabor Deconvolution: Extending Wiener's Method to Nonstationary *CREWES Research Report* **15** 1–27

[2] Margrave G F and Lamoureux M P 2001 Gabor Deconvolution *CREWES Research Report* **13** 241–302

[3] Margrave G F, Henley D C, Lamoureux M P, Iliescu V, and Grossman J P 2002 An Update on Gabor Deconvolution *CREWES Research Report* **14** 1–27

[4] Supriyono 1999 *Prediksi Sifat-Sifat Fisika Lapisan Bawah Permukaan Berdasarkan Analisis Tanggapan Inelastis dan Elastis Batuan Terhadap Gelombang Seismik* (Jakarta: Magister Thesis/Universitas Indonesia)

[5] Oktariena M and Triyoso W 2016 Comparison of Seismic Reconvolution and Gabor Deconvolution in Improving Seismic Images to Detect Local Fluid Trapping *Shifting Paradigms in Indonesia, Supplying Energy in The New Reality-2016 IPA CONVEX Convention Proceeding* **40** 16-102G

[6] Oktariena M 2016 *Dekonvolusi Gabor dan Seismik Rekonvolusi dalam Optimasi Resolusi Seismik sebagai Metode Pre-Conditioning untuk Inversi Seismik Partial Angle-Stack* (Bandung: Magister Thesis/Institut Teknologi Bandung)

Estimation of Existence Geothermal Manifestation Using Very Low Frequency (VLF) Method in the PagerkandangVulcanic, Dieng, Central Java

Asri Wulandari[a)], Ega Asti Anggari, Novi Dwiasih, Imam Suyanto
Gadjah Mada University, Geophysics Department, Yogyakarta, Indonesia

[a)]asriwulandaridari@gmail.com

Abstract. Very Low Frequency (VLF) measurement has been done at Pagerkandang Volcanic, Dieng Volcanic Complex (DVC) to examine the possible existence of conductive zones that related with geothermal manifestation. VLF – EM survey used tilt mode with T-VLF BRGM Iris Instrument operated with two frequencies, they are 22200 Hz from Japan (JJI) and 19800 Hz from Australia (NWC). There are five lines with distance between lines is 50 m, and distance between measure points is 20 m.The parameters measured from VLF method are tilt angle (%) and elliptisity (%). Data processed by tilt angle value with fraser and Karous – Hjelt filter used WinVLF program. Karous – Hjelt filter resulted current density contour to estimate lateral location from conductive and resistive zones. The conductive zone is interpreted as the area which have high current density value. This area located at eastern dan western of Pagerkandang Volcanic. The conductive zone related to geothermal manifestation as like as fumarol that appeared because presenced of normal fault. Whereas the resistive zone is interpreted as the area which have low current density value. This area spread almost in the middle of the Pagerkandang Volcanic. The resistive zone was caused by the high weathering in claystone.

Keywords : Very Low Frequency, Pagerkandang Volcanic, geothermal manifestation

1. Introduction

Dieng Volcanic Complex (DVC) is one of the productive geothermal fields in Indonesia. Dieng geothermal systems have a high temperature and liquid-dominated. High temperatures in geothermal systems can be caused by the age and composition of the heat source (Wohlez and Heiken, 1992; Muffler and Duffield, 1995 in Harijoko A et al. 2016). The total potential of geothermal energy in Dieng is estimated at 355 MW but only 60 MW which has been utilized as a source of power plant. Pagerkandang volcanic is one area that has a geothermal prospects associated with Sileri Crater. Geothermal manifestations can be fumaroles, acid sulfate spring, mud pools, steaming ground and rock alteration (Harijoko A et al. 2016).

VLF method utilizing electromagnetic wave induction can be used to determine the continuity of conductive regions geothermal area based on their surface manifestation. Control of the fault structure largely determines the location of these manifestations. Open fractures formed by the structure used as the hydrothermal fluid flow path and carries with her metallic minerals. In this paper conducted qualitative interpretation of the value of the tilt angle using fraser filter and Karous – Hjelt filter. Karous –Hjelt filter will generate value for the current density equivalent (RAE) to estimate the distribution of the resistive and conductive zones.

2. Geological Setting

Dieng Volcanic complex had being formed by volcanic stratocones, explosion craters, parasitic vent, and also hydrothermally active which originated at the junction of two major fracture zones. The fracture is an east-west trending zone, extending due west from the Dieng Mountains 50 km to the Mount Slamet and the one consist of a row of young cones including Mount Sindoro-Sumbing which extend southeast of the Dieng Mountains for 25-35 km (Zen M T, 1971; Dan Miller C et al., 1983).

Dieng Volcanic Complex also consist of three episodes, there are pre-caldera episode, post- caldera episode and the present episode. The research area located in Pagerkandang Volcanic which included in the post-caldera episode where volcanic activity develops in the caldera. Pagerkandang Volcanic has a crater on the northern part, and there are geothermal manifestations (solfatara and fumarol) scattered along the inside and outside of the crater. Most of the hydrothermal activity are associated with explosion crater, tectonic lineament, or fracture zones in the Dieng Volcanic Complex. According Jacobson (1970) in Zen M T (1971), based on electrical geophysical survey, the thermal areas outlined by surface manifestations and the belt of anomalously low resistivity that related to the northwest - southeast trending lineament. The thermal belt coincides with the zone of volcanic lineament which extends from Pakumadja to Pagerkandang. The research area is dominated by material of post-caldera eruptions, such as olivine andesite, hornblende andesite, and biotite andesite (Dan Miller C et al., 1983). Alteration process changes volcanic rocks into clay minerals which have high conductivity (low resistivity). The alterated rocks are layered by sulfur minerals and are founded surrounding the manifestation. This minerals give important information types and locations of Dieng geothermal heat source system.

3. Very Low Frequency (VLF) Method

VLF method uses electromagnetic waves of radio waves having a frequency ranging from 15 – 30 kHz. Primary electromagnetic field of the transmitter has components vertical electric field and horizontal magnetic that perpendicular to the direction of propagation (the x axis). If a conductive medium exists in the subsurface, the magnetic component of the electromagnetic waves would induce the medium caused induced currents or Eddy currents. Eddy currents would product secondary electromagnetic field (H_s), which has horizontal and vertical components. Interference between the primary magnetic field and secondary magnetic field generates a magnetic field which is resultant polarization ellipse. The components are measured in the VLF method is a tilt angle (α) and ellipticity (ε).

Qualitative interpretation of VLF method data can be performed with fraser filter and Karous -Hjelt filter (linear filter). Data used in the process of filter is the tilt angle (%). In fraser filter large amplitude can be estimated as conductive zones (Sundararajan N et al., 2006). The highest of fraser filter graph shows the increase of the signal conductive medium, where in the conductive anomaly is considered appropriate under the peak graph fraser filter (Santos et al., 2006).The filter divides tilt angle data with 90° to transform zero-crossing into peak, and a low pass smoothing operator to reduce noise, such as topographic effect.

$$f_{2,3} = (M_3 + M_4) - (M_1 + M_2) \qquad (1)$$

$f_{2,3}$ is filter fraser value that plotted midway between $M_2$and M_3 tilt angle data.

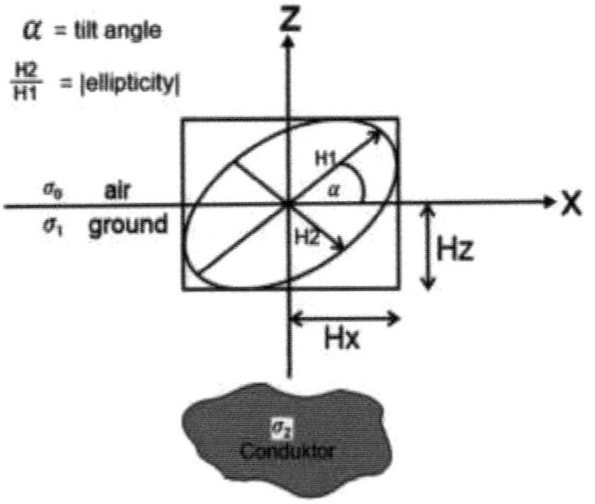

Figure 1. Ellips polarization because of existence conductive object in electromagnetic field (Bahri A S).

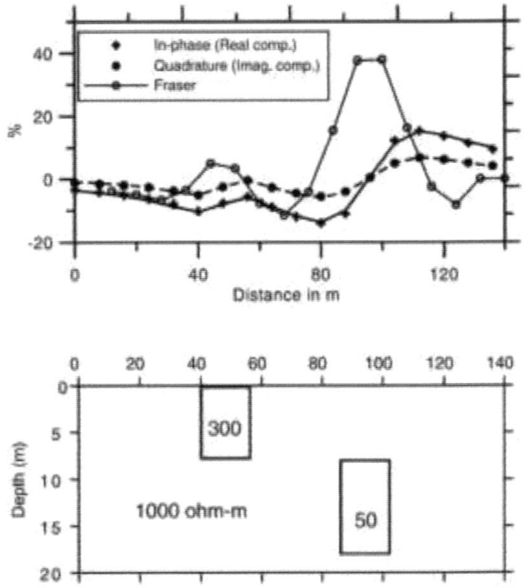

Figure 2. Synthetic Data VLF-EM and fraser filter data that is correlated with the 2-D resistivity models (Monteiro Santos F A et al., 2006).

Karous - Hjelt Filter used to determine the value of the current density equivalent. Lower values of current density shows a resistive zone and vice versa (Benson et al., 1997 in Monteiro Santos F A et al., 2006). Karous - Hjelt filter using linear filter theory to solve the integral equation in the current distribution. The linear equation obtained current density values for various depths along the line from vertical magnetic field each measure point. The formula used is:

$$\frac{\Delta z}{2\pi}\left(\frac{I}{A}\right)\frac{\Delta x}{2} = -\,0{,}102\,H_{-3} + 0{,}059\,H_{-2} - 0{,}561\,H_{-1} + 0{,}561\,H_{1} - 0{,}059\,H_{2} + 0{,}102\,H_{3} \qquad (2)$$

with Δz is the depth (m), $(\frac{I}{A})$ is the current density (%), Δx is spaced measurements (m), and H is the tilt angle of the data (%). Current density $(\frac{I}{A})$ is the flow of the charge current *(I)* in a cross-sectional area *(A)* at a certain point conductor. The position of current dencity can used to interpret the wide and dip of anomaly in certain of depth.

4. Methodology

VLF – EM survey has been done at Pagerkandang Volcanic, DVC. VLF – EM survey used tilt mode with T-VLF BRGM Iris Instrument operated with two frequencies, they are 22200 Hz from Japan (JJI) and 19800 Hz from Australi (NWC). There are five lines with distance between lines is 50 m, and distance between measure points is 20 m. Data processing for Fraser and Karous – Hjelt filter used WinVLF program by Mufaqih A A (2015).

5. Result and Discussion

Based on dissemination of tilt angle at location of VLF measured, the high tilt angles are located at the edge of Pagerkandang Volcanic, with tilt angles are 30 – 80 % (red colors). Whereas in mid Pagerkandang Volcanic has tilt angles ranging from 25 – (-40) % (green colors). The tilt angle related to horizontal and vertical magnetic field, and indirectly can be used to determine conductive zone from high tilt angle value.

Tilt angle values obtained in the field are still influenced by external noise that has a high frequency. Because of the assumption of the waves received by VLF - EM is a wave with a low frequency, the high frequency noise to be removed, one of them uses a moving average filter. Qualitative interpretation on VLF - EM method is using Fraser filter and Karous – Hjelt filter of tilt angle values. Karous – Hjelt filter will generate value for the current density to estimate the lateral location of the resistive and conductive zones. Conductive zones are estimated to be below the chart fraser filter with high amplitude (peak chart). While the linear filter (Karous–Hjelt filter) current density value is a predicted high conductive zone.

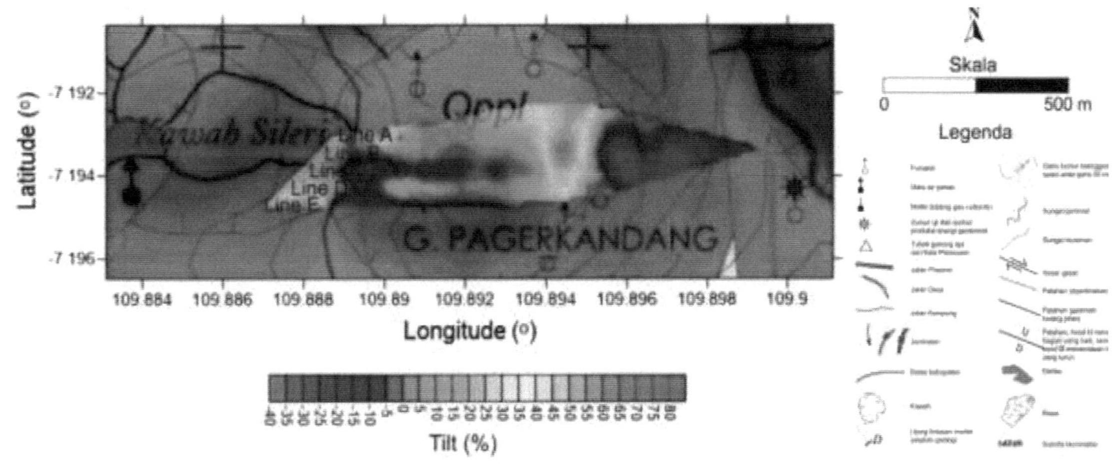

Figure 3. Distribution of tilt angle value at survey area.

On the graph of Fraser filter and Karous-Hjelt filter look anomalous responses that are interrelated. On Line A conductive zones are located at a distance of 120-240 m or on the eastern side of the Pagerkandang Vulcanic assuming depth (skin depth) up to 100 m. Conductive zones on Line A have a value of current density between 8–25%. While the resistive zones on Line A are spread between 300-600 m, with a value of current density between (-5) - (-20)%.

The conductive area on Line B is at a distance 500 m (east) and 680-1180 m or on the western side of the Pagerkandang Vulcanic has a value of current density between 10-50% if the skin depth is up to 200 m. High resistivity area in the middle of the Pagerkandang Vulcanic has a value of current density between -10 - (-30)%.

The conductive zones on Line C is between 160-220 m and 560-800 m which has a continuity with the conductive zones on Line B that located on the western side of the PagerkandangVolcanic. The value of currentdensity on Line C is between 8–40% including a conductive area. High resistivity area is located at 380–480 m which is also located in the central part of the Pagerkandang Vulcanic, with a value of current density between (-12) - (-28)%.

High conductivity zones on Line D are located at a distance of 180-360 m (east) and 600-700 m (west) with current density between 6-20%. High conductivity area in the western side predicted still associated with Line B and Line C in the south. While the resistive area is located at a distance of 120 m - 180 m (east) and 380-560 m (east) with current density between (-4)% - (-20)%, and the skin depth reaches 120 m.

The measurements on Line E was began on the western side of the Sileri Crater bordering the east side of the Pagerkandang Vulcanic. High conductivity zones are located at a distance of 80-260 m or in transition from Sileri Crater and Pagerkandang Volcanic with current density between 20–50%. High

IOP Conf. Series: Earth and Environmental Science **132** (2018) 012023 doi:10.1088/1755-1315/132/1/012023

resistivity areais located at a distance of 270-620 m, or scattered in the central part of the Pagerkandang Vulcanic with a value of current density between 10 - (-10)% and the skin depth reaches 100 m.

The existence of high conductivity area can be influenced by the presence of geothermal manifestations such as fumaroles, fracture structure, and igneous body (intrusion). Fumaroles discovered in the eastern side of Line A and Line C, which is associated with the normal fault as a result the area has high permeabilitythat allows the upwards flow of the thermal fluids, so the geothermal manifestations may appear on the surface. Fumaroles is also located in the western PagerkandangVulcanic, indicated by high conductivity values are constantly on the western side of Line B, Line C, and line D.

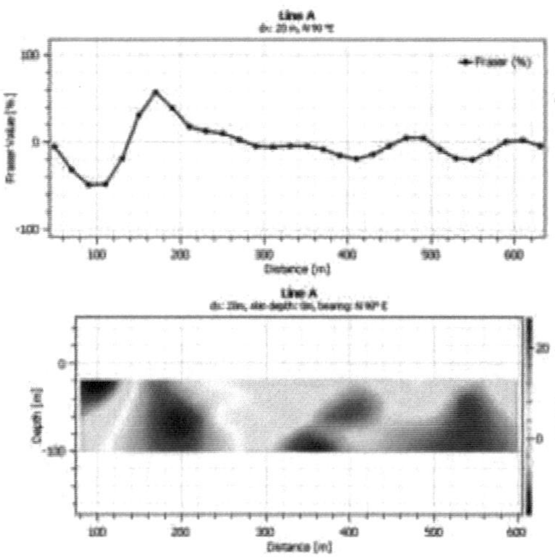

(a)

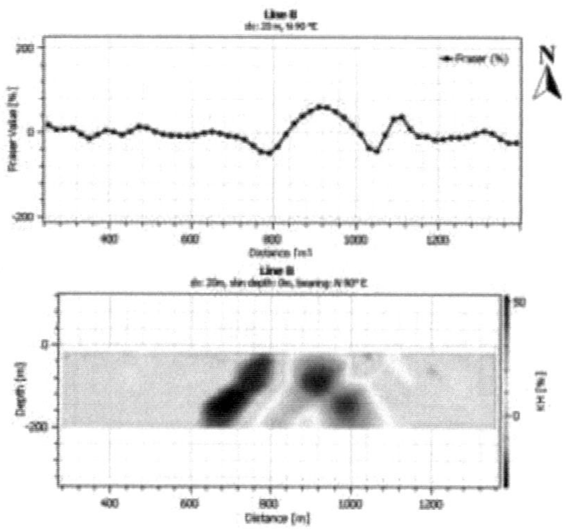

(b)

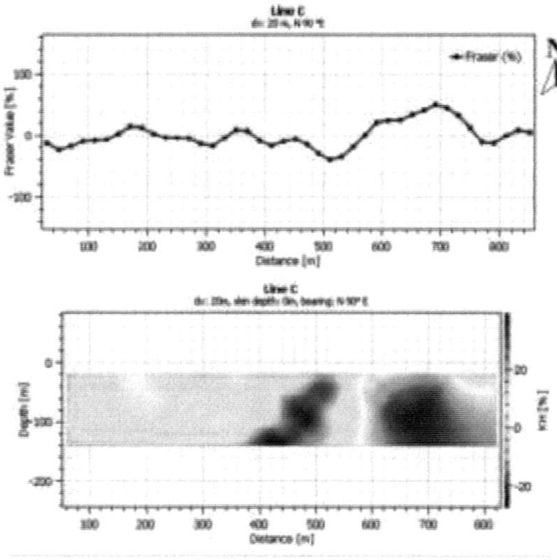

(c)

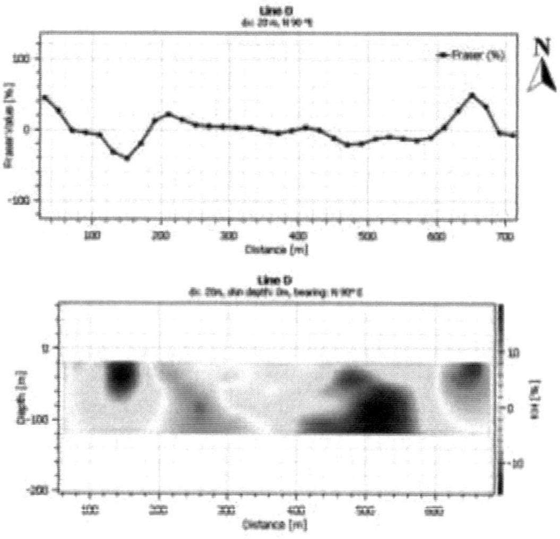

(d)

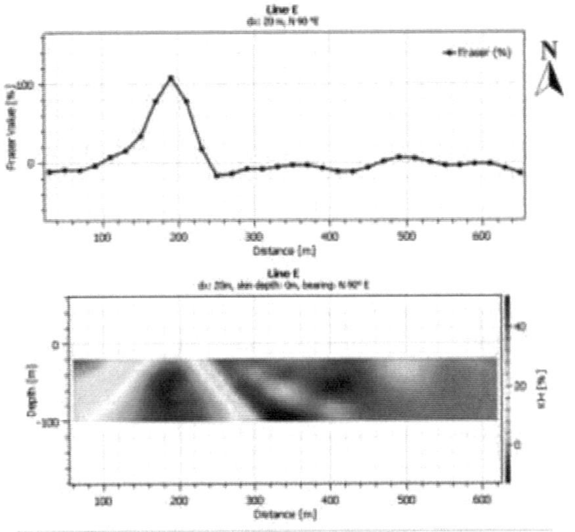

(e)

FIGURE 4. (a) Fraser Filter Result from tilt angle data line A and Pseudosection Result from Karous – Hjelt Filter line A. (b) Fraser Filter Result from tilt angle data line B and Pseudosection Result from Karous – Hjelt Filter line B. (c) Fraser Filter Result from tilt angle data line C and Pseudosection Result from Karous – Hjelt Filter line C. (d) Fraser Filter Result from tilt angle data line D and Pseudosection Result from Karous – Hjelt Filter line D. (e) Fraser Filter Result from tilt angle data line E and Pseudosection Result from Karous – Hjelt Filter line E.

Igneous rocks are found in research area, like outcrop and boulder .Igneous rock outcrops are found andesite and andesite-dasitic with massive structures and porfiro-afanitic texture. The large outcrops of igneous rock are founded on Line E in the eastern side of the PagerkandangVulcanicadjoin with SileriCrater. While the high resistivity area is expected due to the high weathering in claystone. Resistive zonesare spread almost in the middle of the Pagerkandang Volcanic.

(a) (b)

FIGURE 5. (a) Andesite – dasitic rock (b) Fumarol that founded in survey area

6. Conclusion

Based on data analysis, data processing result and interpretation, Very Low Frequency (VLF) method is used to estimate conductive zones and resistive zones are influenced by current density value that related to surface manifestations. The conductive zone is interpreted as the area which have high current density value. This area located at eastern dan western of Pagerkandang Volcanic. The conductive zone related to geothermal manifestation as like as fumarol that appeared because presenced of normal fault. Whereas the resistive zone is interpreted as the area which have low current density value. This area spread almost in the middle of the Pagerkandang Volcanic. The resistive zone was caused by the high weathering in claystone.

7. Acknowledgments

The authors are grateful to PT. Geo Dipa Energy, Department of Dieng Tourism, and Batur, Banjarnegara Government, for their permit to did the research in their area. And Geophysics Department Gadjah Mada University for the support during the fieldwork.

References

[1] Dan Miller C, Sukhyar C, Santoso and Hamidi S 1983 *Eruptive History of The Dieng Mountains Region, Central Java, and Potential Hazard from Future Eruptions Report* (United States : Department of Interior Geological Survey)

[2] Bahri A S *Underground River Mapping in Karst Field Seropan Gunungkidul Using VLF-EM-vGRAD Method (Bahasa)* (Surabaya : Institute of Technology Sepuluh November)

[3] Mutaqih A A 2015 *Data Processing of VLF-EM for Windows Operating System by Graphic User Interface of Practical Work Report (Bahasa)* (Indonesia : Gadjah Mada University)

[4] Fraser D C 1969 *Proc. Contouring of VLF – EM Data* vol 34 (Canada : Geophysics) pp 958– 967

[5] Harijoko A, Uruma R, Wibowo H E, Setijadji L D, Imai A, Yonezu K and Watanabe K 2016 Geochronology and magmatic evolution of the dieng volcanic complex, central java, indonesia and their relationships to geothermal resources *J. Vol.Geo. Res.* **310** 209—213

[6] Karous M and Hjelt S E 1977 *Determination of Apparent Current Density from VLF Measurements: Report* no 89 (Finland : Department of Geophysics, Univeristy of Oulu) p 19

[7] Monteiro Santos F A, Mateus A, Figueiras J and Goncalves M A 2006 Mapping groundwater contamination around a landfill facility using the VLF-EM method — a case study *J. App. Geo.* **6** 115—125

[8] Sundararajan N, Babu V R, Prasad N S and Srinivas Y 2006 VLFPROS—a matlab code for processing of VLF-EM data *Co. Geo.* **32** 1806—1813

[9] Zen M T 1971 *Proc.Geothermal System of The Dieng-Batur Volcanic Complex* vol 6 (Indonesia : Institute of Technology Bandung)

IOP Conf. Series: Earth and Environmental Science **132** (2018) 012024 doi:10.1088/1755-1315/132/1/012024

ERT and Well Data Tie for Nickel Laterite Characterization

Sabrianto Aswad[1a), Difar Mamela Mais[1] Syamsuddin[1] and Wanni[2]

[1]Hasanuddin University, Indonesia
[2]Exploration Department, Vale Indonesia Inc, Indonesia

[a)]sabri@science.unhas.ac.id

Abstract. The need of ERT method in nickel laterites exploration can't be deny. This method have capability to make exploration more effective and efficient. In reality this method still remain ambiguity in its application, especially for geologist and mining expert. These ambiguity related with layer zone determination (limonite, saprolite and bedrock), the same resistivity values in the different zones and determination of bedrock. This paper try to expose interesting fact to overcome this ambiguity by using ERT data and drill data tie. This tie will show characteristic of nickel lateric based on resisitivity value and the contribution of chemistry element for resistivity value. Data ERT was collected by using gradient configuration and well data consist of mayor element and minor element. Tie result showed difference resistivity value in limonite layer influence by Fe, H_2O and Ni, where resistivity value from saprolite layer influenced by Fe, H_2O, $Si0_2$, MgO, Al, Cr, and Ni in certain accumulation. In bedrock layer, almost all drill data did not reach bedrock but only reached the boulder alone, it is supported by the value of the resistivity of rock unserpentinized peridotite which should show a relatively large resistivity value

1. Introduction

Nickel laterites deposit is a product of chemical weathering of ultramafic rocks that Ni-silicate carrier, generally found in areas with tropical climates up to subtropical. Significant laterite deposits range in age from Palaeozoic to Recent and most deposits in the current tropical belts. The most common ways to exploit this deposit by using drill data and test pit, beside to get economic layer also to extract subsurface information. In early stage exploration usually explorationist used sparse drill data to get regional view related with economic distribution of nickel laterites. Before exploitation stages, they will drill more extensively with small distance between one drill hole to others. The more small distance between the drill data the more reliable information they will get, but unfortunately in the same time it needs more time and cost.

The limitation of these two data are narrow coverage area and only show vertically layer variation. Horizontally distribution in 2 and 3 dimension related with continuity done by correlation and geostatistical analysis. Another problem encountered is identification boundary between saprolite and bedrock. Generally this boundary is determined based on drill data, in this case there is a provision that required, for example, if it is assumed to have reached a thickness of 5 meters of fresh rock, the drilling was stopped and the bedrock is assumed to have been obtained. In fact, the characteristic of saprolite layer in west block consist of boulder that has thickness more than 5 meters, it means an error happen in determining the bedrock. This error will impact on the exploitation techniques and calculation of reserves.

Therefore we need another approach that can be used as guidance not only to see horizontal distribution of nickel laterite layer but also can be used for making drilling spacing and how deep drilling must be done. One approach that can be done is by geophysical approach, in this case using the method of geoelectric resistivity. Some results of the research showed satisfactory results using the method of geoelectric resistivity to identify sediment laterite (Robain et al., 1996; Ritz et al., 1999; Beauvais et al, 1999). Especially for the identification of lateritic nickel deposit has been made by Savin et al., (2003) and Robineau et al. (2007). ERT for the correlation between the data and the drill data in the same area but different field have done Aswad et al (2014) even further Aswad et al (2015) have used the data ERT as primary data to determine the location and number of drill point.

This method is done by injecting electricity into the earth, and analyze the response of subsurface materials. This response in resistivity form of rocks subsurface (Telford, 1990). By knowing the subsurface resistivity distribution can make it easier to interpret the material beneath the surface and determine the boundary of the zone from nickel laterite profile.

Results ERT cross-section that has a horizontal resolution either tied with drill data (chemical elements), which has a good vertical resolution. The combination of this data used to make Characteristic of laterite nickel contained in the measurement line.

2. Data and Methodology

The area of research lies in the area of mining concession area of PT. Vale Indonesia, which is administratively located in Luwu Timur. Research area was conducted in an area that includes the area west block. The data used in this study consisted of one line of data ERT (Electrical Resistivity Tomography) and 13 drill point. The line length is 441 meters and the space between the drill points is 25 meters. Configuration used in this study is the gradient configuration. In addition to use of the gradient configuration Wenner configuration has also been used to identify sediment laterite (Ritz et al., 1999; Beauvais et al., 1999; Savin et al., 2003; Robineau et al., 2007). Schlumberger configuration combined with the Wenner configuration has also been used to identify lateritic nickel deposit (Robain et al., 1996).

Configuration gradient is used because it allows generating vast amounts of data with various combinations of electrode potential and current electrodes. This gradient configuration provides better imaging resolution because it is a combination of configuration Characteristic pole-dipole, Wenner and Schlumberger (Dahlin and Zhou, 2006)

3. Result and Discussion

3.1. Prelimanary Interpretation

Based on resistivity cross section (Figure 1) trajectory R can be divided into 3 layers buy seeing trend resistivity value and the initial information about the distribution of nickel laterite profile, threre are limonite, saprolite and bedrock (bedrock). Limonite layer with resistivity values of 50-800 Ωm. Characteristic in this layer has different between the eastern and western parts of the foregoing. The

IOP Conf. Series: Earth and Environmental Science **132** (2018) 012024 doi:10.1088/1755-1315/132/1/012024

eastern part has a resistivity value is relatively much larger than the western part, this shows that the factors that support the eastern part laterisation different from those in the east. Saprolite with 25-450 Ωm resistivity values. In this layer of Ni element is expected enrich. This layer based on preliminary information have characteristic of nickel laterite profile West Block contains many boulder and this can be seen from the cross section of the resistivity (closed curve dotted line). This layer has a low resistivity value relatively small because the water saturation zone, has a small grain size, unless there is a boulder in it. Boulder is very influential in the calculation of reserves. The last layer is a layer of bedrock with resistivity values of more than 400 Ωm. This layer is indicated as a zone of bedrock that has the permeability and porosity is very low. Based on the value of resistivity of this layer belong to ultramafic igneous rock which contains olivine and piroksin (peridotite). A question mark at the beginning of the interpretation of this layer still raises the question whether the error in data collection, data processing, or beneath the surface of this reality is what it is. To answer these questions required additional data is data drill data.

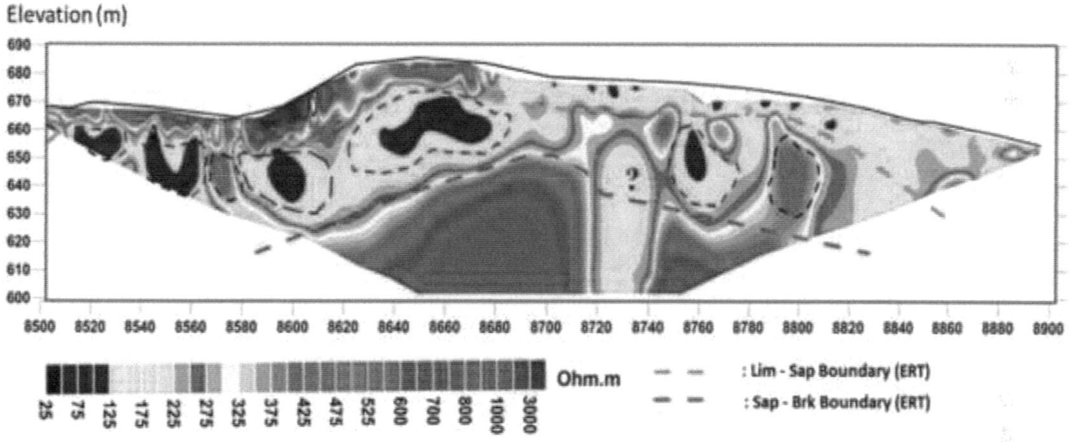

Fig. 1 Resistivity cross section of line R

3.2. Nickel Laterite Profile Based on Drill Data

This cross-section is based on data drill spacing of 25 meters (Figure 2), except for the western part of the drill spacing between R03 and R04 because the topography is very steep. The drilling will be stopped after 5 meters gained massive rock (fresh rock) and annotated have reached bedrock (bed rock). Any data retrieved drill core drilling conducted geological description and examination of the chemical element. From the analysis of chemical elements along with certain criteria determining the limits (cut off) obtained by the boundary zone of limonite-saprolite and saprolite-bedrock boundary for each point of the drill. The entire drill data on the track then correlated to form a cross-section shown in Figure 2. The divisions own profile in accordance with that proposed by Waheed (2009). It was clear enough distinction western and eastern parts of the foregoing. The western part there is a layer of limonite, saprolite and bedrock, while the eastern part is only there is a layer of limonite and immediately met with bedrock without saprolite

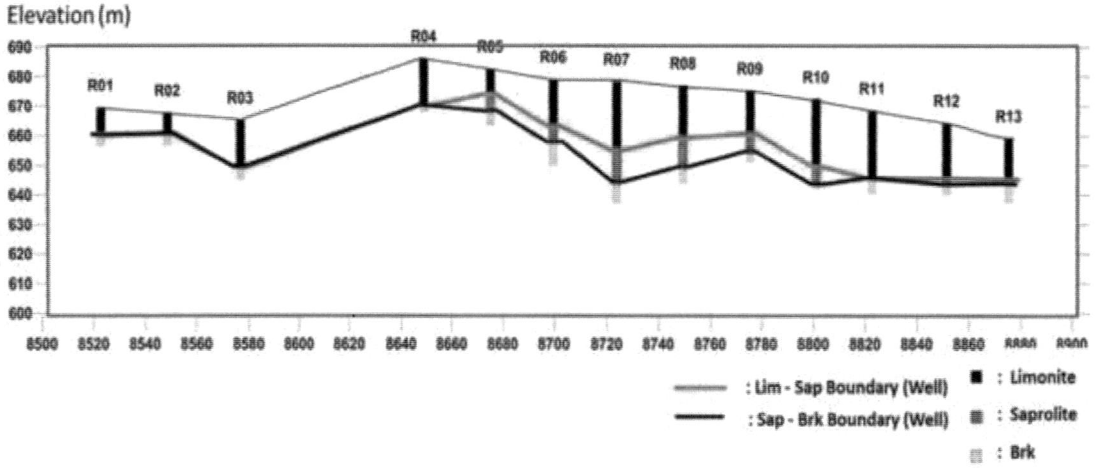

Fig. 2 Nickel laterite Distribution Based on Drill data

3.3. Chemical Mobility of Element

In the ultramafic rocks occur geochemical mobility. Mobility is the ability of an element to be dispersed into the surrounding matrix of another material. Mobility affects the response element to the process disperse. The main factors affecting the mobility geochemistry geochemistry is the chemical stability of the elements (Waheed, 2009).

The mobility of an element that is found in ultramafic rocks or associated with laterite can be classified as follows:

1. Elements that are highly soluble and highly mobile, Easily washed / swept up in the weathering profile and is highly soluble in groundwater flow slightly acidic tropical regions, such as Mg, Si, Ca, Na and K (Figure 3)
2. Elements that are not soluble and unmobile dissolve in ground water at ordinary pH / Eh conditions, these elements make up the bulk or residual soil, such as Fe^{3+} (ferric), Al^{3+}, Cr^{3+}, Ti and Mn^{3+} (Figure 4)
3. Elements with limited solubility and limited mobility (Figure 5), such as: Ni^{2+}, Co^{2+} and Mn^{2+}. Partially soluble in acidic ground water, and insoluble when there are more soluble elements (Si, Mg)
4. ground water, and insoluble when there are more soluble elements (Si, Mg)

IOP Conf. Series: Earth and Environmental Science **132** (2018) 012024 doi:10.1088/1755-1315/132/1/012024

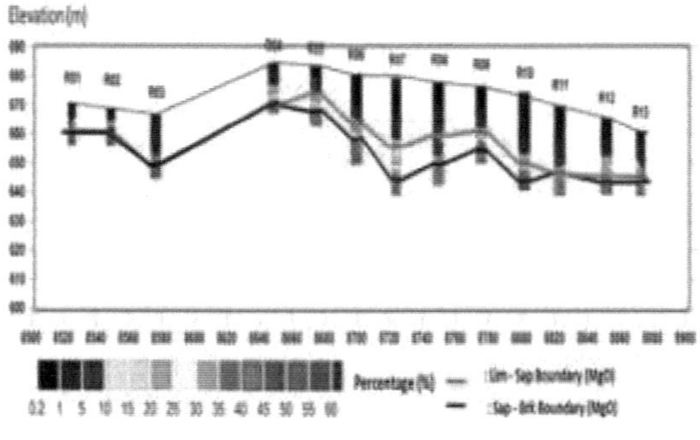

(a)

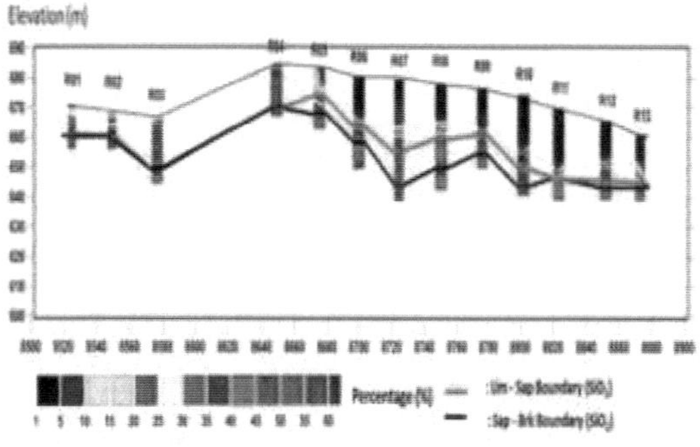

(b)

Fig. 3 Elements that are highly soluble and highly mobile consist of (a) Mgo dan b (SiO_2)

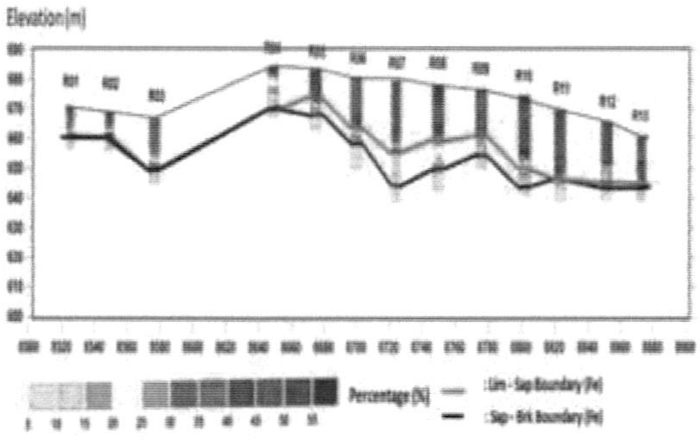

(a)

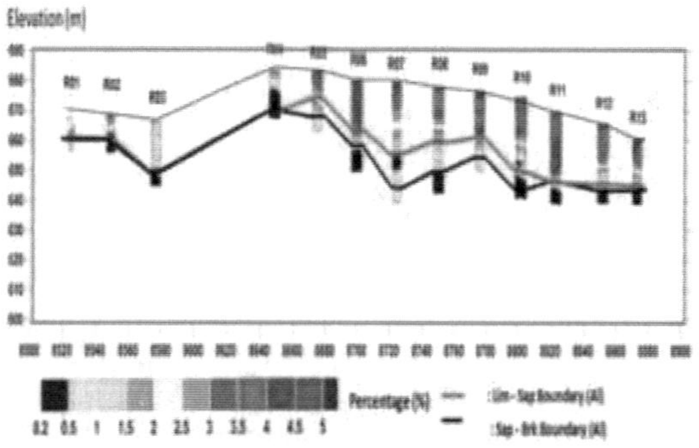

(b)

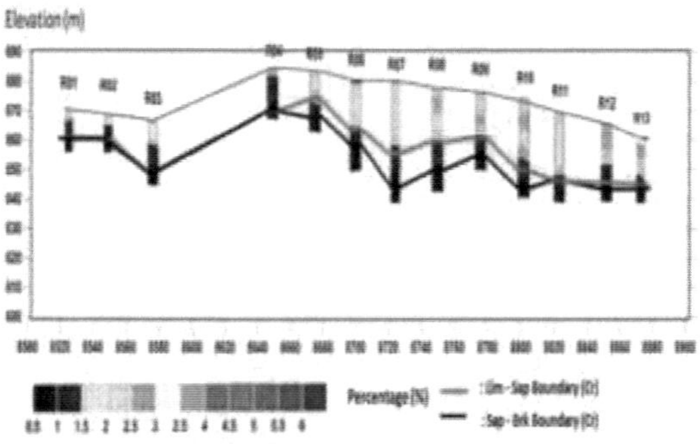

(c)

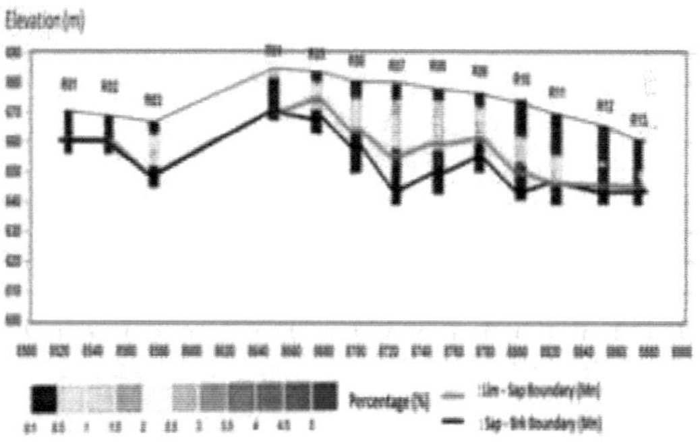

(d)

Fig. 4 Elements that are not soluble and unmobile a)Fe, b) Al, c) Cr dan d) Mn

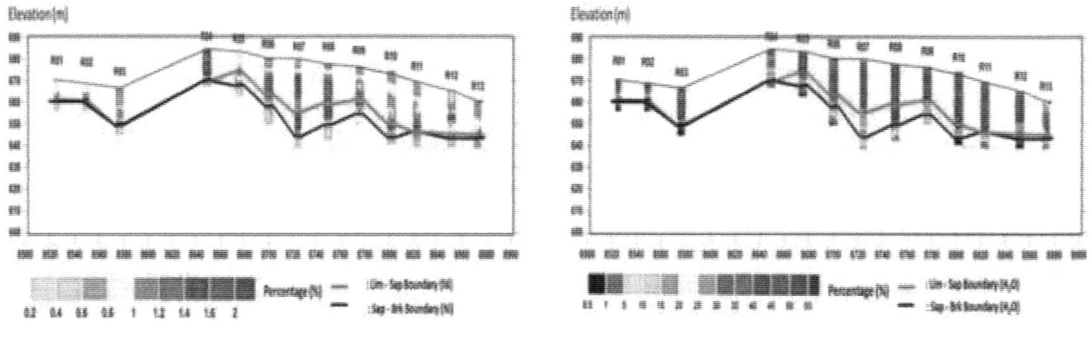

(a) (b)

Fig. 5 Elements with limited solubility and limited mobility Ni (a) and water contain core sample (b)

3.4. ERT and Drill Data Tie

Can be clearly seen the difference between the drill data and resistivity data in determining the upper limit and lower limit of each layer of laterite, especially those boundary of the base layer (bed rock). This has implications for the determination of the difference thickness of the saprolite and ultimately affect the calculation of reserves. The lower limit of the drill data saprolite, in this case the boundary between the saprolite and the bedrock / bed rock is at an average depth of 21.5 meters ±. ERT Data show different things, the depth of the lower limit of saprolite resistivity values are based on average depth of ± 43 meters. This is due to the assumption of the drill data that says after 5 meters found the massive rock (fresh rock) mistakenly based on data ERT. The ERT data showed that the rock still mostly boulder in a large size, which under the bolder they are expected to have a saprolite nickel content which is great. Results from data tie showed limonite zone resistivity difference was more influenced by the content of Fe, Ni whereas H_2O and saprolite resistivity values are influenced by Fe, H_2O, $Si0_2$, MgO, Al, Cr, and Ni with a specific accumulation.

The problem with question marks in initial interpretation from data analysis can be done by comparing anomaly with data vicinity. To strengthen the analysis of geochemical data can be used, particularly by comparing the drill data R06. From the geochemical data that is H_2O, MgO and Al looks a distinct pattern with drill R06. H_2O from R07 possible for smaller if drilled deeper, it is also supported by data MgO were also relatively large. Water as transport will bring the mobile element in this regard MgO deeper. It can be concluded that in this area there is a structure activity, could be fault or fracture (fracture).

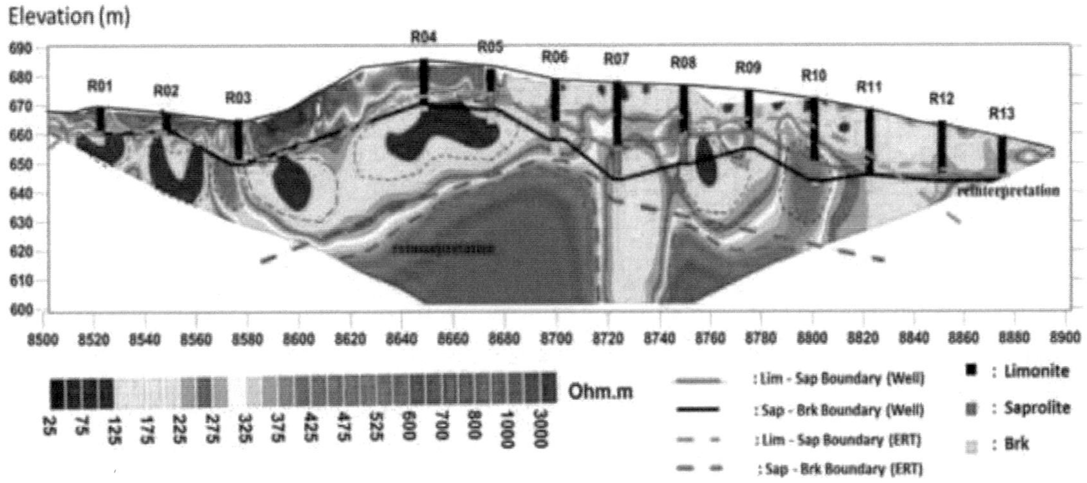

Fig. 6 Reinterpretation using combination ERT and Drill data

4. Conclusion

Interpretation in determination boundary of limonite, saprolite and bedrock would be much better using combination drill data and data ERT, it is seen by the ability of both these data to determine the boundary layers and identify the boulder. Especially for bedrock can be seen clearly in the data ERT with large resistivity values, it is supported by the fact rock bedrock consisting of peridotite unserpentinised. Merging this data also helps in identifying the structures associated with local faults or fractures.

5. Acknowledgments

All the data in this research supported by Exploration Department, PT Vale Indonesia, which is gratefully acknowledged. We wish also thank you to Hasanuddin University for supporting us joint this annual meeting

References

[1] Aswad S., Haerunnisa A.A., Massinai M.A., Syamsuddin., Hadi G.S. 2014. Determination Nickel Laterite Profile using Correlation of ERT (Electrical Resistivity Tomography and Drill Core Data. Proceedings of the 39th Annual Convention and Exhibition, PIT2014-1029

[2] Aswad S., Surawan Y, Massinai M.A., Wanni, 2015, Scenario to Optimize Nickel Laterite Exploration using ERT with Gradien Configuration, . Proceedings of The 12th SEGJ International Symposium

[3] Beauvais A., Ritz M., Parisot J-C., Dukhan M. and Bantsimba C.1999. Analysis of poorly stratified lateritic terrains overlying granitic bedrock in West Africa, using a 2-D electrical resistivity tomography. Earth and Planetary Science Letters173, 413 – 424

[4] Dahlin.T and B. Zhou. 2006. Multiple-gradient array measurements for multichannel 2D resistivity imaging. Near Surface Geophysics. European Association of Geoscientists &

Engineers (EAGE), 113-123

[5] Robain H., Descloitres M., Ritz M, Q Y Atangana. 1996. A multiscale electrical survey of lateritic soil system in the rain forest of Cameroon. Journal of Applied Geophysics, 237-253

[6] Robineau B., Join J.L, Beauvais A., Parisot J-C., Savin C. 2007, Geoelectrical imaging of thick regolith developed on ultramafic rocks : groundwater influence, Austalian Journal of Earth Science, 773-7816

[7] Ritz M., Parisot J-C., Diouf S., Beauvais A., Dione F., Niang M. 1999. Electrical imaging of lateritic weathering mantles over granitic and metamorphic basement of eastern Senegal, West Africa. Journal of Applied Geophysics, 335-344

[8] Savin C., Robineau B., Monteil G., Beauvais A., Parisot J-C., Ritz M. 2003. Electrical imaging of peridotite weathering mantles as a complementary tool for nickel ore exploration in New Caledonia. ASEG 16th geophysical conference and exhibition, Adeleide.

[9] Telford, W.M., 1990. Applied Geophysicst. Cambridge University Press

[10] Waheed. A., 2009. Nickel Laterites-A Short Course: Chemistry, Mineralogy, and formation of nickel laterites (unpublished)

Strike-Slip Fault Deformation and Its Control in Hydrocarbon Trapping in Ketaling Area, Jambi Subbasin, Indonesia

Aldis Ramadhan[1], Alexis Badai Samudra[1], Jaenudin[1], Enik Puji Lestari[1], Julian Saputro[1], Sugiono[1], Yosi Hirosiadi[1], Indi Amrullah[1]

[1]Exploration Department of Pertamina EP, 16th Floor Standard Chartered Building, Jl. Prof. Dr. Satrio No. 164 Jakarta, 12950, Indonesia

Corresponding author: aldis.ramadhan@pertamina.com

Abstract. Geologically, Ketaling area consists of a local high considered as flexure margin of Tempino-Kenali Asam Deep in west part and graben in east part also known as East Ketaling Deep. Numerous proven plays were established in Ketaling area with reservoir in early Miocene carbonate and middle Miocene sand. This area underwent several major deformations. Faults are developed widely, yet their geometrical features and mechanisms of formation remained so far indistinct, which limited exploration activities. With new three-dimensional seismic data acquired in 2014, this area evidently interpreted as having strike-slip mechanism. The objective of this study is to examine characteristic of strike slip fault and its affect to hydrocarbon trapping in Ketaling Area. Structural pattern and characteristic of strike slip fault deformation was examined with integration of normal seismic with variance seismic attribute analysis and the mapping of Syn-rift to Post-rift horizon. Seismic flattening on 2D seismic cross section with NW-SE direction is done to see the structural pattern related to horst (paleohigh) and graben. Typical flower structure, branching strike-slip fault system and normal fault in synrift sediment clearly showed in section. An echelon pattern identified from map view as the result of strike slip mechanism. Detail structural geology analysis show the normal fault development which has main border fault in the southern of Ketaling area dipping to the Southeast-East with NE-SW lineament. These faults related to rift system in Ketaling area. NW-SE folds with reactive NE-SW fault which act as hydrocarbon trapping in the shallow zone. This polyphase tectonic formed local graben, horst and inverted structure developed a good kitchen area (graben) and traps (horst, inverted structure). Subsequently, hydrocarbon accumulation potentials such as basement fractures, inverted syn-rift deposit and shallow zone are very interesting to explore in this area.

1. Introduction

Ketaling area is southeastern part of Jambi Sub-basin. NW-SE seismic cross-section reveals this area geologically consist of horst and graben that formed by paleogene tectonic event. Ketaling and Merang High represent horst whilst Sungai Gelam and East Ketaling Deep depict graben (Fig.1). In general, horst and graben in Jambi Sub-basin have northeast-southwest orientation as different to north-south trend in South Sumatera Basin [1].

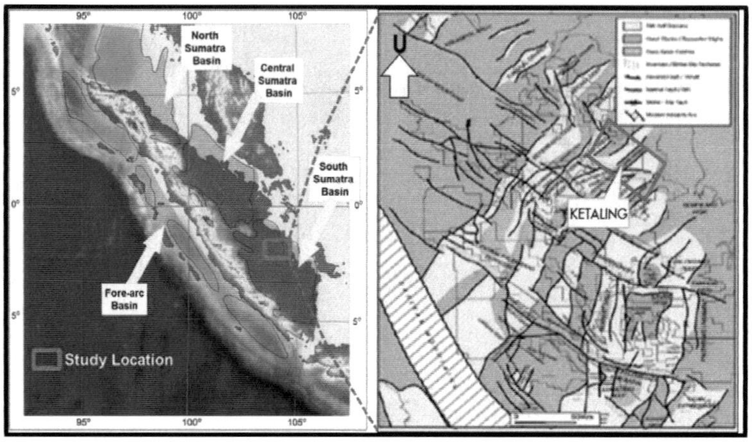

Figure 1. Location Map of Ketaling Area.

Commonly in Jambi Sub-basin, there is no carbonate facies sediment exists yet Ketaling area is unique due to the presence of Middle Miocene carbonate caused by decreasing sediment input southwards. In the west part, this area also experienced the uplift of the Sembilang High as a result of Plio-pleistocene compressional event. This compressional event led to hundreds to thousands feet amount of erosion during the last few millions ago. Furthermore, this tectonic event formed northwest-southeast trending folds that proved to be hydrocarbon-bearing structure (Fig.2).

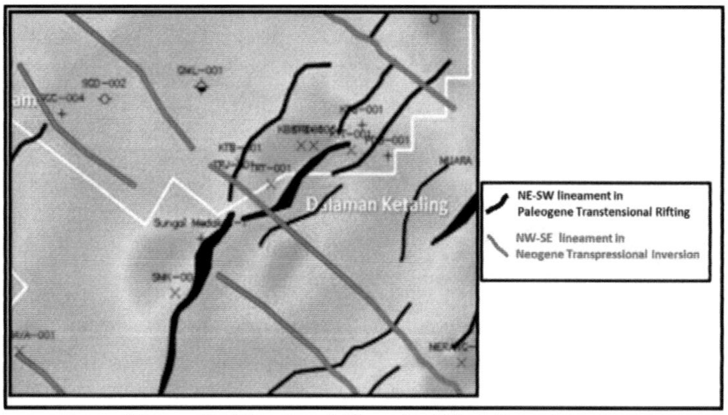

Figure 2. Structural Trend of Ketaling Area.

2. Data and Method

While Global positioning System have been used in many previous studies analysis [2, 3, 4, 5, 6, 7, 8, 9, 10, 11, 12, 13], this study utilize seismic data for further analysis. The structural characterization both geometry and style in this paper was interpreted using New 3D seismic acquired in 2014. The seismic quality is fair to good. The shallow zone shows high amplitude area with much faults shadow effect the seismic quality, whilst the deeper zone shows low amplitude area.

The Mapping of fault and Syn to Post Rift horizon was conducted to examine structural pattern and characteristic of strike slip fault in this area [14]. Moreover, variance attribute analysis was conducted for better analysis and refined interpretation. Seismic flattening with NW-SE direction is done to see the structural pattern related to horst and graben as well as knowing rifting-inversion history of Ketaling area from Early Miocene to Present time (Fig.3).

IOP Conf. Series: Earth and Environmental Science **132** (2018) 012025 doi:10.1088/1755-1315/132/1/012025

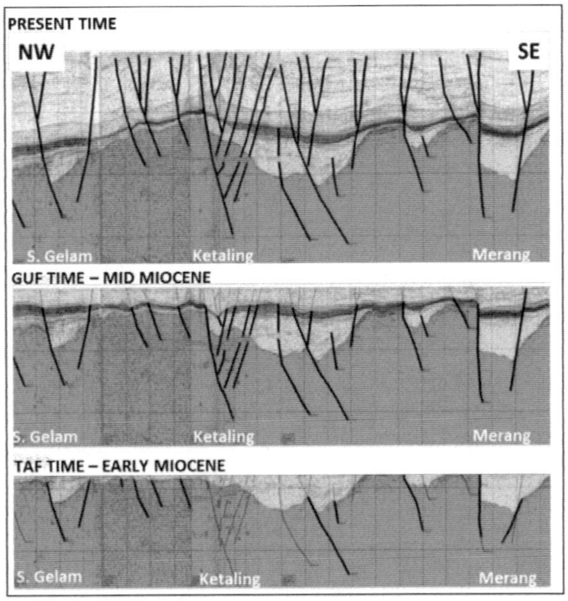

Figure 3. NW-SE Flattening through Gelam-Ketaling-Merang Area.

3. Result and Discussion

3.1. A subsection

Seismic interpretation from new 3D seismic (Fig.4) reveals that the Ketaling Area shows several strike slip features. NW-SE seismic cross section indicates NE-SW main border fault with synthetic and antithetic dipping direction, branching strike slip fault system, positive structural inversion in Post Rift, and typical flower structure features. Whilst SW-NE indicates inverted Syn Rift Deposit, typical flower structure features and development of normal faults in the basement-rift sediment. Flower Structure in this area characterized by relatively planar with high fault-dip angle

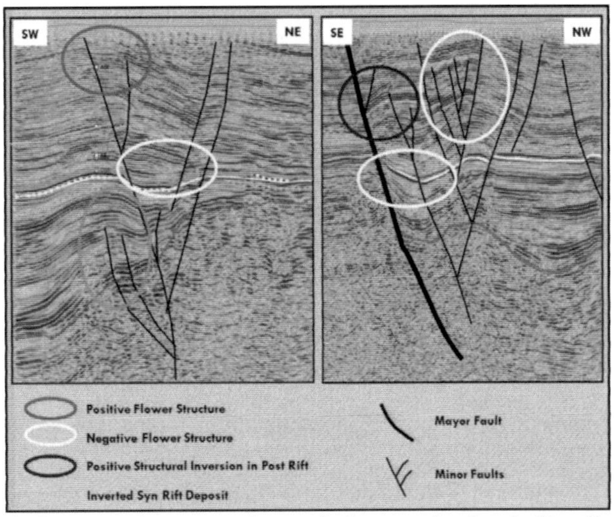

Figure 4. Structure interpretation and characterization in Ketaling area.

Variance seismic attribute is a tool to analyse structures such as faults or micro-faults in the 3D seismic data. This attribute indicates discontinuities in the neighbouring lithology and subsequently in the trace-to-trace variability they become detectable in 3D seismic volumes.

Variance seismic attribute in selected time slice (Fig.5) show Ketaling Area consist of two major NNE-SSW and NW-SE structural trend. NE-SW orientation was formed by transtensional system in Palaeogene yield main border fault with SE dipping. Subsequently, NW-SE structural trend as en echelon fold truncated with NE-SW fault triggered by transpersonal system in middle Miocene to Pliocene. NE-SW fault trend in shallow zone similar to fault trend which formed in Palaeogene hence it showed a reactive fault. This evidence denotes this area experienced polyphase deformation. Those fault features and pattern confirms strike slip mechanism exist in this area.

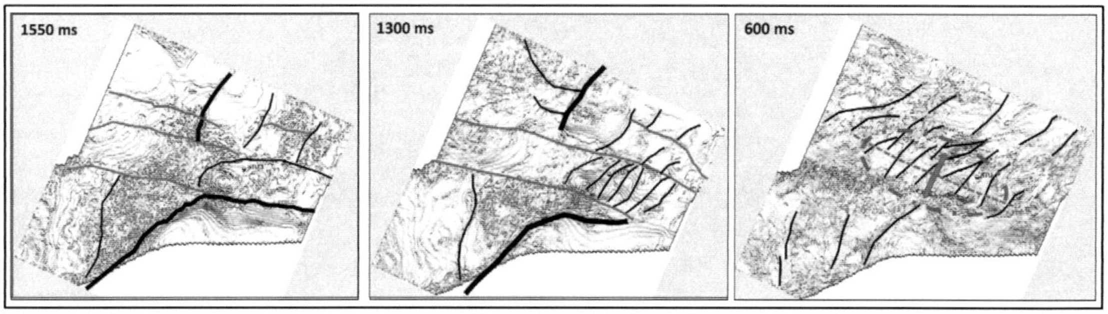

Figure 5. Time slice of variance seismic attributes in - 1550 ms (Basement), -1300 ms (TAF), and -600 ms (Intra ABF) show the changes of structural pattern from Eo-Oligocene to Pliocene time.

3.2. Hydrocarbon Trapping

Eo-Olygocene transtensional tectonic event in Ketaling Area formed graben and horst with NE-SW direction. The graben, Sungai Gelam and East Ketaling Deep, are becomes the kitchen area for Syn Rift organic-rich shale source rock. The Organic-rich sediment in kitchen area consider already mature with regional oil window maturity cut-off at 1700-1800 meter depth.

The horst, Ketaling High, is potential as hydrocarbon trapping especially in basement level. Based on seismic flattening, the Ketaling High was a paleo-high and this area also experienced polyphase deformation that allow for extensive fracture exist in basement level. Moreover, this area directly facing the East Ketaling graben hence increases the chance to be charged by kitchen. The mechanism would magnify geological change in basement level for hydrocarbon accumulation.

Inversion structure in Syn Rift deposit has potential to be home of hydrocarbon accumulation. This structure surrounded by mature source rock , due to its position in kitchen area, which directly charge to reservoir. In shallow zone, proven hydrocarbon trapping occurred mostly on the NW-SE fold follows the major structure trend (East Ketaling Field) formed by middle Miocene to Pliocene transpersonal tectonic event. In this area, the migration pathway through strike slip fault considers as effective mechanism to accumulate HC in the positive structure near the fault system. The existence of hydrocarbon accumulation (East Ketaling Field) proved that hydrocarbon has accumulated and petroleum system has been running in this area.

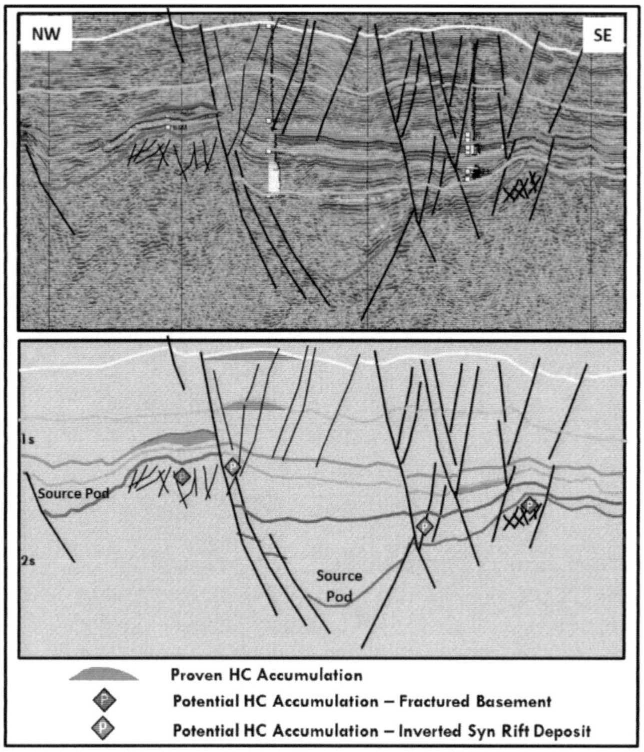

Figure 6. Proven and Potential Accumulation in Ketaling Area.

4. Conclusion

- Transtensional system in Ketaling area formed local graben as kitchen area, while horst developed as traps for hydrocarbon accumulation.
- Transpersonal system in middle Miocene to Pliocene yields NW-SE fold and reactived NE-SW fault. The product of this system act as trap in sediment and potential fractured reservoir in basement.
- Integrated normal seismic and variance seismic attributes interpretation show polyphase deformation from transtensional system with NE-SW orientation to transpersonal system with NW-SE fold and reactived NE-SW fault.
- The change of transtensional to transpersonal system adding the occasion of HC accumulation in Ketaling area.
- Strike slip fault consider as effective migration pathway to store HC in the positive structure near the fault system.
- The existence of hydrocarbon accumulation (East Ketaling Field) proved that hydrocarbon has accumulated and petroleum system has been running in this area.

References

[1] Purwaningsih, M.E.M., Mujihardi, B., Prasetya, L., Suseno, W.A., Sutadiwiria, Y., 2006. Structural Evolution of The Jambi Sub-basin : A Rotated Strike Slip Mechasnism. Proceedings of International Geosciences Conference And Exhibition, Jakarta.

[2] Ito, T., Gunawan, E., Kimata, F., Tabei, T., Meilano, I., Agustan, Ohta, Y., Ismail, N., Nurdin, I. and Sugiyanto, D., 2016. Co-seismic offsets due to two earthquakes (Mw 6.1) along the Sumatran fault system derived from GNSS measurements. Earth, Planets and Space, 68(1), 1, DOI: 10.1186/s40623-016-0427-z.

[3] Anugrah, B., Meilano, I., Gunawan, E. and Efendi, J., 2015. Estimation of postseismic deformation parameters from continuous GPS data in northern Sumatra after the 2004 Sumatra-Andaman earthquake. Earthquake Science, 28(5-6), 347-352, DOI: 10.1007/s11589-015-0136-x.

[4] Alif, S. M., Meilano, I., Gunawan, E. and Efendi, J., 2016. Evidence of Postseismic Deformation Signal of the 2007 M8.5 Bengkulu Earthquake and the 2012 M8.6 Indian Ocean Earthquake in Southern Sumatra, Indonesia, Based on GPS Data. Journal of Applied Geodesy, 10(2), 103-108, DOI: 10.1515/jag-2015-0019.

[5] Ardika, M., Meilano, I. and Gunawan, E., 2015. Postseismic Deformation Parameters of the 2010 M7.8 Mentawai, Indonesia, Earthquake Inferred from Continuous GPS Observations. Asian Journal of Earth Sciences, 8, 127–133, DOI: 10.3923/ajes.2015.127.133.

[6] Ohkura, T., Tabei, T., Kimata, F., Bacolcol, T. C., Nakamura, Y., Luis, Jr., A. C., Pelicano, A., Jorgio, R., Tabigue, M., Abrahan, M., Jorgio, E. and Gunawan, E., 2015. Plate Convergence and Block Motions in Mindanao Island, Philippine as Derived from Campaign GPS Observations (Special Issue on Enhancement of Earthquake and Volcano Monitoring and Effective Utilization of Disaster Mitigation Information in the Philippines). Journal of Disaster Research, 10(1), 59-66.

[7] Gunawan, E., Meilano, I., Abidin, H. Z., Hanifa, N. R. and Susilo, 2016. Investigation of the best coseismic fault model of the 2006 Java tsunami earthquake based on mechanisms of postseismic deformation. Journal of Asian Earth Sciences, 117, 64-72, DOI: 10.1016/j.jseaes.2015.12.003.

[8] Gunawan, E., Maulida, P., Meilano, I., Irsyam, M. and Efendi, J., 2016. Analysis of coseismic fault slip models of the 2012 Indian Ocean earthquake: Importance of GPS data for crustal deformation studies. Acta Geophysica, 64(6), 2136-2150, DOI: 10.1515/acgeo-2016-0106.

[9] Raharja, R., Gunawan, E., Meilano, I., Abidin, H. Z. and Efendi, J., 2016. Long aseismic slip duration of the 2006 Java tsunami earthquake based on GPS data. Earthquake Science, 29(5), 291-298, DOI: 10.1007/s11589-016-0167-y.

[10] Gunawan, E., Kholil, M. and Meilano, I., 2016. Splay-fault rupture during the 2014 Mw7. 1 Molucca Sea, Indonesia, earthquake determined from GPS measurements. Physics of the Earth and Planetary Interiors, 259, 29-33, DOI: 10.1016/j.pepi.2016.08.009.

[11] Gunawan, E., Meilano, I., Hanifa, N. R. and Widiyantoro, S., 2017. Effect of coseismic and postseismic deformation on homogeneous and layered half-space and spherical analysis: Model simulation of the 2006 Java, Indonesia, tsunami earthquake. Journal of Applied Geodesy, DOI: 10.1515/jag-2017-0009.

[12] Pratama, C., Ito, T., Sasajima, R., Tabei, T., Kimata, F., Gunawan, E., Ohta, Y., Yamashina, T., Ismail, N., Nurdin, I., Sugiyanto, D., Muksin, U. and Meilano, I., 2017. Transient rheology of the oceanic asthenosphere following the 2012 Indian Ocean Earthquake inferred from geodetic data. Journal of Asian Earth Sciences, 147, 50-59, DOI: 10.1016/j.jseaes.2017.07.049.

[13] Gunawan, E., Ghozalba, F., Syauqi, Widiastomo, Y., Meilano, I., Hanifa, N. R., Daryono and Hidayati, S., 2017. Field Investigation of the November to December 2015 Earthquake Swarm in West Halmahera, Indonesia. Geotechnical and Geological Engineering, 35(1), 425-432, DOI: 10.1007/s10706-016-0117-4.

[14] Lan, X., Lu, X., Zhu, Y., Yu. H., 2015. The geometry and origin of strike-slip faults cutting the Tazhong low rise megaanticline (central uplift, Tarim Basin, China) and their controlon hydrocarbon distribution in carbonate reservoirs". Journal Of Natural Gas Science and Engineering no.22 p.633-645.

IOP Conf. Series: Earth and Environmental Science **132** (2018) 012026 doi:10.1088/1755-1315/132/1/012026

Preliminary Magnitude of Completeness Quantification of Improved BMKG Catalog (2008-2016) in Indonesian Region

H C Diantari[1], W Suryanto[1], A Anggraini[1], T M Irnaka[1], P Susilanto[2] and D Ngadmanto[2]

[1]Geophysics Laboratory, Department of Physics, Faculty of Mathematics and Natural Sciences, Gadjah Mada University, Yogyakarta, Indonesia

[2]Meteorological Climatological and Geophysics Agency (BMKG)

hastin.chandra.d@mail.ugm.ac.id

Abstract. We present a magnitude of completeness (Mc) quantification based on BMKG improved earthquake catalog which generated from Ina-TEWS seismograph network. The Mc quantification can help us determine the lowest magnitude which can be recorded perfectly as a function of space and time. We use the BMKG improved earthquake catalog from 2008 to 2016 which has been converted to moment magnitude (Mw) and declustered. The value of Mc is computed by determining the initial point of deviation patterns in Frequency Magnitude Distribution (FMD) chart following the Gutenberg-Richter equations. In the next step, we calculate the temporal variation of Mc and b-value using maximum likelihood method annually. We found that the Mc value is decreasing and produced a varying b-value. It indicates that the development of seismograph network from 2008 to 2016 can affect the value of Mc although it is not significant. We analyze temporal variation of Mc value, and correlate it with the spatial distribution of seismograph in Indonesia. The spatial distribution of seismograph installation shows that the western part of Indonesia has more dense seismograph compared to the eastern region. However, the eastern part of Indonesia has a high level of seismicity compared to the western region. Based upon the results, additional seismograph installation in the eastern part of Indonesia should be taken into consideration.

1. Introduction

Indonesian earthquake catalog has experienced a major improvement as implications of significant development seismograph network since 2005 and released by BMKG in 2008. The seismograph network known as Ina-TEWS (Indonesian Tsunami Early Warning System) cover seismograph station in Indonesia and produced improved earthquake catalog. The quality of earthquake catalog is a primary issue and need a special attention because it will be a primary input for seismological earthquake and tectonic analysis.

Magnitude of Completeness (M_c) analysis is one of the effective methode to prove the reliability of earthquake catalog. M_c value show the lowest magnitude which can be recorded complete as a function of space and time. The value is represent sensitivity of seismograph network to detect earthquake complete in space and time. Seismograph density in a region can affect the M_c which is the more dense the network, the more small the M_c of the catalog. Incomplete catalog, generally in lower magnitude can not be used because it have ambiguity so can affect seismicity calculation. Therefore, the analysis of M_c

become essential as a preparation step to makesure earthquake data that will allowed in next calculation. Moreover, identification of seismograph network development affect on M_c variation need a crucial attetion to analyze network performance.

This study is focuse to calculate M_c that affected by the development of the seismograph network in certain time. We used earthquake catalog owned by BMKG which records Indonesian seismicity based on latest earthquake catalogs from 2008 - 2016. Previous research to detect M_c in Indonesia show that M_c is 5,0 and identified as part of seismic hazard analysis [1]. But, this research used compiled catalog from several source, so that, the M_c that they got can not represent the indonesia seismograph network performance. In this research, the analysis is done by correlating BMKG magnitude type with Moment Magnitude (M_w) from GFZ. From that correlation, we obtain Magnitude of Completeness (M_c) and *b*–value distribution around the Indonesian Region. It is a preliminary study to acquire one of many parameters in seismic hazard analysis. We assume that we will get a decreasing M_c in temporal variation due to the development of seismograph and produced vary of *b*-value. Moreover, we will identified the performance density distribution of seismograph network in Western and Eastern of Indonesia by seismicity level.

2. Data and Resources

There are 2 kinds of catalog that used in this study: primary catalogs and reference catalogs. Primary catalog prefers to main catalog that proceed, in this case is BMKG catalog. It covered earthquakes from late 2008 to early 2016, in Indonesia region in area between 95° E to 142° E longitudes and 11° S to 8° N latitudes. It contains about 35675 events with various magnitude type (M, m_b, M_{Lv}, M_w, $M_{w(mB)}$ and M_{wp}) that explained in Table 1.

TABLE 1. Magnitude Type Distribution of BMKG Catalogs 2008-2016.

Magnitude	Abbreviation	Range Data	Number of events
Average Magnitude	M	$1.4 \leq M \leq 6.4$	15342
Short-period body-wave magnitude	m_b	$3.2 \leq m_b \leq 7.2$	3013
Local (Richter) Magnitude	M_{Lv}	$1.4 \leq M_{Lv} \leq 6.6$	16581
Moment Magnitude	M_w	$5 \leq M_w \leq 6.9$	22
Moment Magnitude from P-waves	M_{wp}	$5.5 \leq M_{wp} \leq 6.6$	4
Proxy M_w based on m_B	$M_{w(mB)}$	$3.7 \leq M_{w(mB)} \leq 7.9$	713

Reference catalog used are earthquake catalog from Geo Forschungs Zentrum (GFZ) Potsdam, Germany. We chose this catalog because GFZ have more seismograph installed in Indonesia compare to other foreign seismograph contributor in Indonesia. Moreover, Indonesia has adopted similar method with GFZ to proceed earthquake parameter and known as SeisComp3[2]. Catalog derived in same period and region with BMKG Catalog. We also used seismograph installation document from BMKG to see the additional seismograph that have been installed as shown in Figure 1.

3. Magnitude Homogenization and Declustering

Raw BMKG catalog contains various magnitude type and we have to convert it to become one magnitude type which have similar value by correlating various magnitude type to refference magnitude. Magnitude correlation were proceed by matching earthquake event that occurred in certain time and place that record in test catalog and reference catalog. Next, the correlated events were plotted in a graph (Figure-2) and classified for each BMKG magnitude type. Empirical relationship have derived using least square linear regression method. We correlate it to M_w with consideration that it is a more precise magnitude due to it represent to radiated energy from direct measurement[3].

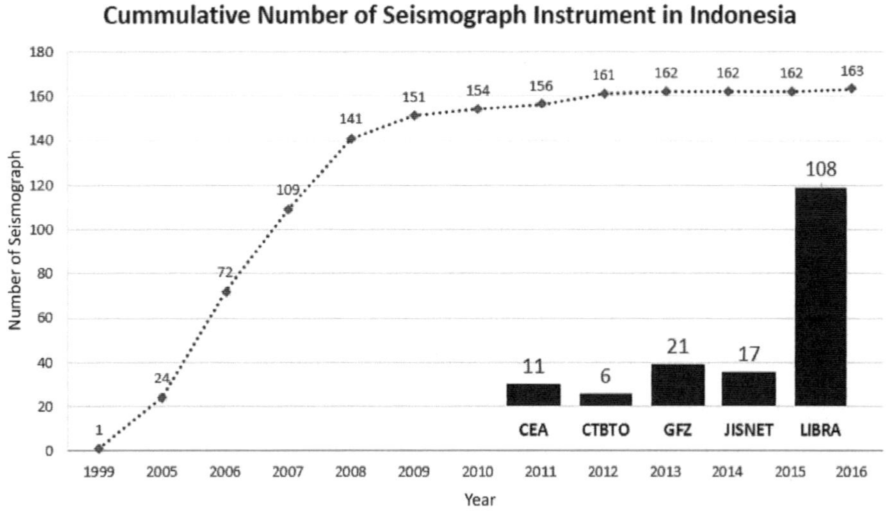

FIGURE 1. Development of seismograph installation in Indonesia from 1999 to 2016. Data source from BMKG. CEA, CTBTO, GFZ, JISNET are foreign seismological institution that contributed seismograph in Indonesia

The acquired empirical relationship above is unique for indonesia region because it converts from specified BMKG catalog. As the result, empirical relationship for each magnitude type are produced that can be seen in Figure-2. Summary of relationship for each magnitude types are listed in Table 2. Those relationship are adopted to get homogen magnitude type in M_w for all earthquake records in BMKG catalog.

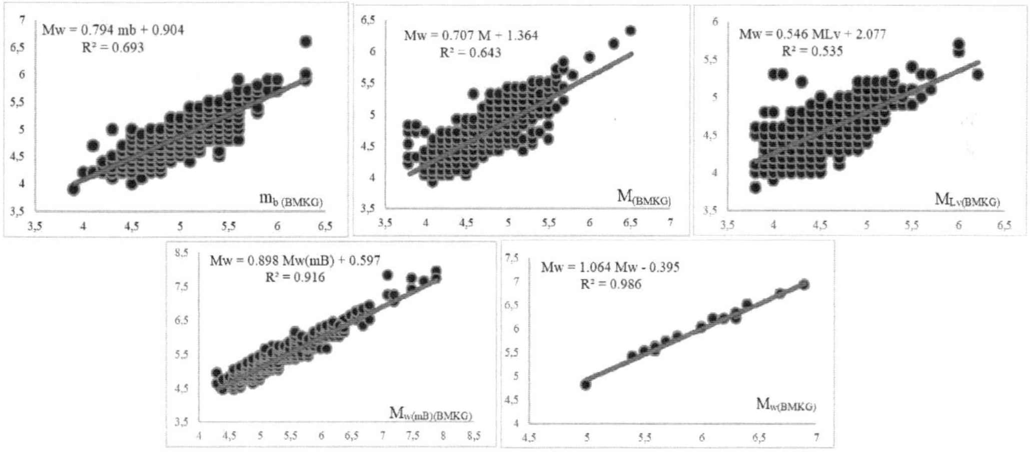

FIGURE 2. Correlating magnitude graph classified into 5 magnitude types (M, m_b, M_{Lv}, M_w and $M_{w(mB)}$). Vertical axis show Mw magnitude from GFZ and horizontal axis show BMKG magnitude. Red line represent linear trend of data distribution

TABLE 2. Correlative Relationship between Moment Magnitude Mw and Various BMKG Magnitude

Empirical Relationship	Number of Events	Determination Coefficient, R^2	Range Data
Mw = 0.707 M + 1.3684	1393	0.6435	$3.8 \leq M \leq 6.5$
Mw = 0.7946 mb + 0.9046	1445	0.6932	$3.9 \leq mb \leq 6.3$
Mw = 0.5468 MLv + 2.076	1144	0.5357	$3.8 \leq Mlv \leq 6.2$
Mw = 1.0646 Mw – 0.3958	16	0.9861	$5.0 \leq Mw \leq 6.9$
Mw = 0.8981 Mw(mB) + 0.5978	476	0.9164	$4.7 \leq Mw(mB) \leq 7.9$

To remove aftershock, mainshock, and foreshock effect, we applied seismicity declustering to separate from natural earthquake occurrence. Declustering followed conjugate methode from Urhammer (1986) and Gardner-Knopoff (1974) method provide by Openquake. It reduced earthquake event data plotted and remained 25689 that classified as natural earthquakes events of mainshocks.

4. Methodology

Seismicity rate of certain region expressed in Frequency Magnitude Distribution (FMD) with follow Gutenberg-Richter (GR) relationship.

$$\log n(M) = a - bM \tag{1}$$

with $n(M)$ is earthquake quantity with magnitude greater than or equal to M in a certain region with time interval T; a and b value are positive parameters indicating the level and seismicity characteristics[4]. a-value, describe the productivity in given region and time period and b-value that shows the relative size distribution of earthquake.

FMD is expressed in cumulative and increment data plot. The cumulative summing all frequencies above the lowest magnitude. Increment curve counts the number of event with magnitude in each bin with a certain range. GR relationship assume that principally earthquake occurrence have to be distribute linier. It describes that earthquake with higher magnitude should be more infrequently happened than low magnitude earthquake, and earthquake will linearly distributed. But due to limitless seismic network, instrument characteristics, low magnitude earthquake sometime not recorded well, so make the distribution is incomplete.

We analyzed magnitude of completeness (Mc) by estimating point that starts to deviate out of GR trend[5]. It aims to limiting FMD to be conform with conform with GR law. Besides, Mc can determine by maximum curvature method, that select point of maximum curvature in increment curve[6]. Next, FMD below Mc have to be removed to get perfect slope. Data processing of earthquake occurrence distribution was using Openquake Project with based on Python Language. Openquake Project is an open source code for seismic hazard and risk calculation. It provide us to calculating all seismic hazard parameter with the right data input. The b-value is calculated using maximum likelihood method using this following equation.

$$b = \frac{log_{10}^{e}}{M - Mc} \tag{2}$$

With $\boldsymbol{M}$ is the average magnitude and M_c is the Magnitude of completeness.

5. Seismicity Analysis

FMD distribution have to contain the homogenized and mainshock event only [7]. We visualized earthquake catalog in cumulative and increment curve of FMD distribution. FMD proceed in temporal variation start from 2009 to 2016 in Indonesian region. It used to compare and find out variations of Mc and b-value each year, due to difference magnitude distribution density and numbers of earthquake occurred.

Figure 6 expressed that there are different FMD type in each temporal variation that affect to produce different Mc and b-value, due to various earthquake occurrence density. Table 3 showed a summary of difference parameter in each year. It conclude that Mc was decrease with no significance change and parameter seismicity a and b-value is fluctuative. Mc of BMKG Catalog vary from 4.9 and decrease to 4.6. This result compare with development of seismograph installation (Figure 1) in Indonesia had affected Mc of Indonesia, although development were not significance. Temporal seismicity were observed with applied the Mc and b-value each year and showed by Figure 4 and conclude that seismicity level in Western Indonesia higher than Eastern of Indonesia.

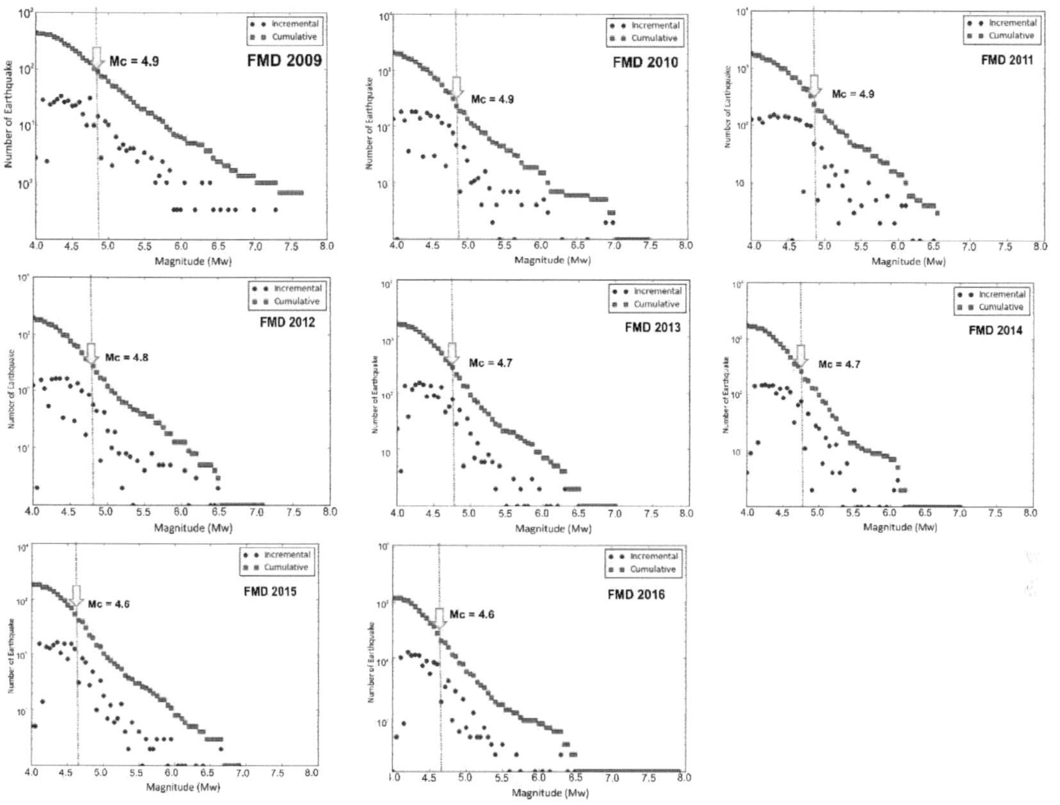

FIGURE 3. Variation of FMD from 2008 to 2016. Different distribution of FMD in each year results in different variation of Mc (marked by black dotted line)

Temporal seismicity were observed with applied Mc and b-value each year and showed by Figure 4. Colour scales shows the normalized earthquake occurence in certain years. As we can see, Eastern region of Indonesia have a higher level of seismicity compare to western region. If we matched this assumption with seismograph distribution in Figure 5, we can see that western part of Indonesia has more dense seismograph compared to the eastern region. However, the eastern part of Indonesia has a high level of seismicity compared to the western region.

TABLE 3. Summarize of analysis seismicity temporal variation

Year	Number of Events	Magnitude Completeness	b-value	a-value
2009	2460	4.9	0.75 ± 0.02	11.82 ± 3.52
2010	3168	4.9	0.76 ± 0.01	0.11 ± 0.00
2011	2336	4.9	0.74 ± 0.03	13.06 ± 4.46
2012	2812	4.8	0.76 ± 0.02	15.84 ± 4.35
2013	2130	4.7	0.79 ± 0.02	24.84 ± 5.30
2014	2415	4.7	0.81 ± 0.02	31.43 ± 6.11
2015	3173	4.6	0.78 ± 0.02	65.50 ± 11.33
2016	1528	4.6	0.79 ± 0.02	136.65 ± 36.54

IOP Conf. Series: Earth and Environmental Science **132** (2018) 012026 doi:10.1088/1755-1315/132/1/012026

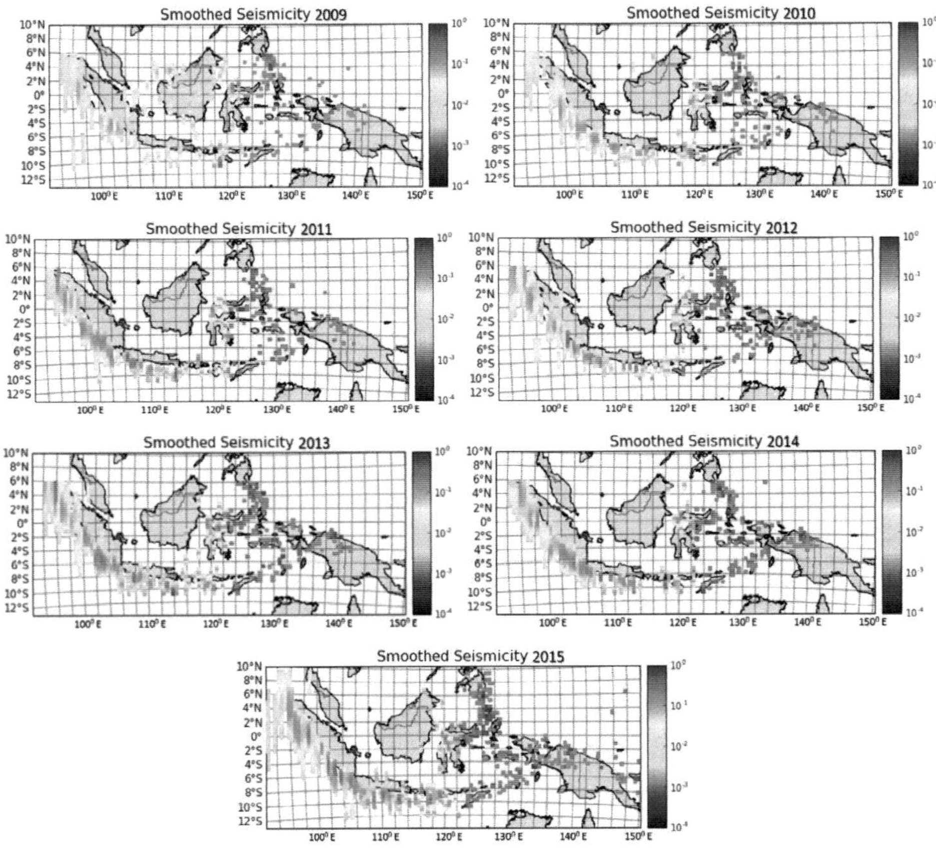

FIGURE 4. The seismicity temporal variation from 2008 to 2015. It has been plot depend on amount of earthquake that happened in certain time period. High seismicity level tendency in east part of Indonesia, and moderate seismicity level distribution disseminated from Sumatra to Java.

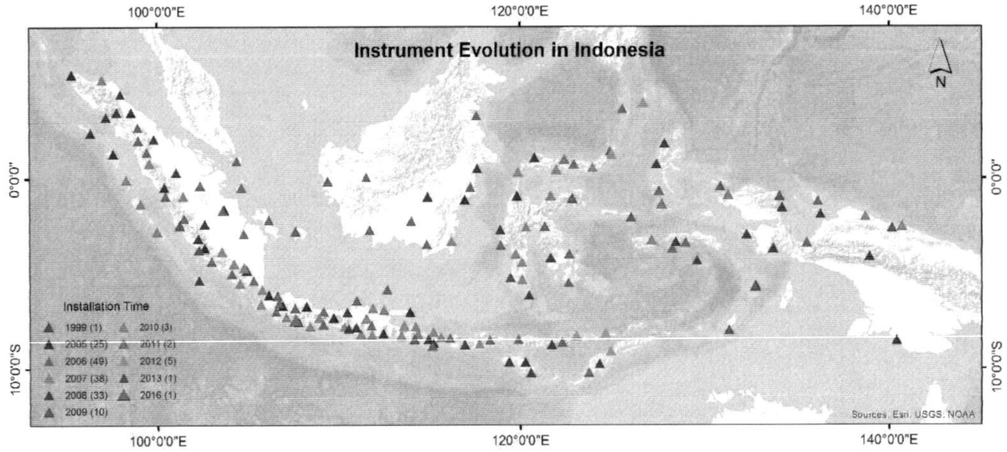

FIGURE 5. Seismograph distribution in Indonesia shows that western region have more dense seismograph installed compare to eastern region of Indonesia. Basemap source derived from ESRI, USGS, and NOAA

6. Summary

Seismicity Analysis for determining b-value parameter as preliminary quantifying have proceed based on latest earthquake catalog in Indonesian region from 2008 to 2016. Some products that produced are empirical relationship between various BMKG magnitude type correlate to Moment Magnitude (Mw) of GFZ-Potsdam and Mc variation of Ina-TEWS seismograph network. Mc are decreasing annually proved that the development of seismograph during 2008-2016 has affect Mc although it is not really significant. Consider to seismograph network distribution, it showed that the western part of Indonesia has more dense seismograph compared to the eastern region. However, the eastern part of Indonesia has a high level of seismicity compared to the western region. Based upon the results, additional seismograph installation in the eastern part of Indonesia should be taken into consideration. Another method will be improved to get more reliable quatification and seismicity parameters.

7. Acknowledgements

The authors gratefully acknowledge to Indonesian Agency for Meteorology, Climatology, and Geophysics (BMKG) and Lab. Geophysics UGM squad for their support and kindness during this research.

References

[1] Asrurifak M 2010 *Peta Respon Spektra Indonesia untuk Perencanaan Struktur Bangunan Tahan Gempa Berdasaran Model Sumber Gempa Tiga Dimensi dalam Analisa Probabilitas* Ph.D thesis (Bandung : Intitut Teknologi Bandung)
[2] Hanka W, Saul J, Weber, Becker J, Harjadi P, and Fauzi, 2010 *Nat. Hazards Earth Syst. Sci*, **10** pp 2611–2622
[3] Scordilis E M 2006 *Journal of Seismology*. **10** pp 225-236
[4] Wells D L and Coppersmith K J 1994 *Bulletin of the Seismological Society of America*. **84** pp 974-1002
[5] Shater A B A, Shater A and Mahmoud S M 2012 *African Journal of Basic & Applied Sciences*. **4(3)** pp 77-82
[6] Duni L L, Kuka S H and Kuka N 2010 *Acta Geod. Geoph. Hung*. **45(3)** pp 317-323
[7] Al-Heety E 2014 *Arab J. Geosci*, **7** pp 4727–4732

Strain Variation along Cimandiri Fault, West Java Based on Continuous and Campaign GPS Observation From 2006-2016

A A Safitri[1,a], I Meilano[1,2], E Gunawan[2], H Z Abidin[1], J Efendi[3], and E Kriswati[4]

[1] Geodesy and Geomatics Engineering, Faculty of Earth Sciences and Technology, Bandung Institute of Technology, Jalan Ganeca 10, Bandung, Indonesia
[2] Graduate Research on Earthquake and Active Tectonics, Faculty of Earth Sciences and Technology, Bandung Institute of Technology, Jalan Ganeca 10, Bandung, Indonesia
[3] Geospatial Information Agency of Indonesia, Jalan Raya Jakarta-Bogor 46, Cibinong, West Java, Indonesia
[4] Center for Volcanology and Geological Hazard Mitigation, Geological Agency, Jalan Diponegoro 57, Bandung, Indonesia

[a] corresponding author: alwidya13@gmail.com

Abstract. The Cimandiri fault which is running in the direction from Pelabuhan Ratu to Padalarang is the longest fault in West Java with several previous shallow earthquakes in the last 20 years. By using continues and campaign GPS observation from 2006-2016, we obtain the deformation pattern along the fault through the variation of strain tensor. We use the velocity vector of GPS station which is fixed in stable International Terrestrial Reference Frame 2008 to calculate horizontal strain tensor. Least Square Collocation is applied to produce widely dense distributed velocity vector and optimum scale factor for the Least Square Weighting matrix. We find that the strain tensor tend to change from dominantly contraction in the west to dominantly extension to the east of fault. Both the maximum shear strain and dilatation show positive value along the fault and increasing from the west to the east. The findings of strain tensor variation along Cimandiri Fault indicate the post seismic effect of the 2006 Java Earthquake.

Keywords: The Cimandiri Fault, GPS, Strain Tensor

1. Introduction

The convergent motion of the Australian and the Eurasia Plates results subduction at the Sunda Arc. This subduction zone process imposes tectonic stresses on the fore arc region offshore and on land of Java thus causes the formation of earthquake fault zones to accommodate the plate movement [1]. There are faults in-land Java to accommodate that plate movement; Cimandiri fault, Lembang fault, and Baribis fault.

The Cimandiri fault is located in Sukabumi, West Java, which runs in the direction from Pelabuhan Ratu, passing Sukabumi, Cianjur and Padalarang [1]. According to [2], Cimandiri is a sinistral strike-slip fault with the axial direction of some N150°E trending folds in the southeast turn clockwise while those to the northwest turn anticlockwise (**Figure 1**). Geologic evidence indicates that Cimandiri fault is active fault. Several previous earthquakes such as Pelabuhan Ratu earthquake (1900), Cibadak earthquake (1973), Gandasoli earthquake (1982), Padalarang earthquake (1910), Tanjungsari earthquake (1972), Conggeang earthquake (1948), and Sukabumi earthquake (2001), occurred along the Cimandiri fault [1].

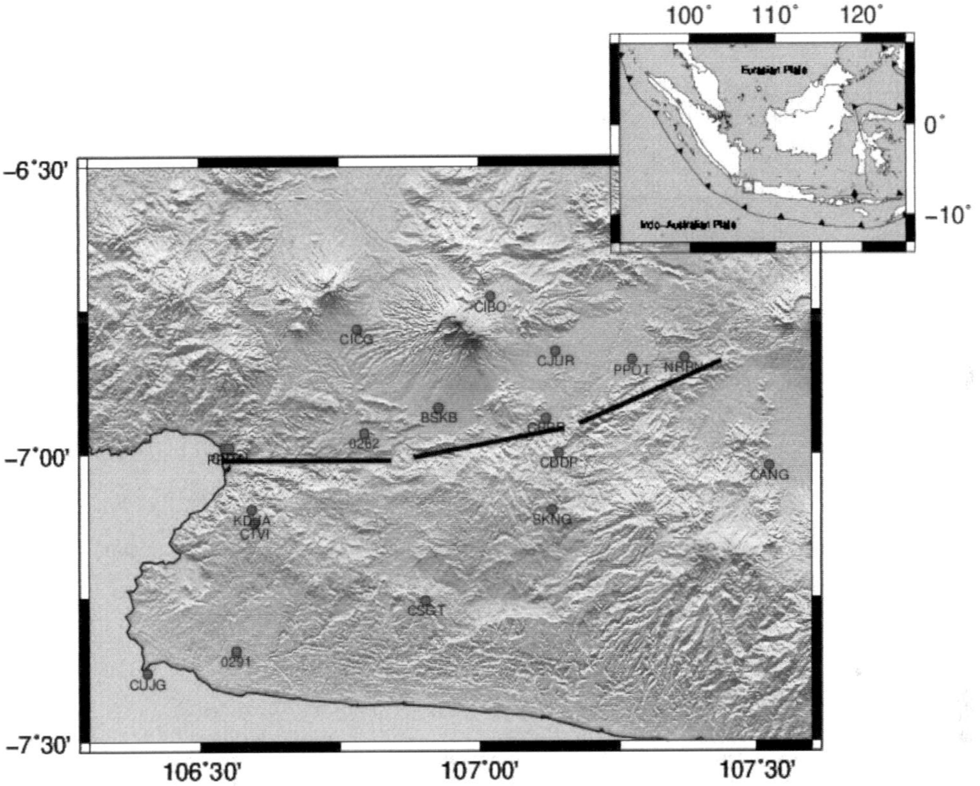

Figure 1. Location of Cimandiri Fault and the surrounding GPS stations indicates by red point. Solid black line indicates the main fault line.

By using continues Global Positioning System (GPS) and campaign GPS measurement, the strain is estimated along the Cimandiri fault. The surrounding GPS stations are very important to show the tectonic activities which influenced to the research area since the campaign GPS data was taken soon after the 2006 Java earthquake. The campaign GPS data were taken by Geodesy Research Division of Bandung Institute of Technology showed a significant signal of post seismic deformation of the 2006 Java earthquake [1]. According to Gunawan et al (2016), the post seismic of the 2006 Java earthquake also affected BAKO station which is located in distance about 440 km from the earthquake epicenter. Therefore, the aim of this research is to obtain the activity of the Cimandiri fault indicates by the strain pattern around the fault and the other tectonic activities which might influence the surrounding fault area.

2. Data and Method

This research combined two kinds of GPS monitoring. Continuous GPS stations cover entire fault zone which located in far distance of the fault. Campaign GPS stations cover area near fault line. The continuous GPS data are provided by Geospatial Information Agency of Indonesia (BIG) existing data from 2008 to 2015 whilst campaign GPS data were observed by Geodesy Research Group of Institute

IOP Conf. Series: Earth and Environmental Science **132** (2018) 012027 doi:10.1088/1755-1315/132/1/012027

of Technology Bandung in collaboration with Center for Volcanology and Geological Hazard Mitigation in year 2006 to 2016.

The GPS data processing is done using GAMIT/GLOBK software [3, 4]. We used surrounding IGS stations around the research area to fix the preliminary coordinate results in ITRF2008 within 95% confidence level. Since the GPS stations located in Sundaland block, the effect of the Sundaland movement should be removed. We applied reference frame transformation from ITRF2008 to ITRF2000 using transformation parameters by [5]. The calculation of Sundaland block motion and rotation is applied by using parameters by [6].

The strain tensor is calculated within a GPS Delaunay triangulation which every triangle defined as discrete rigid body. The continuum model is applied into calculation to obtain the result of principal strain rate as the representation of tectonic activities within the area [7].

3. Results and Discussion

To produce the strain tensor within the area, we used the result of the velocity field of related GPS stations. The velocity field (**Figure 2**) clearly shows the movement of related GPS station to the southeast direction with varying magnitude around 25 to 30 mm/year which is fixed into stable reference ITRF2008. The result of the velocity field in ITRF2000 has the similar pattern of the velocity field which fixed in ITRF2008.

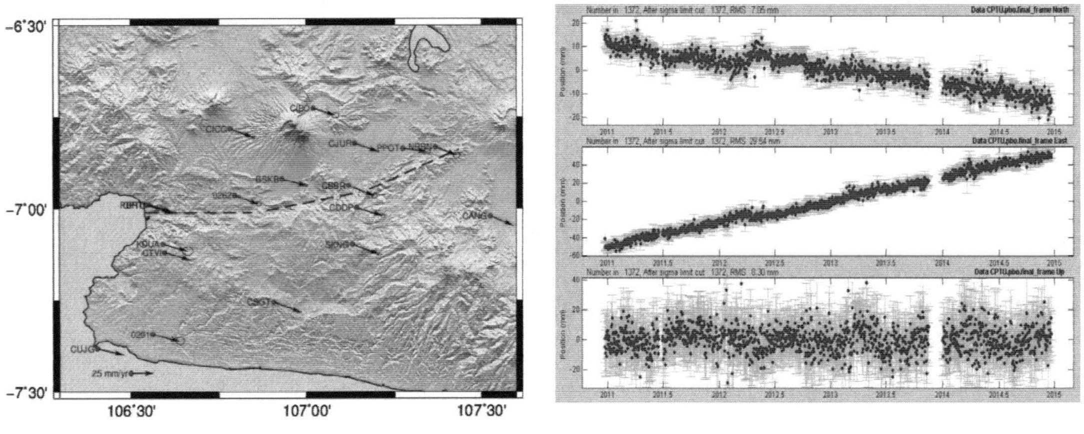

(a) **(b)**

Figure 2. (a) Velocity field of surrounding GPS stations. The blue dash line indicates the Cimandiri fault. (b) The time series of one of GPS station (CPTU) breaks down into its component.

The strain tensor which is calculated from the velocity field shows through the principal strain rate in **Figure 3(a)** below. The principal strain indicates the significant extension movement from the west to the east. The west part of the research area tends to dominated by the shortening whilst in the east of the area the strain gradually changes into extension.

Our result of the principal strain rates indicates that the area dominantly influenced by the post seismic activity of the 2006 Java earthquake. The displacement of the post seismic displacement of the 2006 Java earthquake is also calculated according to afterslip model by [8] using STATIC 1D code analyzed in a layered, spherical earth model [9]. The post seismic displacement (**Figure 3(b)**) in each GPS station clearly shows the changes of the post seismic magnitude increasing from west to east which is agree with the result of the principal strain rates. The influence of the post seismic process get higher in east area since the east area located relatively near to the epicenter of the 2006 Java earthquake rather than the west area (**Figure 4**).

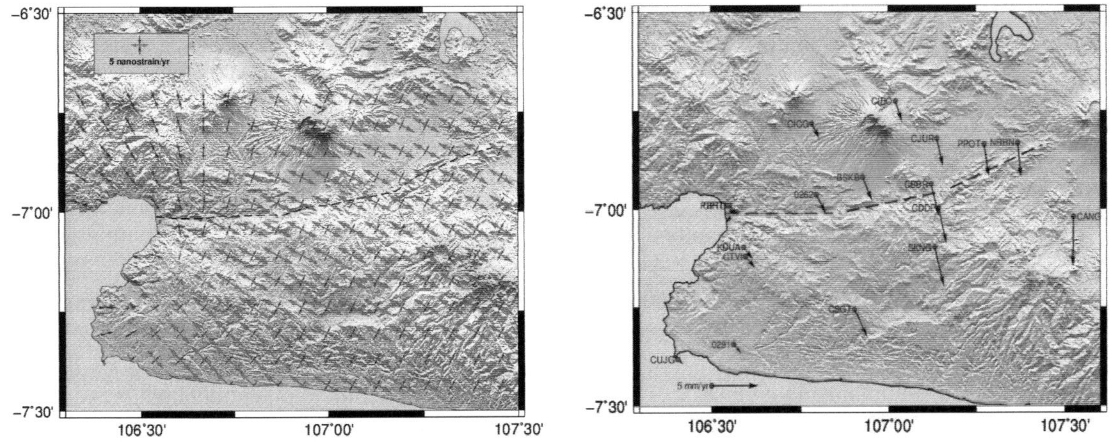

(a) (b)

Figure 3. (a) Horizontal principal axes of strain rates. The red arrow indicates extension while the blue arrow indicates shortening. (b) Afterslip displacement in GPS stations.

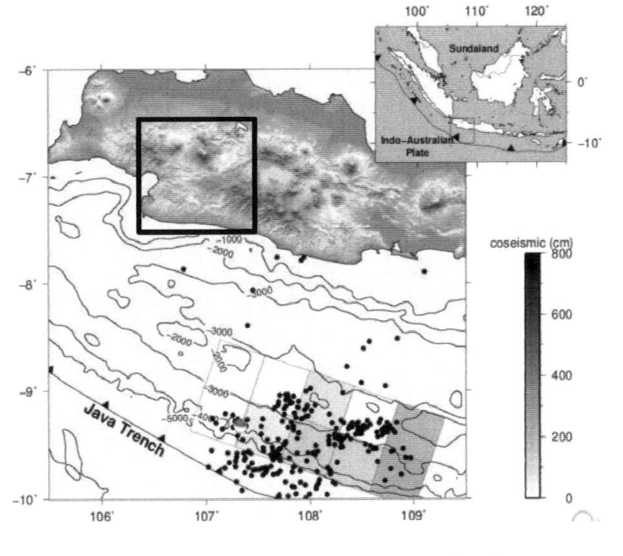

Figure 4. Location of research area indicated by black rectangular in Java map. The epicenter of the 2006 Java earthquake is about hundred kilometers away from GPS station which is used in this research (Modification from [10]).

4. Conclusion

We used combination of continues and campaign GPS data observed from 2006 to 2016 to produce the velocity field within the research area around the Cimandiri fault. The strain rates which is produced from the GPS velocity field indicates the dominant post seismic effect in the east area through the dominant patter of extension in the east whilst the shortening in the west area. These findings support the previous findings that the post seismic of the 2006 Java earthquake affect to the distance of 440 km away from the earthquake epicenter.

Acknowledgements
This research was supported by the Indonesia Endowment Fund for Education (LPDP). We would like to thanks PVMBG for the support during the GPS data acquisition. Figures were generated using GMT software [11].

References

[1] Abidin, H. Z., Andreas, H., Kato, T., Ito,T., Meilano, I., Kimata, F., Natawidjaya, D.H., and Harjono, H. 2009. Crustal deformation studies in Java (Indonesia) using GPS. *Journal of Earthquake and Tsunami, Vol.3, No.2*, 77-88.

[2] Dardji, N., Villemin, T., and Rampnoux, J. P. 1994. Paleostress and strike-slip movement: The Cimandiri Fault Zone, West Java, Indonesia. *Journal of Southeast Asian Earth Sciences, Vol.9, No.1*, 3-11.

[3] Herring, T. A., King, R .W., and McClusky, S. C. 2010a. *GAMIT - Reference Manual GPS Analysis at MIT, Release 10.4.* Cambridge: Massachusetts Intitute of Technology.

[4] Herring, T. A., King, R. W., and McClusky, S. C. 2010b. *GLOBK Global Kalman Filter VLBI and GPS analysis program, Release 10.4.* Cambridge: Massachusetts Institute of Technology.

[5] Altamimi, A., Collilieux, X., and Metivier, L. 2011. ITRF2008: an improved solution of the international terrestrial reference frame. *Journal of Geodesy, Vol. 85, Issue 8*, 457-473.

[6] Simons, W. J. F., Socquet, A., Vigny, C., Ambrosius, B. A. C., Haji Abu, S., Promthong, C., Subarya, C., Sarsito, D. A., Matheussen, S., Morgan, P., and Spakman, W. 2007. A decade of GPS in Southeast Asia: Resolving Sundaland motion and bounderies. *Journal of Geophysical Research: Solid Earth (1978-2012), Vol.112, No. B6*, doi: 10.1029/2005JB003868.

[7] Calais, E. 2015. *Strain Analysis.* ecalais@purdue.edu.

[8] Gunawan, E., Meilano, I., Abidin, H. Z., Rahma, N. R., and Susilo. 2016. Investigation of the best cosesimic fault model of the 2006 Java tsunami earthhquake based on mechanism of postseismic deformation. *Journal of Asian Earth Sciences, Vol. 117*, 64-72.

[9] Pollitz, F. F., 1996. Coseismic deformation from earthquake faulting on a layered spherical Earth. *Geophysical Journal International*, *Vol. 125 No. 1*, 1–14.

[10] Fujii, Y., and Satake, K. (2006). Source of the July 2006 West Java tsunami estimated from tide gauge records. *Geopysical Research Letters. Volume 33*, 24.

[11] Wessel, P., and Smith, W. H. F. (1998). New improved version of the generic mapping tools released. *Eos Trans. AGU. Volume 79(47)*, 579.

IOP Conf. Series: Earth and Environmental Science **132** (2018) 012028 doi:10.1088/1755-1315/132/1/012028

Free Surface Downgoing VSP Multiple Imaging

Fahdi Maula and The Nguyen Dac

[1]Schlumberger Data Services, South East Asia, Jalan P Ramlee, Kuala Lumpur, Malaysia

Corresponding author: fmaula@slb.com

Abstract. The common usage of a vertical seismic profile is to capture the reflection wavefield (upgoing wavefield) so that it can be used for further well tie or other interpretations. Borehole Seismic (VSP) receivers capture the reflection from below the well trajectory, traditionally no seismic image information above trajectory. The non-traditional way of processing the VSP multiple can be used to expand the imaging above the well trajectory. This paper presents the case study of using VSP downgoing multiples for further non-traditional imaging applications. In general, VSP processing, upgoing and downgoing arrivals are separated during processing. The up-going wavefield is used for subsurface illumination, whereas the downgoing wavefield and multiples are normally excluded from the processing. In a situation where the downgoing wavefield passes the reflectors several times (multiple), the downgoing wavefield carries reflection information. Its benefit is that it can be used for seismic tie up to seabed, and possibility for shallow hazards identifications. One of the concepts of downgoing imaging is widely known as mirror-imaging technique. This paper presents a case study from deep water offshore Vietnam. The case study is presented to demonstrate the robustness of the technique, and the limitations encountered during its processing.

1. Introduction

Borehole seismic or vertical seismic profiling (VSP) plays an important role in oil and gas exploration. Throughout the years, the technology has developed beyond providing time-depth relations into providing high-resolution images in the reservoirs. The traditional subsurface imaging from VSP data consists of extracting the up-going Wavefield (reflection Wavefield) from below the well trajectory.

With the receiver in a borehole and its source on the surface, borehole seismic data will record both downgoing and up going waves, along with their multiples. In normal processing, multiples have been considered unwanted arrivals, and several processing methods are normally proposed to eliminate multiples. For certain environments (such as deep-water) a free surface multiple can be generated and differentiated for imaging, since the free surface multiple will have larger time separation with direct arrival.

This free surface multiple travels a longer path in the subsurface and has the possibility to illuminate certain areas better than the primary source. Downgoing VSP multiples provide a broader coverage area and enhanced illumination over the use of primary arrivals only [1]. Reflectors above receivers can be imaged with downgoing multiples [2].

The wave recorded in the borehole seismic configuration can be schemed in figure 1a below, where downgoing and upgoing can be differentiated based on their apparent velocity. The multiple is indicated by their time separation with the primary. Figure 1b shows the schematic of the possible free surface multiple and their ability to illuminate the reflection above the trajectory.

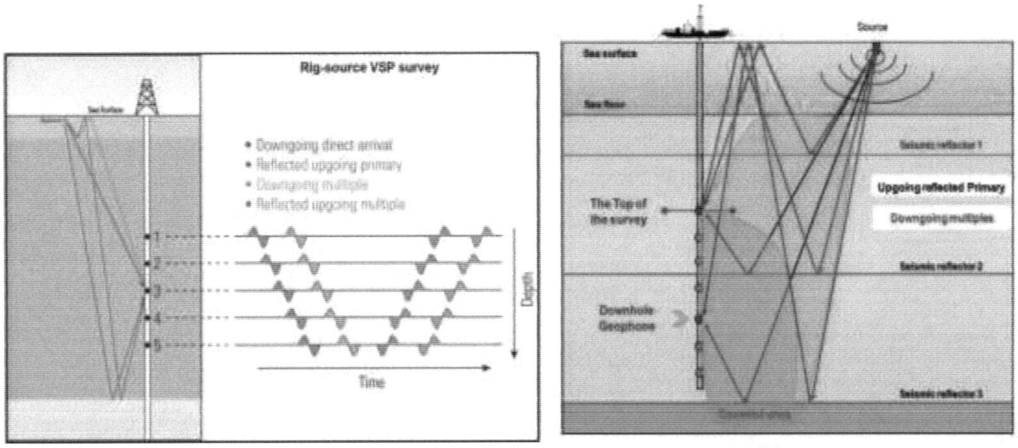

(a) (b)

Figure 1. (a) Schematic of basic wave recorded in borehole seismic – upgoing and downgoing wavefield, (b) illustration of free surface multiple for borehole seismic configuration. [3]

2. Methodology

2.1. VSP Mirror Imaging

As depicted in the schematic above, multiple is basically repetition of the primary arrival. In certain environments (such as deep water) and relatively noncomplex subsurface layering (i.e. flat layers), the multiple can be repositioned as if it's recorded by virtual receivers created by the free surface mirrors [2]. It has been proposed that lateral subsurface coverage in VSP surveys be extended [4]. In this study, the authors pose a real-life example of a barber shop having mirrors on opposite walls, allowing one to see the reflection repeatedly between two mirrors.

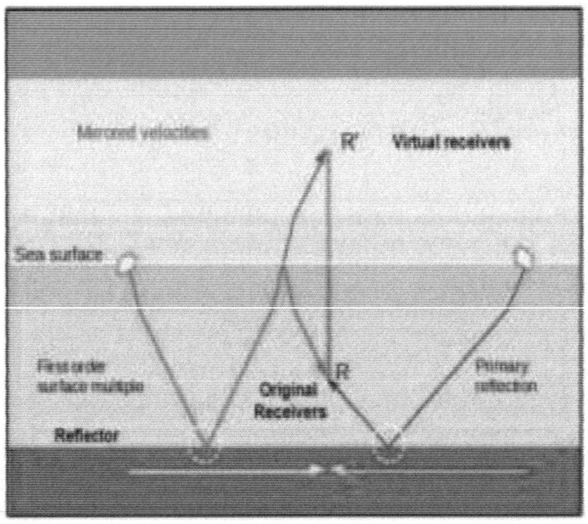

Figure 2. Illustration of VSP mirror imaging, by mirroring the velocity model and projecting the receiver for the free-surface multiple [2].

IOP Conf. Series: Earth and Environmental Science **132** (2018) 012028 doi:10.1088/1755-1315/132/1/012028

2.2. VSP Processing Example (Traditional and Multiple Imaging)

The example data below shows the VSP record from deep water off the coast of Vietnam [5]. From this data, the sea surface multiple is comprised of a wave train of reflections. It started with the sea floor reflection and continued to the reflection below the seafloor. In deep water, the wave train is easily distinguishable, as the seafloor has low acoustic impedance in contrast with the overlying water.

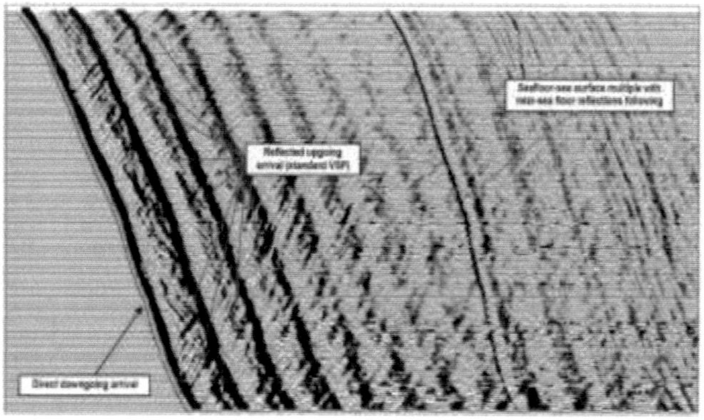

Figure 3. A vertical component of VSP data (offshore deepwater Vietnam), where the free surface multiple has a large time separation with the direct downgoing arrival.

The objective of this VSP acquisition is a seismic tie for a deviated well. Vertical incidence configuration was proposed to achieve the objective using the Schlumberger Versatile Seismic Imager (VSI*) as the downhole receiver. A vertical incidence survey will record vertical time and vertical reflection in deviated wells (maximum well deviation is 55 deg). Subsurface layering in the area is relatively not complex, and the dip is relatively gentle. These conditions are ideal for downgoing free surface multiple imaging.

For traditional imaging, standard VSP processing was proposed using a median velocity filter to separate the downgoing and upgoing wavefields. Deconvolution operators have been designed on downgoing wavefields and applied to the upgoing wavefield to collapse the multiple. Since this is vertical incidence imaging, the 2D reflection image below the well trajectory can be constructed to give a high resolution VSP image. This image represents the primary reflection from below the trajectory, captured by the VSP receiver within the borehole. This is the traditional way of looking at VSP data.

IOP Conf. Series: Earth and Environmental Science **132** (2018) 012028 doi:10.1088/1755-1315/132/1/012028

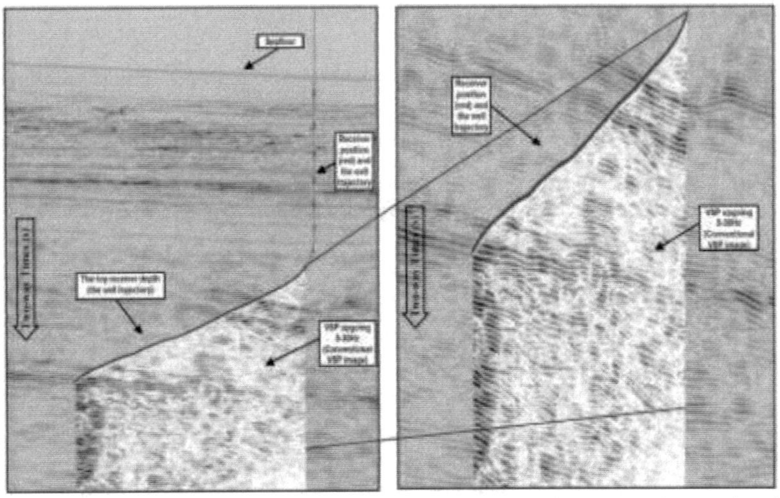

Figure 4. Upgoing VSP image (traditional VSP image) overlaid on surface seismic image; the traditional VSP image only gives correlation below the well trajectory.

To extract the downgoing free surface multiple from VSP, the multiple reflection (identified by a later arrival in the data) was isolated. Waveshaping deconvolution using operators designed from direct downgoing wavefields were also applied to the free surface multiple to get a zero-phase image. To produce an image from these VSP multiples, a velocity model was constructed. These velocity models consist of the mirror image of the actual subsurface. Virtual receivers were placed in the mirrored region, and multiples were simulated using ray trace modeling. The ray trace modeling has been shown with mirror imaging, as the multiple can be used to image reflection above the well trajectory and the traditional VSP imaging can only give the image below well trajectory.

Using the mirrored velocity model, the multiple image can be migrated (VSP-CDP Transform) into a 2D seismic image. This will give a 2D VSP image above well trajectory with reasonably high frequency, which can then be used for seismic calibration up to the seabed.

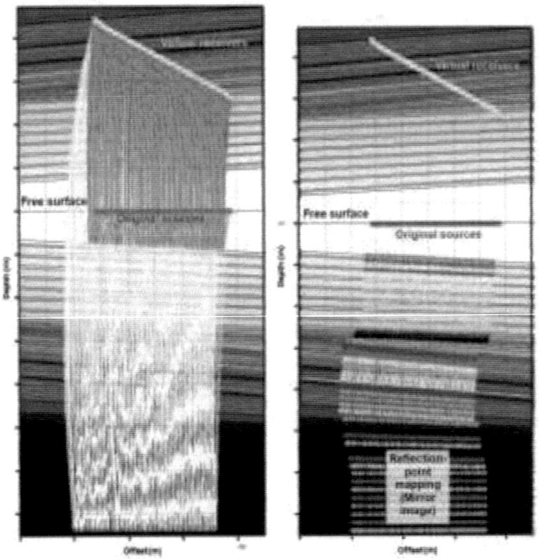

Figure 5. Velocity model for mirror imaging (left), and ray trace modeling has shown the reflection points from mirror image cover the reflection up to seabed.

IOP Conf. Series: Earth and Environmental Science **132** (2018) 012028 doi:10.1088/1755-1315/132/1/012028

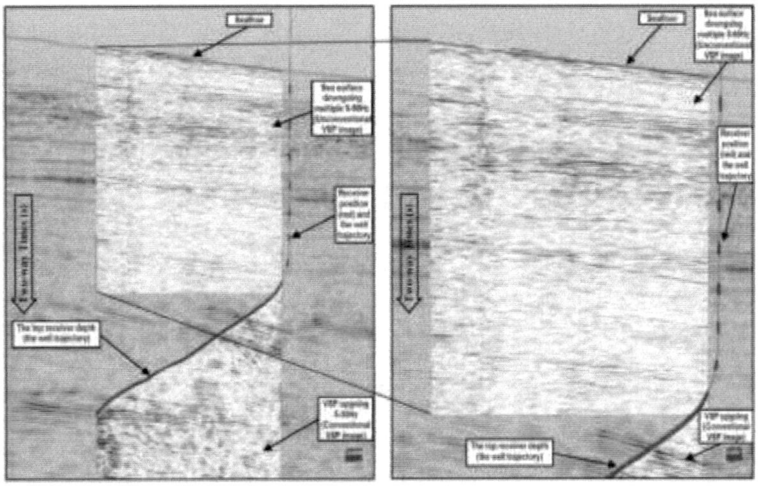

Figure 6. VSP upgoing image and VSP multiple image (left), zoomed in on the multiple image show reasonably good well tie with the surface seismic

3. Discussion

Mirror imaging is an alternative and simpler way to use free-surface multiple reflections from VSP data. By using virtual receivers created by free-surface mirrors, the illuminations can be extended to the surface. This methodology works best for receivers near the seafloor, because the seismic rays have traversed the shortest distance through the formations. If the receivers are placed deeper, the seismic multiple energy will suffer more attenuation due to the fact that the multiple will have traveled more than the primary (multiple rays traversing the formation three times).

The seafloor dip will also have an effect on the multiple image. If the seafloor is dipping, the multiple reflections will also move updip, and the data will require migration with proper velocity model to get the correct image. Since this is based on a mirroring technique, a relatively non-complex subsurface is preferred.

The mirror imaging technique works best in a deep-water environment, where the free surface multiple can be identified easily and have a larger separation from the direct arrival. Several case studies of using the method in shallow water have also been published [6], but the processing need to be carried out with caution when separating the free surface multiple.

References

[1] Leung, V., Wong, M., and Zhang, R. 2013. Up- and Down-going Wave Multiples RTM Imaging for VSP. *Presented at the SEG Annual Meeting*, Houston, Texas

[2] Farmani, M.B., Atkinson, B., Muzi, J., Brooks, N., Hilton, T., and O'Brien, J. 2012. Imaging First Order Multiples in VSP Data: A Detailed Field Study. *Presented at the SEG Annual Meeting*, Las Vegas, Nevada

[3] Pereira, A.M., and Jones,M., *Fundamental of Borehole Seismic Technology*, 2010, (Schlumberger)

[4] Alam, A.M., Method for Extending The Lateral Subsurface Coverage in Vertical Seismic Profiling Surveys. 1988. *European Patent, EP 0 317 288 B1*, submitted 16.11.88, issued 28.04.93

[5] Dac, T.N., Sanders., M., 2015a. Using Free Surface Multiples to Obtain Unconventional VSP

Image Above the Well Trajectory – A Case Study in Offshore Vietnam. *IPTC Conference*. IPTC-18305-MS
[6] Dac, T.N., Sanders., M., 2015b. Obtaining a Seismic Image Above the Top VSP Level Using Sea Surface Multiple – A Case Study in Nam Con Son Basin, Offshore Vietnam. *SPE Conference*. SPE-177322-MS

Acknowledgments
The author would like to thank PetroVietnam (PVN) and Talisman Energy Vietnam for the support and permission to publish this paper. We also would like to thank Michale Lawson, Foo Say Jeow, Khanh Tran Van from Talisman Energy Vietnam for the discussion, and Michael Sanders (Schlumberger) for the extensive discussion.

IOP Conf. Series: Earth and Environmental Science **132** (2018) 012030 doi:10.1088/1755-1315/132/1/012030

Geotourism of Batu Dinding Kilo Tiga, Amurang District, South Minahasa Regency, North Sulawesi - Indonesia

V.A. Elvien Kristian[a), Ahmad Ridho[b), and Febriany Diaz Alffi[c)

[a) asprilavincentius@gmail.com, [b)ahmdrdo@gmail.com, [c) febridiazalffi@gmail.com

Abstract. Start your abstract Batu Dinding Kilo Tiga or famously called Batu Dinding is one of tourist destinations in Kilo Tiga village, Amurang district, South Minahasa regency, North Sulawesi. In terms of Geology, Batu Dinding is a rock cliff with columnar structure and constituted in the form of andesitic thrakit rocks which is one of intermediate igneous rocks of volcanic aged Late Tertiary until Early Quarter, precisely during the Pleistocene. The structure and appearance of Batu Dinding was like a ladder, make Batu Dinding in demand as a destination for wall climbing or cliff hanger. The area around the Batu Dinding, there are Pamurapa River that at a time when it was high tide, can be used for rafting. Unfortunately, the accommodation and accessibility of the Amurang Batu Dinding are poor, causing Batu Dinding rarely to visit. The aim of the paper is to introduce as well as media publications of Batu Dinding tourism in order to increase local income and in addition it can be as a geological site for learning purposes.

1. Introduction

According to Farsani et al (2013), geotourism is nature tourism with unique geological and geomorphological landscape. The geotourism as country tourism asset could be a solution in order to support local businesses and products. North Sulawesi has many potential tourism sites that not well explored yet even it has high attractiveness. One of the tourism resources that has not been explored is geotourism features (Mangindaan et al, 2014). Geotourism defined as tourism which focuses on a landscape (or others geological features) as the basis for providing visitor engagement, learning, and enjoyment. It has links with adventure, cultural, and ecological tourism that focus on creating geotourism product that protect geoheritage and build communities (Robinson, 2015).

Geographically, North Sulawesi located between 00°15’51” N – 05°34’06” N and 120°07’00” E - 127°10’30” E. based on territory administration, North Sulawesi province has 11 districts and four cities, namely; Bolaang Mongondow, Minahasa, Sangihe archipelago, Talaud archipelago, South Minahasa, North Minahasa, North Bolaang Mongondow, Sitaro archipelago, Southeast Minahasa, South Bolaang Mongondow, East Bolaang Mongondow, Manado, Bitung, Tomohan, and Mobagu.

Batu Dinding as Geotourism site is located in Kilo Tiga village, Amurang district, South Minahasa regency, North Sulawesi province (**Error! Reference source not found.**), precisely 5 km northwest of Amurang, and about 60 km southwest of Manado. According to Sompotan (2012), Sulawesi has complex geological conditions and complex litotectonic. This complexity is due to the convergence of three tectonic plates, the Indo-Australian plate, Pacific plate, and Eurasian plate.

Batu Dinding as a natural rock cliff with columnar structure that has high up to 90 meters, composed of andesitic thrakit (**Error! Reference source not found.**). Batu Dinding used as cliff hanger and wall climbing arena. As tourist destination, Batu Dinding has poor supporting infrastructure. The nearest lodgment and restaurant are located in Amurang, about 5 km from Kilo Tiga village.

Beside to wall climbing, the beauty of Kilo Tiga village also interesting to visit. In fact, in this village there is Pamurapa River that usually used for rafting. Little publication and poor supporting tourist infrastructure make Batu Dinding and Kilo Tiga village rarely visited. The aims of this paper are to publicize the tourist attractions in Kilo Tiga village, especially Batu Dinding and enhances the usability of Batu Dinding as a tourist destination or as a field study.

2. Methodology

Mouriki and Charalampos (2009) explained that certain methodology required to assess the geotopes (geological site). The methodology using model to evaluate the touristic value and conservation status of landform and geomorphosite. In this research, after conducting field observations and interviews directly with Komunitas Pecinta Alam (KPA) Sulawesi Utara and Kilo Tiga villagers, the authors evaluated two Geotourism site in Kilo Tiga village, Batu Dinding and Pamurapa River. Scoring method used to evaluate both Geotourism sites. The scoring method based upon paper "Geoheritage and Geotourism Potential of the Strzelin Hills (Sudetic Foreland, SW Poland)", written by Solarska Anna and Jarry Zdislaw (2010). The scoring method used four aspects as a reference assessments, consists of accessibility, state of preservation, scientific worth, and education significance (**Error! Reference source not found.**).

Table 1. Criteria of assessment for Kilo Tiga's geosite evaluation (according to Anna and Zdislaw, 2010)

Criterion	Traits	Points
Accessibility	Site clearly visible, located directly on the touristic trail or nature's path	5
	Site clearly visible, located on the road or path	4
	Site barely visible, located more than 250 m away from the path or road	3
	Site difficult to access for tourist (ex. significantly overgrown or difficult to access)	2
	Site unavailable for tourists	1
State of preservation	Well preserved site with no visible signs of degradation	5
	Site in slight violation of its structure	4
	Partially destroyed	3
	Site heavily modified by human	2
	Site destroyed - loss character of geosites	1
Scientific worth	Very high: one site in the region, unique in a wider scale	10
	High: very important for regional studies	8
	Average: significant for regional research	6
	Low: Common site with average values	4
	Very low: no partikular distinctive features	2
Education significance	Very high: number of represented issues: 5 and more	10
	High: number of represented issues: 4	8
	Average: number of represented issues: 3	6
	Low: number of represented issues: 2	4
	Very low: number of represented issues: 1	2

3. Geology of Study Area

The research area located in Kilo Tiga village, Amurang district, South Minahasa regency, North Sulawesi. This paper also discussed the geological conditions and litotectonic system of Sulawesi Island, particularly in the northern arm of Sulawesi Island.

3.1. Regional Geology of Sulawesi

Geologically, Sulawesi and the surrounding area is a complex area. Complexity is due to the convergence of three lithospheric plates, the Indo-Australian plates (northward), Pacific plate (westward), and Eurasian plate (south-southeast) (**Error! Reference source not found.**). In the northern part of Sulawesi, there is North Sulawesi Trench which is a subduction zone of Sulawesi Sea (Celebes Sea) and northern arm of Sulawesi (Darman, 2011).

Based on lithologic and tectonic conditions, Sulawesi and the surrounding area is divided into five tectonic provinces (**Error! Reference source not found.**), the North Sulawesi Magmatic Arc, West Sulawesi Plutono-Volcanic Arc, Central Sulawesi Metamorphic Belt as part of Australian block, East Sulawesi Ophiolite which is a segment of Trias – Miocene oceanic crust, and Banggai-Sula Microcontinental fragments (Sompotan, 2012).

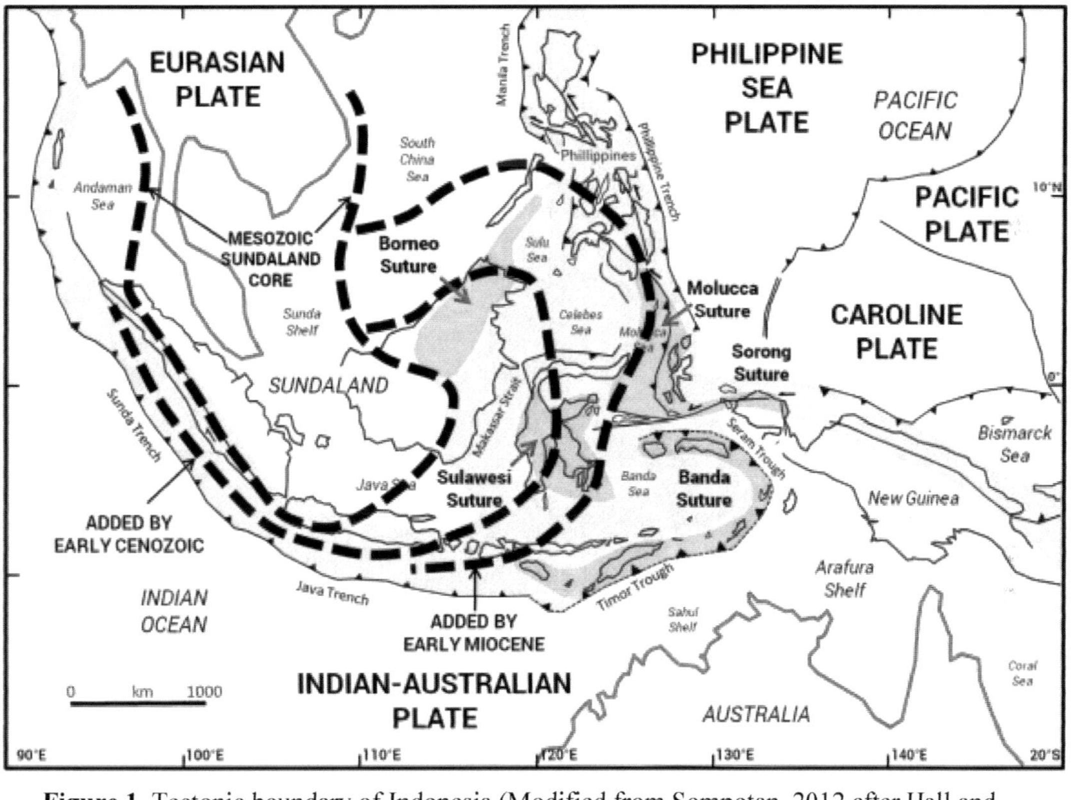

Figure 1. Tectonic boundary of Indonesia (Modified from Sompotan, 2012 after Hall and Smyth, 2008)

3.2 North Sulawesi Magmatic Arc

The North Sulawesi Magmatic Arc (**Error! Reference source not found.**) extending from Buol to Manado, covering North Sulawesi Province and Gorontalo. North Sulawesi Arc extends about 500 km from 121° E to 125°20' E with relatively constant width of 50 – 70 km. This area reaches altitude 2065 m and 3255 m in western Gorontalo (area around neck of Sulawesi) (Darman, 2000).

Evolution of North Sulawesi Arc could be divided into two main stages (Sompotan, 2012):

1. West-trending subduction during Early Miocene (22 – 16 Ma)
2. Post-collison rifting and the appointment of arc, with the onset of subduction along North Sulawesi Trench during Late Miocene to Quaternary (9 Ma).

North Sulawesi region dominated by limestone as a constituent lithology of Ratatotok sedimentary basin. The others lithologies that compile the basin are (Sompotan, 2012):

1. Breccia-sandstone group which consists of breccia to coarse conglomerate, fine and coarse sandstone cross bedding, siltstone, claystone (found in Ratatotok – Basaan), and pyroxene-andesite breccia.
2. Tondano Tuff group (Pliocene) which consists of coarse volcanic rock fragments containing fragments of andesite, pumice, tuff, ignimbrite breccia, and andesitic thrakit rock.
3. Quaternary rocks group that composed of andesite-basalt lava, bombs, and ash.
4. The youngest group consist of coral reef limestone, lake sediment, stream sediment, and alluvium.

Pliocene and Quaternary volcanicity belonging to the Sangihe Arc (**Error! Reference source not found.**) conceals much of the Early Miocene geology near Manado (**Error! Reference source not found.**). Small exposures of andesite and diorite below Quaternary volcanic cover on the Sangihe island (north of Manado), suggest that older arc volcanics form the basement of the present-day Sangihe Arc (Darman, 2000).

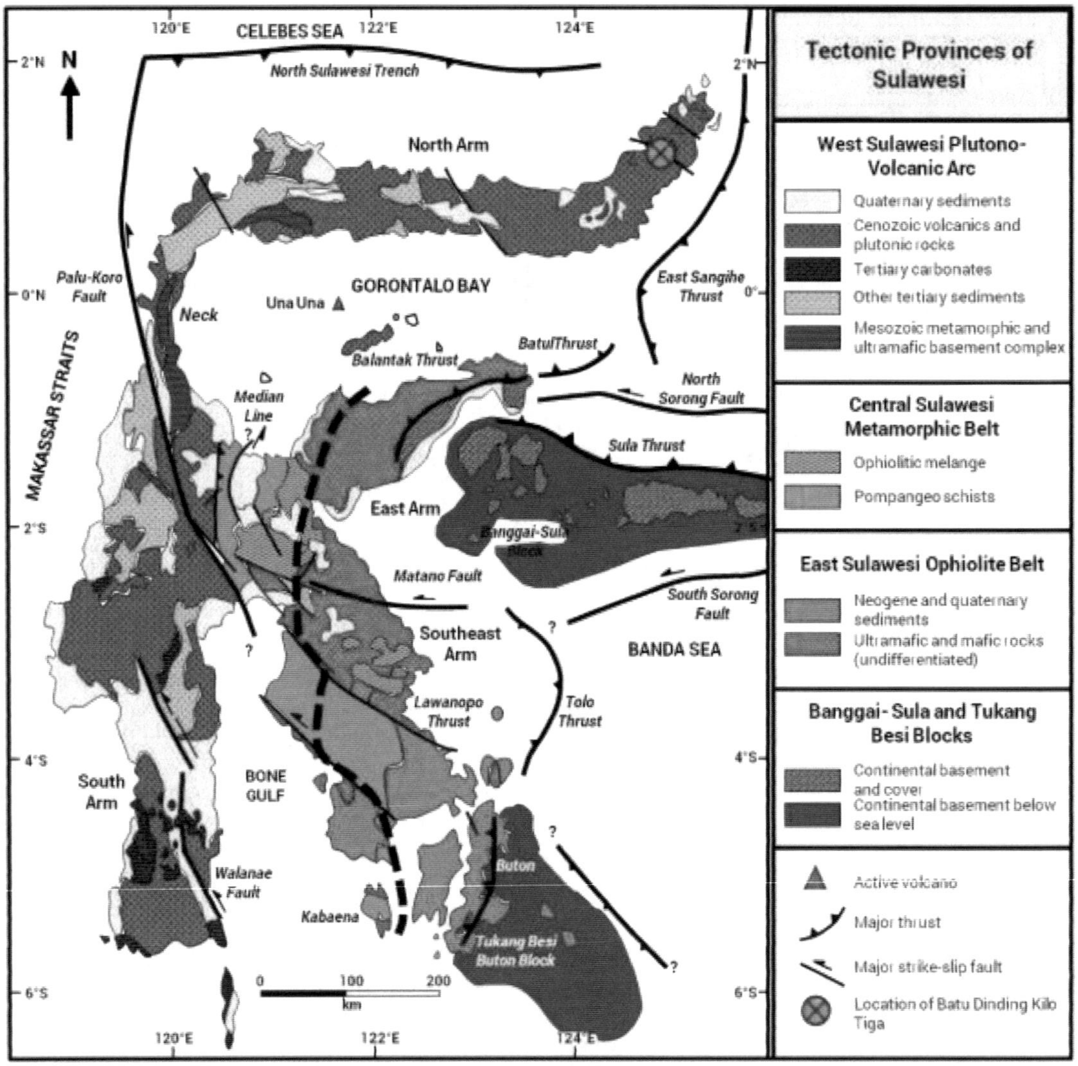

Figure 2. Tectonic provinces of Sulawesi (Modified from Sompotan, 2012 after Hall and Wilson, 2000). Location of Batu Dinding Kilo Tiga located in northern arm of Sulawesi.

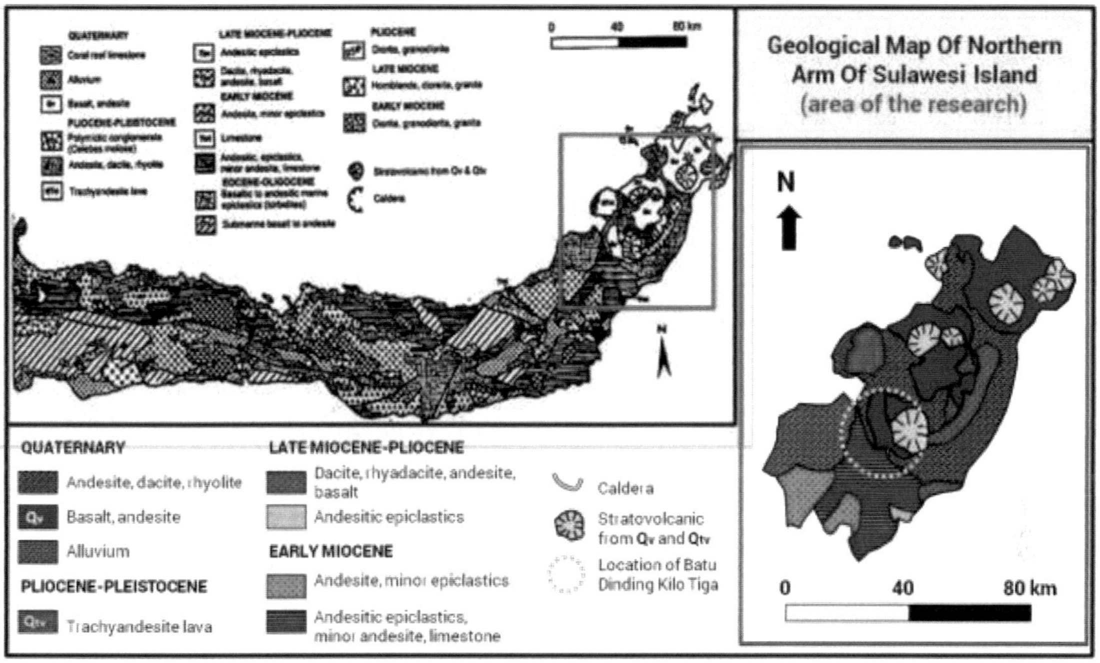

Figure 3. Simplified geological map of northern arm of Sulawesi Island (Modified from Darman, 2000 after Kavalieris et al, 1992).

4. Batu Dinding As Geotourism Site

Batu Dinding is located in Kilo Tiga village, Amurang district, South Minahasa regency, North Sulawesi. The village itself is housed in the route between Amurang and Tombatu. Batu Dinding can be reached from Amurang or Manado by private vehicles or public transportations. However, Batu Dinding can't be achieved with one trip, visitors have to use taxibike from public transportation drop-off point to Kilo Tiga village, and then walk to Batu Dinding.

Batu Dinding is a natural rock cliff that has 90 meters high. This cliff consists of andesitic thrakit with columnar structure. The unique shape with difficult level for climbing, make Batu Dinding used for cliff hanger and wall climbing (**Error! Reference source not found.**).

The configuration of columnar rocks were reversed, make the climbing became difficult and challenging. There are several climbing routes that have been installed by Kelompok Pecinta Alam (KPA) Sulawesi Utara. The climbing routes are:

1. Ofu route (bee in English), the route where the climbing path through some beehives which are some of them inhabited.
2. Spider route, the route where climbing path disjoint, so to finish this route, climbers have to jump from one path to another.
3. Hang Dog route, the route where to pass this route, climbers have to hang for several times.
4. Climb or Swim (COS) route, the route where the path crosses and passing water flow that flows in one side of the cliff.
5. Tragedy route, the route to commemorate the team members who were killed because of swept by Ranoyapo River when the flooding occurs.
6. Malaria route, the second hardest climbing route after Ratapan route.
7. Ratapan route (lament in English), the hardest climbing route in Batu Dinding. Stamina, mind, and strength of climbers will be drained during climbing, make the climbers complain or lamenting.

Until now, only a few people who are able to finish Malaria and Ratapan route. Because it is still in the stage of the development plan, Batu Dinding is free from charge, but it is recommended for visitors to leave a voluntary donation for maintenance and cleanliness management that conducted independently by villagers. For visitors who want to climb, it is recommended to bring their own climbing equipment and contact KPA Sulawesi Utara first.

Near to Batu Dinding, there is Pamurapa River that usually used for Rafting (**Error! Reference source not found.**). Pamurapa River is a short river that carries volcanic and fluvial sediment from volcanoes around Amurang. The river also used as traditional sand quarry and suitable as a sedimentology and hydrology learning site. Score for both Batu Dinding and Pamurapa River are shown in **Error! Reference source not found.**.

Figure 4. Batu Dinding in Kilo Tiga village with its unique structure used for wall climbing or as a place to take pictures.

Table 2. Score (evaluation result) for Kilo Tiga's geosite

Geosite	Criterion				Total
	Accessibility	State of preservation	Scientific worth	Education significance	
Batu Dinding	3	5	8	10	26
Pamurapa River	3	5	6	5	19

5. Conclusions

Batu Dinding is a site which consist of vertical colmnar joint stands about 90 meters from the ground. The Batu Dinding columnar joint classified as andecite-basaltic rocks which undergone cooling period. During the period of cooling, lava started to crack and the crack grows perpendicular to the surface. Because of the uniqueness and the history of forming, Batu Dinding is suitable as Geotourism destination and structural geology educational place.

Pamurapa river is also suitable as geotourism destination. The river flows in the north of Batu Dinding site. During the rain season, the river will cover the flood plain peripheral. After this season, the materials that brought by the flood will cover the flood plain and its suitable as educational place for sedimentology.

Accessibility to Batu Dinding and Pamurapa River fairly easy, less than 250 meters from main road and only be reached by walk. Both sites are well preserved and still rarely visited. Management and maintenance of Batu Dinding and Pamurapa River conducted independently by the local community. In terms of the usefulness of field studies, Batu Dinding and Pamurapa River are very important for geological study of Sulawesi, especially North Sulawesi.

Lodgment and restaurant only exist in the nearby town of Kilo Tiga village, Amurang and Manado. For the convenience of visitors, it should be established tourist supporting infrastructure such as rental climbing equipment, restaurant, and lodgment in the area of Kilo Tiga village.

6. Acknowledgements

Thanks to Drs. Adi Susilo, M.Si, Ph.D as advisor of this paper. And also thanks to Amurangers, Kelompok Pecinta Alam (KPA) Sulawesi Utara, and villagers Kilo Tiga for friendliness and assistance.

References

[1] A. F. Sompotan, *Struktur Geologi Sulawesi*, Perpustakaan Sains Kebumian Institut Teknologi Bandung, Bandung, 2012.

[2] Farsani, N.T., Celesta O.A., & Carlos M. 2013. *Rural Goutourism: A New Tourism Product. Acta Geotoristica, volume 4, number 2*. Portugal.

[3] Herman Darman, et al., *The Geology of Indonesia*, Ikatan Ahli Geologi Indonesia, Jakarta, 2000.

[4] Herman Darman, *Seismic Expression of North Sulawesi Subduction Zone*, in *Berita Sedimentologi Number 22*, Ikatan Ahli Geologi Indonesia, 2011.

[5] Mangindaan, H.,Djabir, H., Rahman, K., & Indriyanti, S. 2014. *Assessment of Tourism Development In North Sulawesi Based On The Holistic Geotourism Perspective*, in *IOSR Journal of Business and Management (IOSR-JBM)*. e-ISSN: 2278-487X, p-ISSN: 2319-7668. Volume 16, Issue 1. Ver. V (Feb.2014), PP-06.

[6] Mouriki, D., and Charalampos, F. 2009. *Quantitative Assessment of Psiloritis' Geotopes with Emphasis on Protection and Geotourism (Crete, Greece)*. University of Crete, Heraklion. Proceedings of The VII European Geoparks Conference.

[7] Robinson, A.M. 2015. *Geotourism and Geoparks A Tourism Development Opportunity for Australia*. Ecotourism Australia Ltd. Australia.

[8] Solarska Anna and Jary Zdizlaw, *Geoheritage and Geotourism Potential of The Strzelin Hills (Sudetic Foreland SW Poland)*, in *Geographica Pannonica Volume 14*, 2010.

IOP Conf. Series: Earth and Environmental Science **132** (2018) 012031 doi:10.1088/1755-1315/132/1/012031

A Comparison between Model Base Hardconstrain, Band-limited, and Sparse-Spike Seismic Inversion: New Insights for CBM Reservoir Modelling on Muara Enim Formation, South Sumatra

Faris Mohamad Noor* and Agra Adipta

Geophysical Engineering Department, Universitas Pembangunan Nasional "Veteran" Yogyakarta, SWK 104 Northern Ringroad, Yogyakarta, Indonesia

*farismohamadnoor@gmail.com

Abstract. Coal Bed Methane (CBM) as a newly developed resource in Indonesia is one of the alternatives to relieve Indonesia's dependencies on conventional energies. Coal resource of Muara Enim Formation is known as one of the prolific reservoirs in South Sumatra Basin. Seismic inversion and well analysis are done to determine the coal seam characteristics of Muara Enim Formation. This research uses three inversion methods, which are: model base hard-constrain, bandlimited, and sparse-spike inversion. Each type of seismic inversion has its own advantages to display the coal seam and its characteristic. Interpretation result from the analysis data shows that the Muara Enim coal seam has 20 (API) gamma ray value, 1 (gr/cc) – 1.4 (gr/cc) from density log, and low AI cutoff value range between 5000-6400 (m/s)*(g/cc). The distribution of coal seam is laterally thinning northwest to southeast. Coal seam is seen biasedly on model base hard constraint inversion and discontinued on band-limited inversion which isn't similar to the geological model. The appropriate AI inversion is sparse spike inversion which has 0.884757 value from cross plot inversion as the best correlation value among the chosen inversion methods. Sparse Spike inversion its self-has high amplitude as a proper tool to identify coal seam continuity which commonly appears as a thin layer. Cross-sectional sparse spike inversion shows that there are possible new boreholes in CDP 3662-3722, CDP 3586-3622, and CDP 4004-4148 which is seen in seismic data as a thick coal seam.

1. Introduction

Coal Bed Methane (CBM) as a newly developed resource in Indonesia is one of the alternatives to relieve Indonesia's dependencies on conventional energies. Indonesia has approximately 453.3 Trillion Cubic Feet (TCF) of CBM reserve that covers 11 basins over Sumatra, Java, Borneo, and Celebes. The biggest resource is in the South Sumatra Basin (183 TCF or 40.37% of national resources), followed by the Barito Basin and Kutai Basin. (Hasyim, 2010).

Coal-Bed Methane (CBM), or *Coal Seam Gas* (CGS), or *Coal-Mine Methane* (CMM) is an unconventional resource of natural methane gas which founded in coal deposits or coal seams (coal-bed). Subsurface explorations on coal seams are definitely identified by well-logging data. But the presence of coal seams on conventional seismic data may not easily determine. By practicing seismic inversion we could achieve the coal seam distribution known from its low impedance value. Various inversion methods are available in the industry and each will characterize the data differently. This research was purposed to emphasize which inversion that most suites on a CBM field.

2. Geological Setting and Stratigraphy of South Sumatra Basin

The South Sumatra basin was formed by three major tectonic phases: 1) extension during late Paleocene to early Miocene forming north-trending grabens that were filled with Eocene to early Miocene deposits; 2) relative quiescence with late normal faulting from early Miocene to early Pliocene; and 3) basement-involved compression, basin inversion, and reversal of normal faults in the Pliocene to Recent forming the anticlines that are the major traps in the area (Pulunggono, 1986). South Sumatra Basin is prolific reservoirs which contain oil, gas, and coal bed methane reserves. The stratigraphy of South Sumatra Basin has been massively discussed by the petroleum expertise or geoscientist since last decades. According Ginger, 2005, the stratigraphy is given as follow (Fig 1).

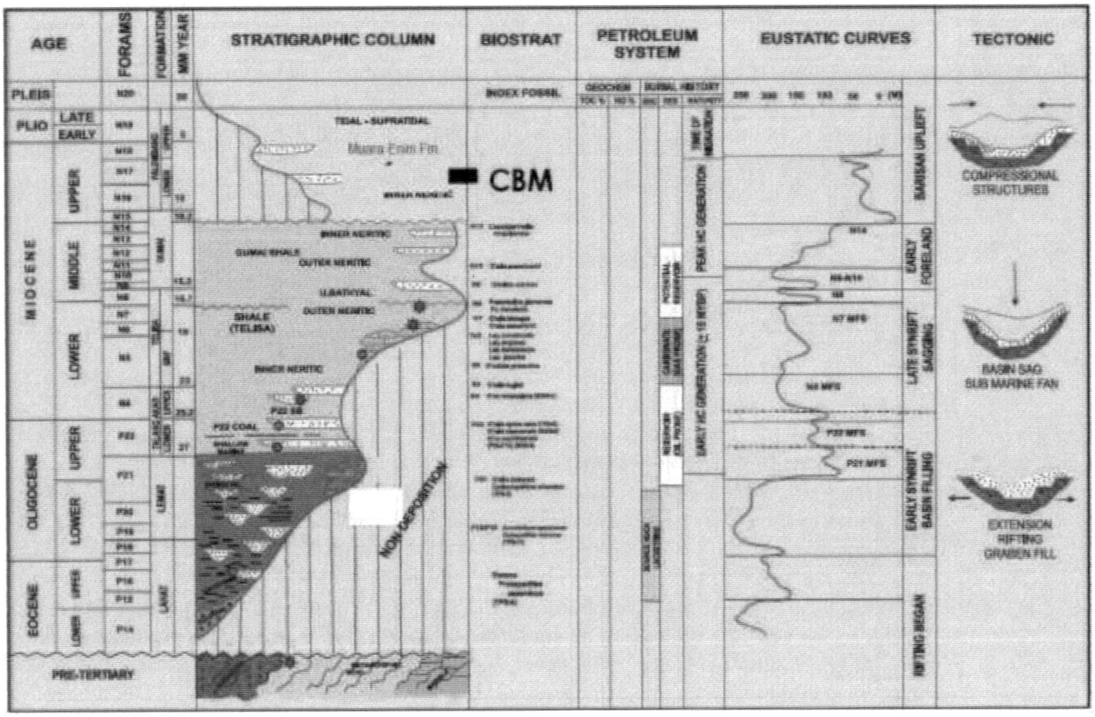

Fig 1. Stratigraphy of South Sumatra Basin (Ginger, 2005)

South Sumatra Basin is distinguished to three group of rocks, which are pre-tertiary rocks, tertiary rocks, and Quaternary rocks. Pre-tertiary rocks act as a basin floor which appears in igneous, metamorphic, and sedimentary rocks. During the middle to late Cretaceous, the pre-tertiary rocks intensively had had folding and faulting and being intruded by igneous rocks since the orogenesis of middle Mesozoic (De Coster, 1974). Tertiary rocks consist of Lahat Formation, Talang Akar Formation, Baturaja Formation, Gumai Formation which is deposited during transgressive phase in Early Eocene to Middle Miocene (Telisa Group) and Air Benakat Formation, Muara Enim Formation, and Kasai Formation which is deposited during regressive phase in Middle Miocene to Pliocene (De Coster, 1974). Quaternary rocks consist of the decomposition of igneous, sedimentary, and metamorphic rocks in which appear as alluvial deposits. The alluvial deposits have various thickness and deposited spreads along the river as a meander both in the middle and the edge of the river.

The presence of coal seam exists in Muara Enim Formation and distinguishes as coal bearing formation which is being our target (Fig 2). Muara Enim Formation represents the last phase of tertiary

IOP Conf. Series: Earth and Environmental Science **132** (2018) 012031 doi:10.1088/1755-1315/132/1/012031

regression which comfortably overlain above Air Benakat Formation in shallow marine, delta plain, paluda, and non-marine. The thickness of Muara Enim Formation is 500 to 1000 m, consists of sandstone, claystone, siltstone, and coal. Coal seam of Muara Enim Formation appears as a lignite which is deposited during Late Miocene to Early Pliocene.

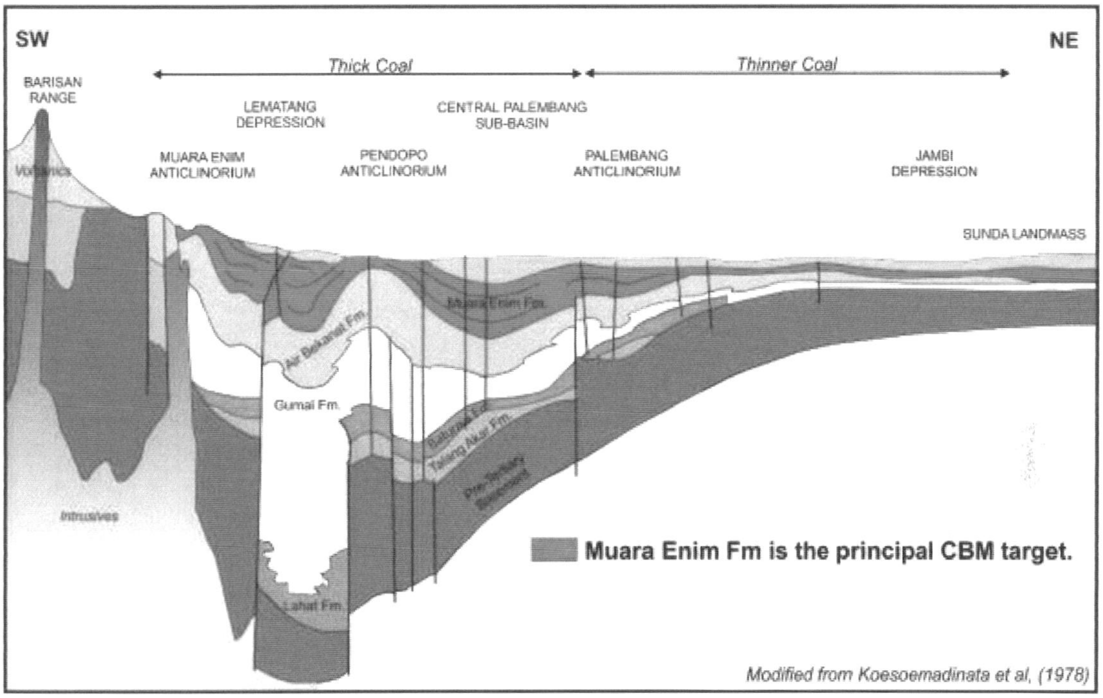

Fig 2. Muara Enim Formation as a CBM prone formation in South Sumatra Basin (Koesoemandinata, 1978)

3. Method

According to Sukmono, 2000, Seismic Inversion is a sub-surface modeling process which requires seismic and well data input as a control. There are few types of post-stack inversions which are model base, bandlimited, and sparse-spike inversion. Model Based Inversion generally adopts GLI method (Generalized Linear Inversion), a process to provide a seismic model (impedance) which eventually compared with seismic data and iteratively repeated until obtaining lowest error count. Model Based Inversion is divided into two types, which are Model-Based Soft-Constrain and Hard-Constrain. Soft-Constrain Inversion doesn't set any absolute bounds on how far the impedance can move from the initial guess, while Hard-Constraint Inversion imposes fixed boundaries on the calculated impedance. Band-Limited inversion is a modification of discrete recursive inversion. This inversion purposed to re-obtain low frequencies which are lost in the process of recursive inversion. Low frequencies are achieved from low-filtered well-log data. Band-Limited inversion provides simple Acoustic Impedance section. Nevertheless, band-limited are depended on the acoustic impedance of the first layer, lacked with lateral resolution and noisy data are also being inverted. Sparse Spike Inversion basically seeks minimal series of high-value reflection coefficient. This value directly impacts the appearance geological event such as unconformity, lithology boundaries, etc. Iteratively, sparse spike inversion applies additional reflectors which are dimmed on each iteration. Iteration process will be halted if RC model represents seismic data. RC obtain from such process of deconvolution. Sparse Spike also able to provide higher resolution model by the increasing of bandwidth reflectivity and less dependent on initial model (Hampson and Russel, 2005).

IOP Conf. Series: Earth and Environmental Science **132** (2018) 012031 doi:10.1088/1755-1315/132/1/012031

On this paper, the authors only rely the linear programming which relies on sparseness, frequency, windows length, impedance output option, and time range. Linear programming is a recursive inversion with a sparse spike assumption. If high-resolution deconvolution is applied then it will be assumed as the real value of reflectivity (Hampson and Russel, 2005). This average of reflectivity value will lose low-frequency component. The absence of low frequency will be replaced by the frequencies of the geological model (Oldensburg, 1983).

4. Pre-Inversion Analysis

The pre-inversion analysis is one of the important step before determining the proper inversion to display the coal seam layer beneath the surface. The initial model is made based on log data and seismic horizons. The impedance value is obtained from two wells to aim the interpolation pattern of impedance value (Fig 3). Based on this initial model we could able to determine a quick-look of a probable final inversion. The initial model expressed a distribution of low impedance coal from SW-NE.

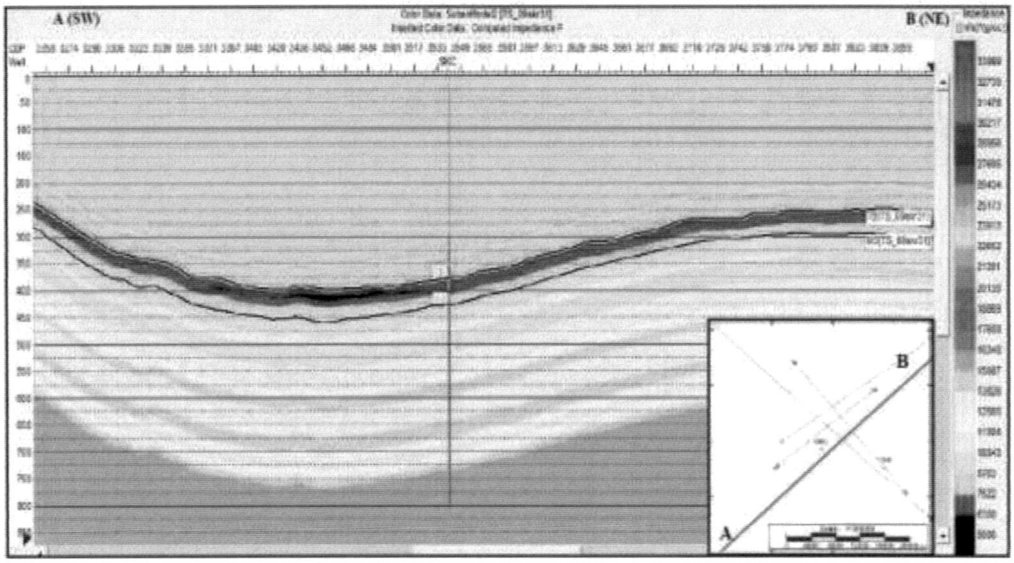

Fig 3. Pre-Inversion initial model on determined seismic section.

Based on those three inversions, it will be chosen the best inversion depending on the highest correlation value on the cross plot analysis and the inversion which is able to display the best coal seam. All the inversions add 40ms both the top and the bottom horizon. The Model Based Inversion shows that hard constraint is better than soft-constraint inversion based on the number of iteration (Table 1) and error comparison graphic (Fig 4).

Table 1. Iteration and error ratio of Hard-Constrains and Soft-Constrains

Hard-Constrain											
Iteration/Error	Constrain (%)										
	0	5	10	15	20	25	30	35	40	45	50
5	1785	1089	1160	1214	1310	1359	1359	1359	1359	1359	1359
10	1780	1118	1249	1273	1352	1411	1420	1420	1420	1420	1420
15	1806	1221	1296	1383	1417	1473	1495	1553	1465	1465	1465
20	1817	1395	1352	1393	1450	1421	1514	1553	1553	1553	1553

IOP Conf. Series: Earth and Environmental Science **132** (2018) 012031 doi:10.1088/1755-1315/132/1/012031

Soft-Constrain

Iteration/Error	Constrain (%)										
	0	0.05	0.1	0.15	0.2	0.25	0.3	0.35	0.4	0.45	0.5
5	2778	2570	2118	2024	1956	1502	1463	1463	1399	1367	1338
10	3038	2374	2174	2037	1948	1489	1489	1451	1390	1364	1337
15	3965	2375	2191	2064	1541	1492	1492	1455	1392	1364	1339
20	4045	2743	2216	1616	1549	1499	1499	1460	1394	1367	1342

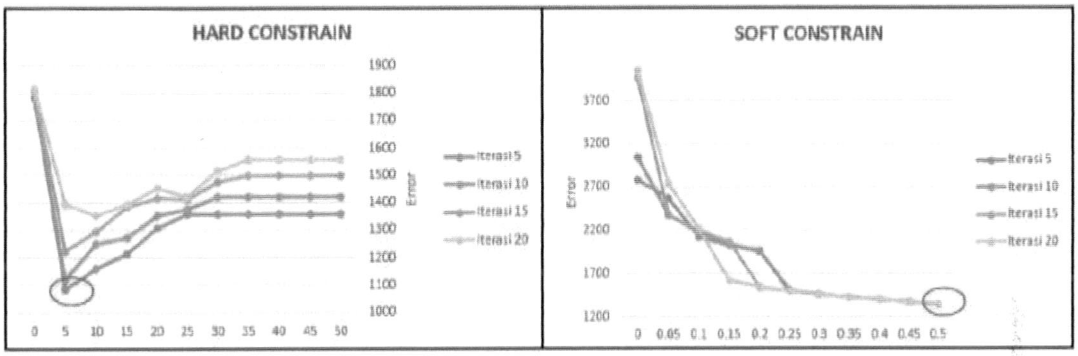

Fig 4. Error comparison chart of Hard-Constraint and Soft-Constraint.

In Bandlimited Inversion, the author only applied the constraint high cut frequency that appears on the frequency 65 Hz and the lowest error (fig 5). In Sparse Spike, Inversion was obtained error value based on sparseness and frequency, from the graphic shows that sparseness at 10 and frequency at 65 Hz has the lowest error (fig 6). Based on the error analysis previously and the inversion cross plot of each method, the model based hard constraint appears as the best inversion. Crossplot inversion is done based on trial and error for each type of inversion in which the highest correlation belongs to spars spike inversion (a= 0.884757). It shows the result of the correlation between inversion trace and seismic data (fig 7). Among Model Base Hardconstrain, Bandlimited, and Sparse Spike Seismic Inversion, the best inversion to display the continuity of coal seam is Sparse Spike Inversion.

Bandlimited

f	40	45	50	55	60	65	70	75
Error	1635	1476	1476	3252	3252	1219	1219	1219

Sparse-Spike

f	Sparseness									
	5	10	15	20	25	30	35	40	45	50
55	540	579	616	635	619	748	676	692	657	741
60	514	560	535	535	566	544	606	606	671	672
65	482	445	523	523	494	570	578	547	576	617

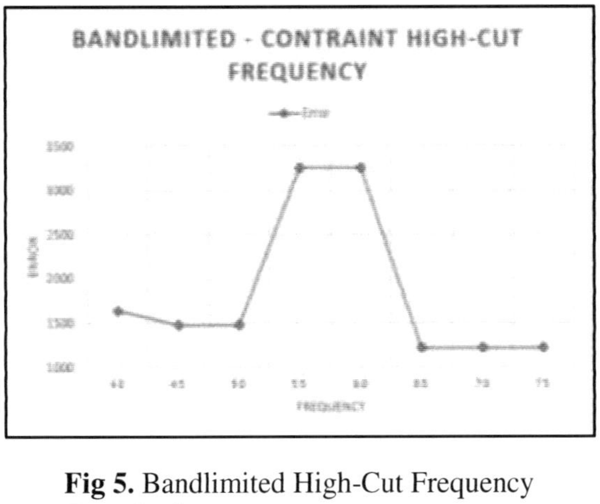

Fig 5. Bandlimited High-Cut Frequency Error Ratio

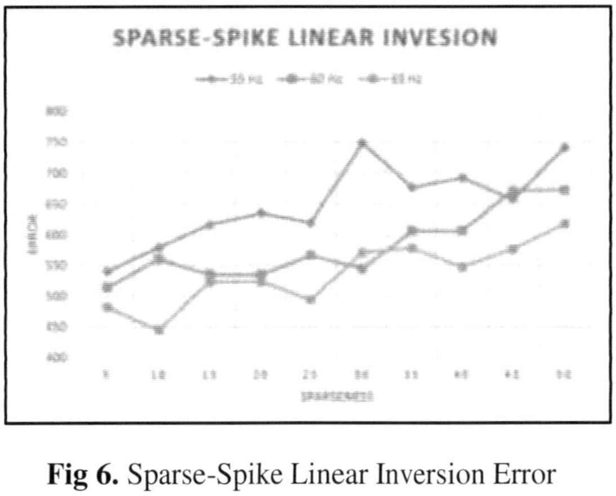

Fig 6. Sparse-Spike Linear Inversion Error Ratio

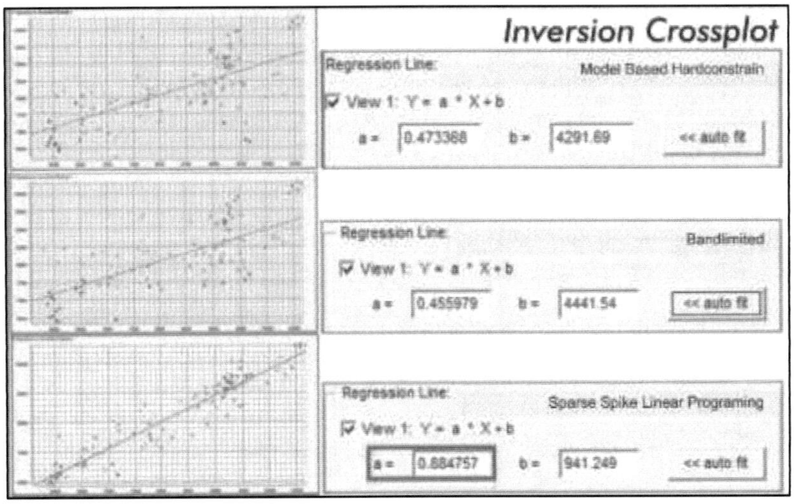

Fig 7. Inversion cross-plot comparison that concludes Sparse-Spike has the highest correlation value among others

5. Result and Discussion

Reservoir, cap rock, and source rock of coal bed methane are the coal seam its self. Our main interest is located at Muara Enim Formation as coal bearing formation. Coal has specific properties, which low gamma ray, low density, and low p-wave as appear in cross-plot analysis to recognize the cut off value between coal and other lithologies such as sand and shale (Fig 8 & 9). Based on the analysis data the Muara Enim Coal Seam has 20 (API) gamma ray value, 1 (gr/cc) – 1.4 (gr/cc) from density log, and low AI cutoff value range between 5000-6400 (m/s)*(g/cc). Based on Sparse Spike Inversion, the coal seam is quite thick based on SKC-well and it shows its continuity among the others (Fig. 10, 11, & 12). The horizon slice at Top Coal (Fig 10) shows that the coal seam is thinning from northwest to southeast.

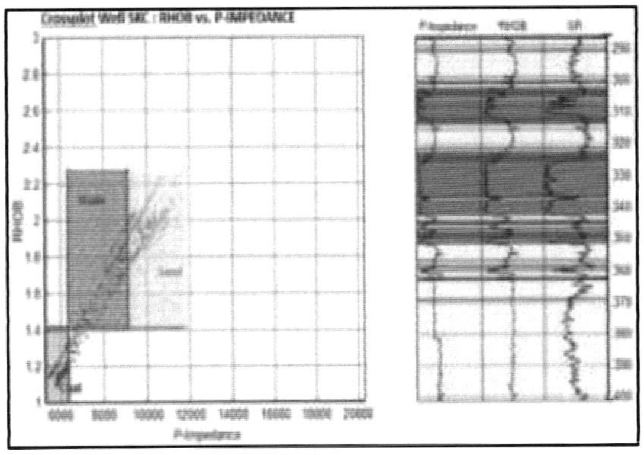

Fig 8. Well SKC

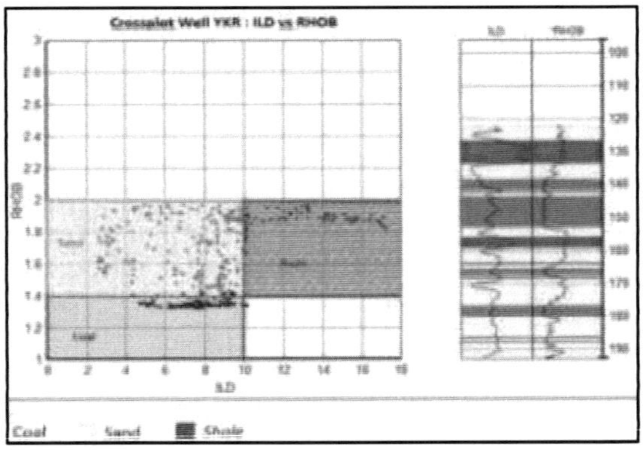

Fig 9. Well YKR

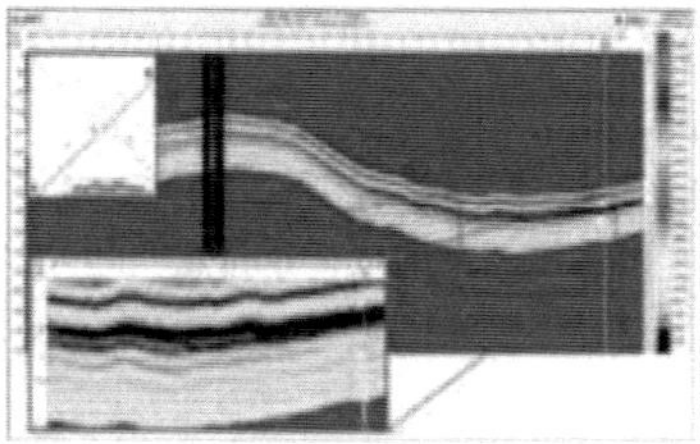

Fig 10. Sparse-Spike

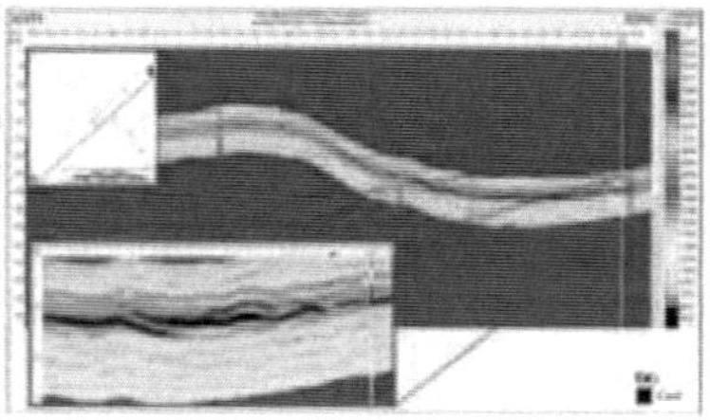

Fig 11. Bandlimited

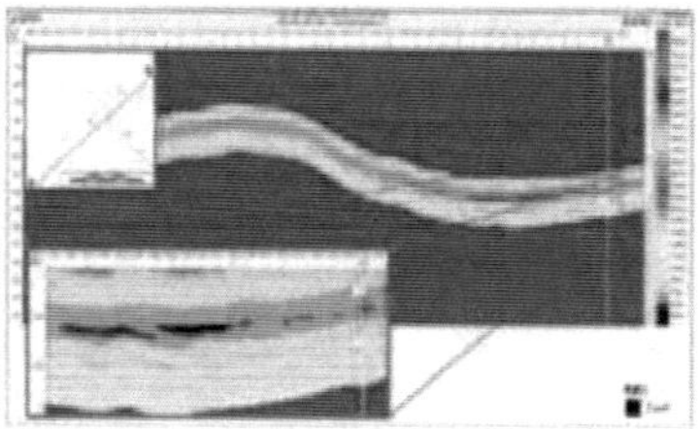

Fig 12. Modelbased

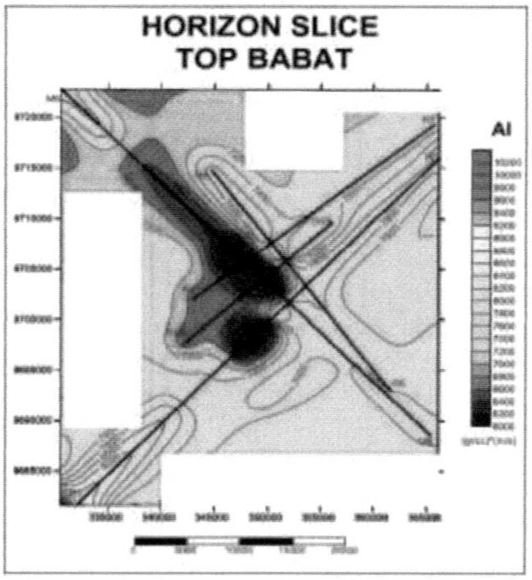

Fig 13. Top Coal overlaid with Base map

Theoretically, Sparse Spike Inversion has high amplitude to do the inversion, and the coal in the research area is lignite to bituminous which has low density and Vp. Hence, the impedance contrast will give giant peak as high amplitude Further drillings are purposed by identifying coal seams on the inverted seismic trace, preferred due lack of well data to do well correlation. Based on "Rule of Thumb" which regulate CBM to only exploit coal seams located over than 300m below the surface. Other criteria for consideration are structures which occurred on the coal. CBM drilling targets anticline flanks or plain seams. Structures consequently impact coal characteristics. Anticline seam will develop fast dewatering causing cleat shrinkage which diminishes pore permeability and syncline seam will lead over-emphasized water production. Most likely thick seams is an ideal target. Based on considered parameters, drilling planning will be purposed at CDP 3586-3622 (fig 14), CDP 4004 – 4148 (fig 15), and CDP 3661 – 3722 (fig 16.)

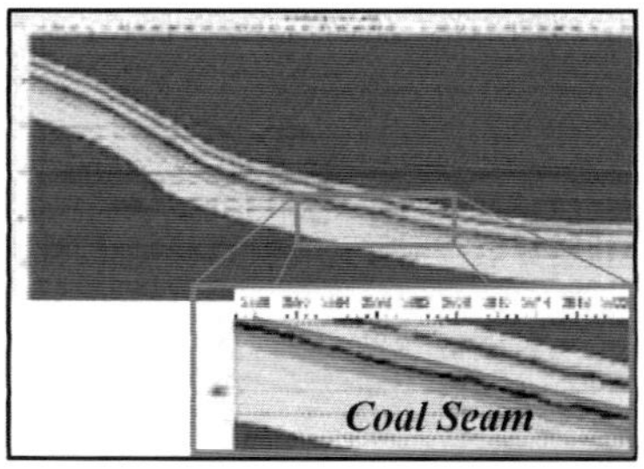

Fig 14. CDP 3586 – 3622

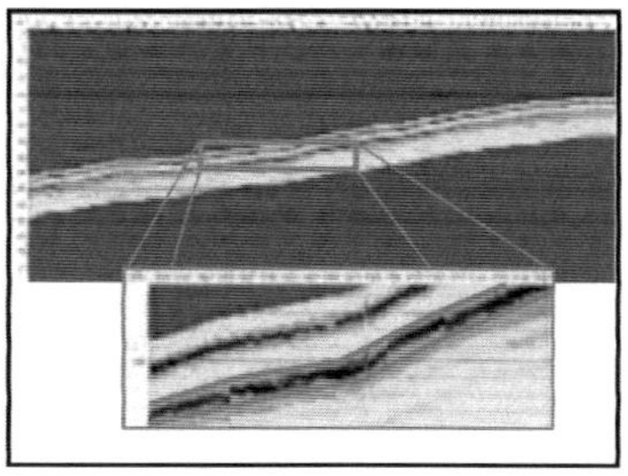

Fig 15. CDP 4004 – 4148

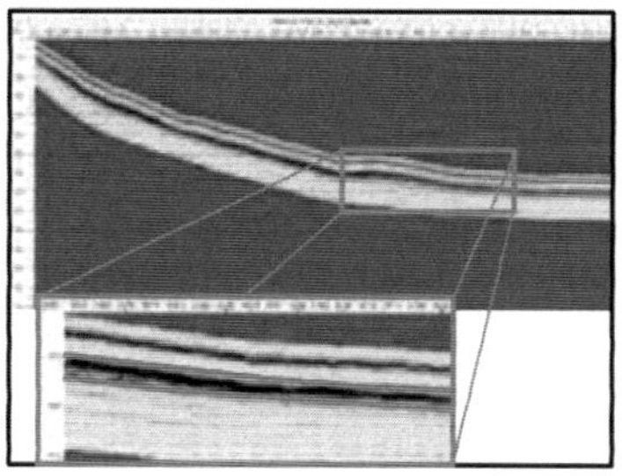

Fig 16. CDP 3661 – 3722

IOP Conf. Series: Earth and Environmental Science **132** (2018) 012031 doi:10.1088/1755-1315/132/1/012031

6. Conclusion

Each seismic inversion will express coal seams differently. Sparse-Spike is considerably the most proper seismic inversion to display thick coal seams within its wide distribution area. Therefore, dark areas on top coal horizon map results are considered as coal thickness. Lateral coal distribution is gradually thinning from NW to SE. Moreover, anticline identified formed from SW to NE trapping two seams. Consequently, a number of three purposed wells are adequate to cover this field's future production.

Acknowledgements

Authors would like send our highest gratitude to PT.PHE Metana Suban II management for providing support and permission to publish this paper, particular recognition to Akhmad Sukoco as authors mentor to establish this paper, and best regards to Indonesia Associaton of Geophysics for conducting The 41th Scientific Annual Meeting and giving us the opportunities to publicly present this paper.

References

[1] De Coster G L 1974 The Geology of Central and South Sumatra Basin, *Proceedings Indonesia Petroleum Association*, 3rd Annual Convention, p77-110.
[2] Ginger D and Fielding K. 2005 The Petroleum System and Future Potential of The South Sumatra Basin. *Proceedings Indonesian Petroleum Association*, 30th Annual Convention, p39-51.
[3] Hampson D and Russell B 2005 *STRATA: Seismic Inversion Workshop*, Hampson-Russel Software Services Ltd, Canada.
[4] Hasjim Mahmud 2010 Dillema of Multi Energy Resources In South Sumatra Province *Journal of International Conference of Coal Bed Methane*, Palembang.
[5] Koesoemadinata R P 1980 Geologi Minyak dan Gas Bumi *2nd Edition ITB Publishing*, Bandung
[6] Oldenburg D W 1983 Funnel Functions in Linear and Nonlinear Appraisal, *Journal of Geophysic Research,* Vol 88 p7387-7398
[7] Pulunggono A 1986 Tertiary Structural Features Related to Extensional and Compressive Tectonics in the Palembang Basin, South Sumatra *Proceedings Indonesian Petroleum Association*, 15th Annual Convention p67-89.
[8] Russel Brian 1988 Introduction to Seismic Inversion *Society of Exploration Geophysicists* p60-71
[9] Sukmono S 2000 Seismik Inversi Untuk Karakterisasi Reservoir, *ITB Publishing*, Bandung

Identifying Intraplate Mechanism by B-Value Calculations in the South of Java Island

Ida Bagus Suananda Y[a]., Irfan Aufa, Ulvienin Harlianti

Geophysical Engineering Major, Bandung Institute of Technology, Indonesia

[a]suanandayogi@gmail.com

Abstract. Java is the most populous island in Indonesia with 50 million people live there. This island geologically formed at the Eurasia plate margin by the subduction of the Australian oceanic crust. At the south part of Java, beside the occurrence of 2-plate convergence earthquake (interplate), there are also the activities of the intraplate earthquake. Research for distinguish this 2 different earthquake type is necessary for estimating the behavior of the earthquake that may occur. The aim of this research is to map the b-value in the south of Java using earthquake data from 1963 until 2008. The research area are divided into clusters based on the epicenter mapping results with magnitude more than 4 and three different depth (0-30 km, 30-60 km, 60-100 km). This location clustering indicate group of earthquakes occurred by the same structure or mechanism. On some cluster in the south of Java, b-value obtained are between 0.8 and 1.25. This range of b-value indicates the region was intraplate earthquake zone, with 0.72-1.2 b-value range is the indication of intraplate earthquake zone. The final validation is to determine the mechanism of a segment done by correlating the epicenter and b-value plot with the available structural geology data. Based on this research, we discover that the earthquakes occur in Java not only the interplate earthquake, the intraplate earthquake also occurred here. By identifying the mechanism of a segment in the south of Java, earthquake characterization that may occur can be done for developing the accurate earthquake disaster mitigation system.

Keywords: intraplate, interpolate, b-value, Java Island

1. Introduction

Java Island, which has population of 50 million people, is the most populous in Indonesia. This island geologically formed at the Eurasia plate margin by the subduction of the Australian oceanic crust. At the south part of Java, beside the occurrence of 2-plate convergence earthquake (interplate), there are also the activities of the intraplate earthquake. Indo-Australia plate subduction occurs until 100-200 km beneath the surface, which made Java has high seismicity (Rohadi et al., 2007). In southern part of Java there are two types of earthquake, the first is earthquake that happened in boundary of the plates (interplate), the other one is earthquake that happened inside the plate (intraplate).

Based on previous researches, b-value on the frequency-magnitude distribution depends on tectonic characteristic and stress level or structure material of an area (Scholz, 1968). The b-value variation is related to structure heterogeneities and stress distribution of that area. The b-value for intraplate area is approximately 0.72-1.20 (Al-Hetty, 2011). Nugraha et al. (2016), have done b-value mapping for Sunda arc, based on this research b-value of Java has both high b-value (>1.2) and low value (0.6-0.8).

2. Basic Theory

2.1. Tectonic Settings

Seismicity in southern Java Island occurs because of two-plate interaction those are Indo-Australia and Eurasia. The Indo-Australia plate is relatively moved to the north 7 cm/year to the Eurasia plate. These tectonic processes have produce Indonesia Island and its volcanoes; beside that there are many big and small earthquakes that have been happened.

Overall, the slab subducted below Java Island was continuous, except the southern part of east java where there was a slab hole. Based on Widyantoro et al. (2011), using seismic velocity tomography, there is a slab hole below the Eurasian Plate. After the trench stepped back, subduction resumed behind the plateau causing the hole to develop (Hall et al., 2009). Only a few of big earthquakes occurred in southern Java Island. Some of big earthquakes are the 1921, 1937, and 1943 Java earthquakes that classified as intraplate earthquakes based on past research. (Okal, 2012). Meanwhile, earthquakes that generate tsunami such as Banyuwangi Earthquake in 1994 and Pangandaran earthquake in 2006 were classified as interplate earthquakes (Okal, 2012; Gunawan et al., 2016a; Raharja et al., 2016; Gunawan et al, 2017).

2.2. Intraplate & Interplate Earthquakes

Kanamori and Anderson (1975) classified the earthquake into two groups, interplate and intraplate earthquakes. The interplate earthquakes refer to those that occur along, or parallel to, the major plate boundaries with a large slip rate. Major thrust earthquakes along subduction zones (Anugrah et al., 2015; Ardika et al., 2015; Alif et al., 2016) and along transform faults (Ito et al., 2016; Gunawan et al., 2016b; Pratama et al., 2017) are classed into' this group. Earthquakes that occur clearly within the plate are intraplate earthquakes. In the present paper, earthquakes whose fault planes are not along the plate boundary are classed as intraplate earthquakes, even though they occur near the plate boundary. By this definition, the San Fernando and Kern County events are intraplate earthquakes (Kanamori and Anderson, 1975).

2.3. Gutenberg-Richter

Frequency-Magnitude Distribution Relation is one from many methods to determine the seismicity in a region. This relation firstly stated by Gutenberg-Richter (1944). Frequency-Magnitude Distribution are power relation which modified in logarithmic equation. The method to determine the seismic and tectonic parameter of a region using Gutenberg-Richter relation written as:

$$\log n\,(M) = a - bM$$

where a and b are the regression parameter model, and n(M) is the number of earthquake with magnitude M. A-value is seismic parameter which the value depends on the amount of event for particular region depend on volume determination and time window. B-value shows the gradient from relationship of earthquake frequency and magnitude on linear equation.

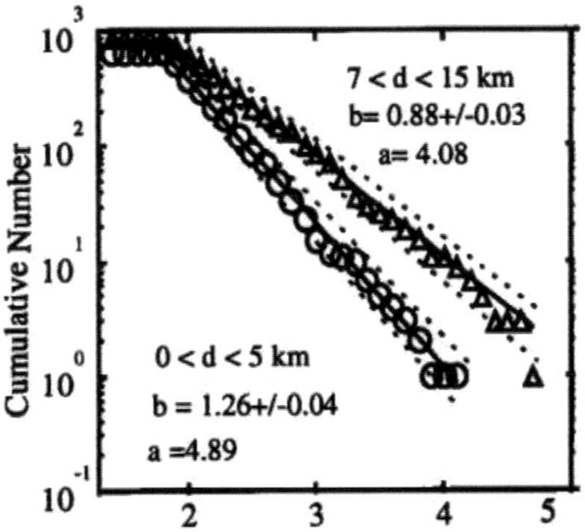

Figure 1. Example of Frequency-Magnitude Distribution Relationship for certain depth of Calaveras Fault Earthquake (Wiemer and Wyss, 1997).

3. Method

Data that we used in this research were Engdahl earthquake catalog data (EHB) between 1963-2008 with depth 0-100 km, the number of events recorded are 1033 events. The catalog we used still consists of every type of earthquake, the main shock, foreshock, and aftershock, so we filtered the events to get only the main shock. Furthermore, we transform all the magnitude with different scale to moment magnitude scale before the data inputted in b-value processing. We filter the main shock data using de-clustering method. If we use all the earthquake data without de-clustering, the b-value we obtain probably less accurate because not all of the events are main shock events

After we get main shock catalog then we applied smoothing distance in calculating the grid rates. Smoothing distance used to calculate the value in each grid especially grid with less earthquake (Petersen et al., 2008). Earthquake above M 4 are counted in each grid cell and used Gaussian operator to smoothing, this study used grid 0.1x0.1 degree of longitude and latitude as in Figure 2. When smoothing distance we tried some distance as 10 km, 30 km, 50 km, and 75 km to get great model. The catalog divided into three-depth range 0-30 km, 30-60 km, and 60-100 km. Generally, the amount of data decreases with increasing depth. From the trial distance we get result that smoothing distance 50 km has more accurate model than the other distance, so the b-value calculated based on this result. Smoothing distance will produce patterns of earthquake distribution, from the pattern we will get earthquake cluster and we can calculate a-value and b-value of each cluster.

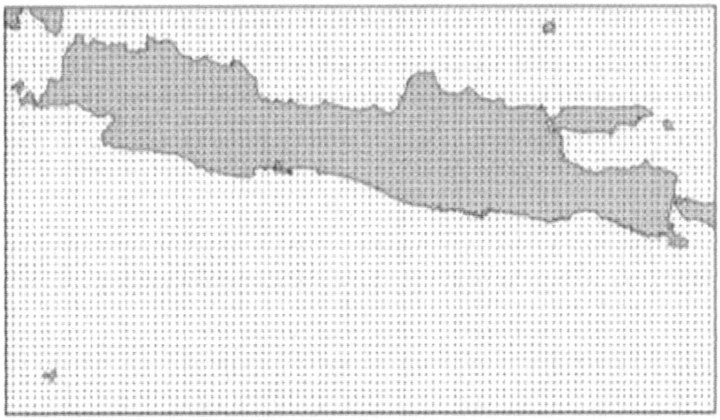

Figure 2. The grid used in the processing of smoothing distance.

4. Result and Discussion

The initial earthquake distribution map is shown in Figure 3. This map has foreshock, main shock, and aftershock. The earthquake distribution map after applied de-clustering is shown in Figure 4. There is different amount of data in both figures, its mean that in the southern Java occurred many foreshock and aftershock. By using main shock data only, the obtained the b-value becomes more accurate.

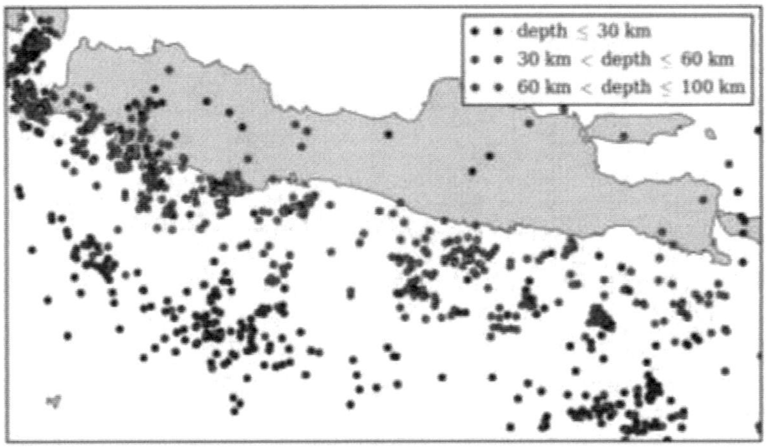

Figure 3. Earthquake distribution before de-clustering (foreshock, main shock and aftershock).

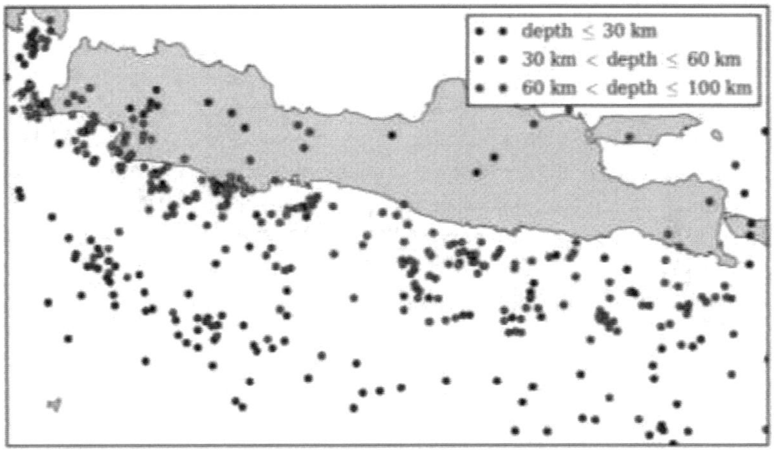

Figure 4. Earthquake distribution after de-clustering.

4.1. Smoothing Distance

The result of smoothing distance 50 km to the earthquake at depth 0-30km is shown in Figure 5. From the image obtained two clusters of earthquake with b-value in cluster 1 is 1.181 and cluster 2 is 1,164. The both of clusters are located in south of Java subduction zone. In cluster 1 was occurred Pangandaran earthquake on July 17, 2006.

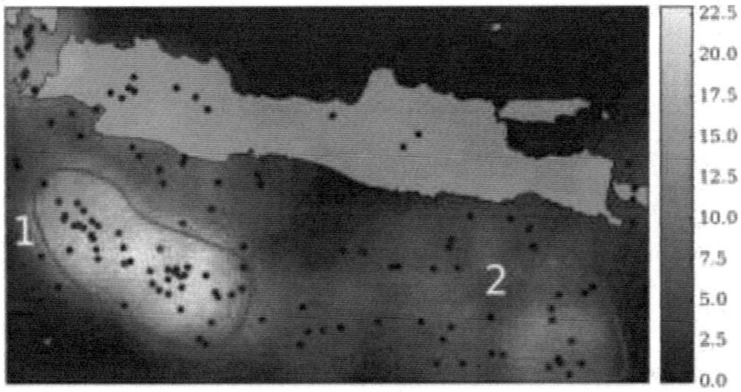

Table 1. a-value and b-value at depth 0-30 km

cluster	a-value	b-value
1	7.406	1.181
2	6.936	1.164

Figure 5. Earthquake cluster at depth 0-30 km.

The result of smoothing distance 50 km to the earthquake at depth 30-60 km is shown in Figure 6. From the image obtained three clusters of earthquake with b-value in cluster 1 is 0.884, cluster 2 is 1.253, and cluster 3 is 0.853. Cluster 1 is located in the south of West Java, cluster 2 is located in the south of Central Java, and cluster 3 is located in the south of East Java. The location of all cluster in this group are closer to Java than clusters at depth 0-30 km.

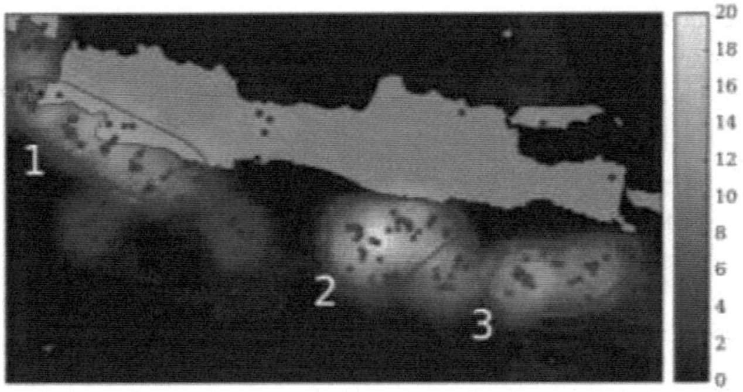

Table 2. a-value and b-value at depth 30-60 km

cluster	a-value	b-value
1	5.99	0.844
2	7.816	1.253
3	5.582	0.853

Figure 6. Earthquake cluster at depth 30-60 km.

The result of smoothing distance 50 km to the earthquake at depth 60-100 km is shown in Figure 7. From the image obtained three clusters of earthquake with b-value in cluster 1 is 1.214 clusters 2 is 0.893, and cluster 3 is 0.994. Cluster 1and 2 are located in the south of West Java and cluster 3 is located in the south of Central Java. The location of earthquake in this group is tending to be more to the north than the both of depth group before.

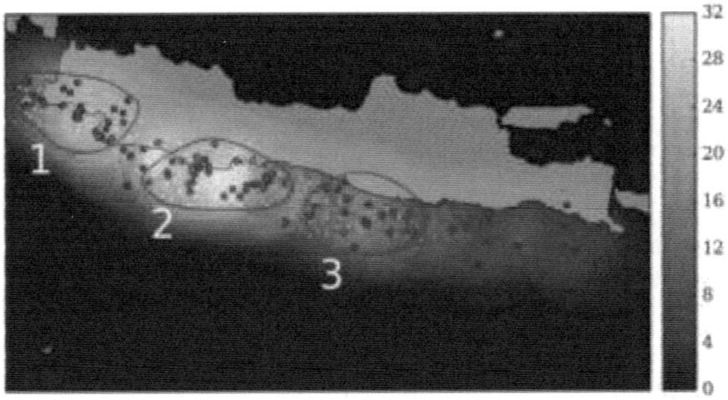

Table 2. a-value and b-value at depth 60-100 km

cluster	a-value	b-value
1	7.842	1.214
2	5.764	0.893
3	6.165	0.994

Figure 7. Earthquake cluster at depth 60-100 km.

5. Conclusion

From our study, we conclude:

- Based on the distribution of earthquake from each group depth, the tendency of the hypocenters is deeper towards the north. These earthquakes follow the direction of subduction in southern Java.
- The b-value of the southern Java regions from Engdahl (1963-2008) ranges from 0.85 to 1.25.
- Referring to Al-Hetty (2011), states that the region with b-value range of 0.7-1.2 is an intraplate zone, hence we can say that the earthquakes derived from Engdahl (1963-2008) in Southern Java were intraplate earthquake. However, many reports define Banyuwangi Earthquake 1994 and Pangandaran Earthquake 2006 as interplate earthquake, hence there should be more consideration that not all of earthquake zone with b-value 0.7-1.2 is an intraplate earthquake.

References

[1] Al-Hetty, E. A. M., 2011. Variation of b – Value In The Earthquake Frequency – Magnitude Distribution With Depth In The Intraplate Regions, International Journal of Basic & Applied Sciences IJBAS-IJENS, Vol: 11 No:66, 29-37.

[2] Alif, S. M., Meilano, I., Gunawan, E. and Efendi, J., 2016. Evidence of Postseismic Deformation Signal of the 2007 M8.5 Bengkulu Earthquake and the 2012 M8.6 Indian Ocean Earthquake in Southern Sumatra, Indonesia, Based on GPS Data. Journal of Applied Geodesy, 10(2), 103-108, DOI: 10.1515/jag-2015-0019.

[3] Anugrah, B., Meilano, I., Gunawan, E. and Efendi, J., 2015. Estimation of postseismic deformation parameters from continuous GPS data in northern Sumatra after the 2004 Sumatra-Andaman earthquake. Earthquake Science, 28(5-6), 347-352, DOI: 10.1007/s11589-015-0136-x.

[4] Ardika, M., Meilano, I. and Gunawan, E., 2015. Postseismic Deformation Parameters of the 2010 M7.8 Mentawai, Indonesia, Earthquake Inferred from Continuous GPS Observations. Asian Journal of Earth Sciences, 8, 127–133, DOI: 10.3923/ajes.2015.127.133.

[5] Gunawan, E., Meilano, I., Abidin, H. Z., Hanifa, N. R. and Susilo, 2016a. Investigation of the best coseismic fault model of the 2006 Java tsunami earthquake based on mechanisms of postseismic deformation. Journal of Asian Earth Sciences, 117, 64-72, DOI: 10.1016/j.jseaes.2015.12.003.

[6] Gunawan, E., Maulida, P., Meilano, I., Irsyam, M. and Efendi, J., 2016b. Analysis of coseismic fault slip models of the 2012 Indian Ocean earthquake: Importance of GPS data for crustal deformation studies. Acta Geophysica, 64(6), 2136-2150, DOI: 10.1515/acgeo-2016-0106.

[7] Gunawan, E., Meilano, I., Hanifa, N. R. and Widiyantoro, S., 2017. Effect of coseismic and postseismic deformation on homogeneous and layered half-space and spherical analysis: Model simulation of the 2006 Java, Indonesia, tsunami earthquake. Journal of Applied Geodesy, DOI: 10.1515/jag-2017-0009.

[8] Gutenberg, R. and Richter, C. F., 1944. Frequency of Earthquakes in California, Bulletin of the Seismological Society of America, 34, 185-188.

[9] Hall, R., Cross, L, Clements, B. & Spakman, W., 2009. Neogene subduction beneath Java: slab tearing, arc deformation and arc magmatism, and their causes. In: Southeast Asian Gateway Evolution Conference (abstract). Royal Holloway University of London, United Kingdom.

[10] Ito, T., Gunawan, E., Kimata, F., Tabei, T., Meilano, I., Agustan, Ohta, Y., Ismail, N., Nurdin, I. and Sugiyanto, D., 2016. Co-seismic offsets due to two earthquakes (Mw 6.1) along the Sumatran fault system derived from GNSS measurements. Earth, Planets and Space, 68(1), 1, DOI: 10.1186/s40623-016-0427-z.

[11] Kanamori, H. and Anderson, D. L., 1975. Theoretical basis of some empirical relations in seismology, Bulletin of the Seismological Society of America, 65 (5), 1073-1095

[12] Nugraha, A. D., Shiddiqi, H. A., Widyantoro, S., Sutiyono, Handayani, T., 2016. Analysis of Spatiotemporal Variation in b-value for the Sunda Arcusing High Precision Earthquake Location

[13] Okal, E. A., 2012. The South of java earthquake of 1921 September 11: A negative search for a large interplate trust event at Java Trench. Geophysical Journal International, 190(3), 1657-1672.

[14] Petersen, M. D., Mueller, C. S., Frankel, A.D., Zeng, Y., 2008. Spatial Seismicity Rates and Maximum Magnitudes for Background Earthquakes, USGS Open-File Report.

15] Pratama, C., Ito, T., Sasajima, R., Tabei, T., Kimata, F., Gunawan, E., Ohta, Y., Yamashina, T., Ismail, N., Nurdin, I., Sugiyanto, D., Muksin, U. and Meilano, I., 2017. Transient rheology of the oceanic asthenosphere following the 2012 Indian Ocean Earthquake inferred from geodetic data. Journal of Asian Earth Sciences, 147, 50-59, DOI: 10.1016/j.jseaes.2017.07.049.

[16] Raharja, R., Gunawan, E., Meilano, I., Abidin, H. Z. and Efendi, J., 2016. Long aseismic slip duration of the 2006 Java tsunami earthquake based on GPS data. Earthquake Science, 29(5), 291-298, DOI: 10.1007/s11589-016-0167-y.

[17] Rohadi, S., Grandis, H., Ratag, M. A., 2007. Studi Variasi Spatial Seismisitas Zona Subduksi Jawa, Jurnal Meteorologi Dan Geofisika, Vol. 8, 42 - 47

[18] Scholz, C. H., 1968. Experimental study of the fracturing process in brittle rock, Journal of Geophysical Research, vol. 73, 1447–1454

[19] Widyantoro, S., Pesicek, J. D., Thurber, C. H., 2011. Subducting Slab Structure Below The Eastern Sunda Arc Inferredfrom Non-Linear Seismic Tomographic Imaging. Geological Society, London, Special Publications, vol. 355, 139-155

[20] Wiemer, S. and Wyss, M., 1997. Mapping The Frequency-Magnitude Distribution In Asperities: An Improved Technique To Calculate Recurrence Times?, Journal of Geophysical Research, vol. 102, 15115-15128

Completed Ensemble Empirical Mode Decomposition: a Robust Signal Processing Tool to Identify Sequence Strata

H Purba[1], J T Musu[1], S A Diria[1], W Permono[1], O Sadjati[2], I Sopandi[2], F Ruzi[2]
[1] Research and Development Center for Oil and Gas Technology "Lemigas", Jl. Raya Ciledug, Jakarta Selatan, 12230, Indonesia
[2] BOB PT. Bumi Siak Pusako - Pertamina Hulu Energi, Jakarta, Indonesia

Abstract. Well logging data provide many geological information and its trends resemble non-linear or non-stationary signals. As long well log data recorded, there will be external factors can interfere or influence its signal resolution. A sensitive signal analysis is required to improve the accuracy of logging interpretation which it becomes an important thing to determine sequence stratigraphy. Complete Ensemble Empirical Mode Decomposition (CEEMD) is one of non-linear and non-stationary signal analysis method which decomposes complex signal into a series of intrinsic mode function (IMF). Gamma Ray and Spontaneous Potential well log parameters decomposed into IMF-1 up to IMF-10 and each of its combination and correlation makes physical meaning identification. It identifies the stratigraphy and cycle sequence and provides an effective signal treatment method for sequence interface. This method was applied to BRK-30 and BRK-13 well logging data. The result shows that the combination of IMF-5, IMF-6, and IMF-7 pattern represent short-term and middle-term while IMF-9 and IMF-10 represent the long-term sedimentation which describe distal front and delta front facies, and inter-distributary mouth bar facies, respectively. Thus, CEEMD clearly can determine the different sedimentary layer interface and better identification of the cycle of stratigraphic base level.

1. Introduction

The sequence stratigraphy method of stratigraphic interpretation combines the stratal architecture (geometric relationship) of the sedimentary facies within a framework of geometries assumed to be the result of the repeated fill of accommodation by successions of sedimentary strata [1] and the chronological order of their accumulation to enhance the interpretation of depositional setting and prediction of stratal continuity [2]. Interface identification and sequence partition are the basic of sequence stratigraphy which well log data, core data, outcrop samples and another supporting data (e.g seismic) are used for its object. A problem may arise when information from core data and outcrop are limited and seismic data has poor quality of temporal resolution. The availability of well logging data becomes more important for the partition of sequence stratigraphy [3]. However, during well log recording, various disruptions can interfere its resolution and quality so that logging data can be non-stationary signal in space.

Complete Ensemble Empirical Mode Decomposition (CEEMD), a robust signal processing approach can handle non-stationary and non-linear process [4, 5, 6]. CEEMD calculates local mean curve which decomposes the complex signal to a series of component function for instantaneous frequency with apparent physical meaning. In this paper, the CEEMD method is applied for well log data processing and identifying sequence interface in logging.

2. Subsurface Geology

Study area is located on Beruk field, Central Sumatra Basin which consists of 36 wells with the inter-well distance 100 meters up to 2 kilometers (Figure 1).

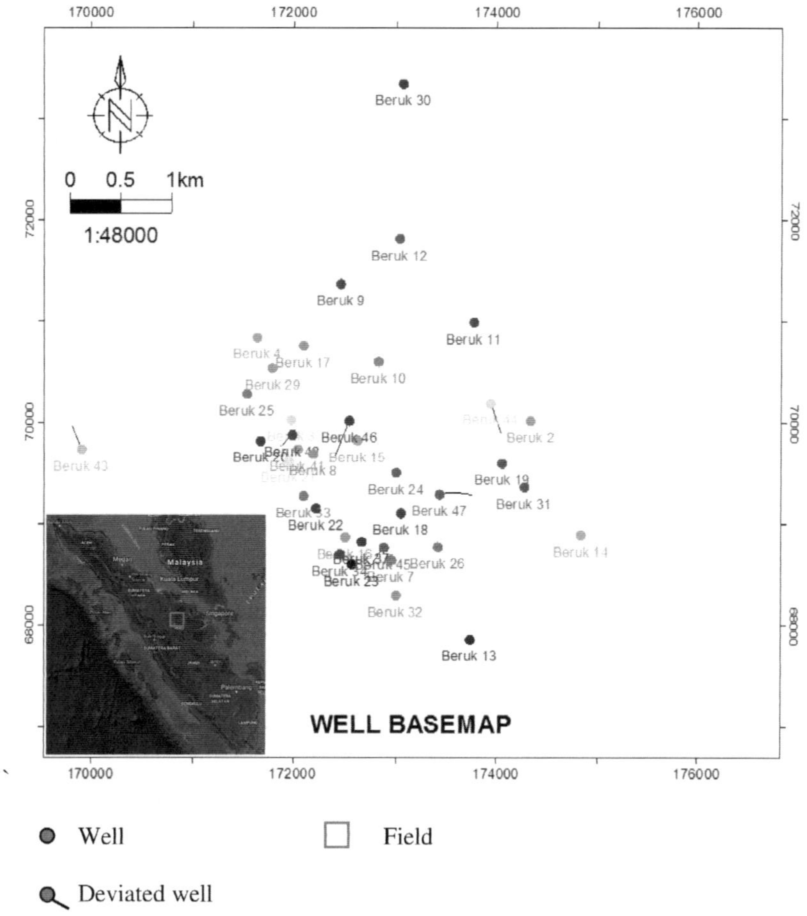

Figure 1. Study location on Beruk field Central Sumatra Basin (shown by red rectangle) with massive production wells. Inter-well distance varies from 100 meters up to 2 kilometers

Correlation model from well log describes eight zones (Zone 1 to 8) of delta front facies model from west to south-east (Figure 2a) and north to south traverse (Figure 2b) which interspersed by distal delta front facies, shale break, and prodelta facies.

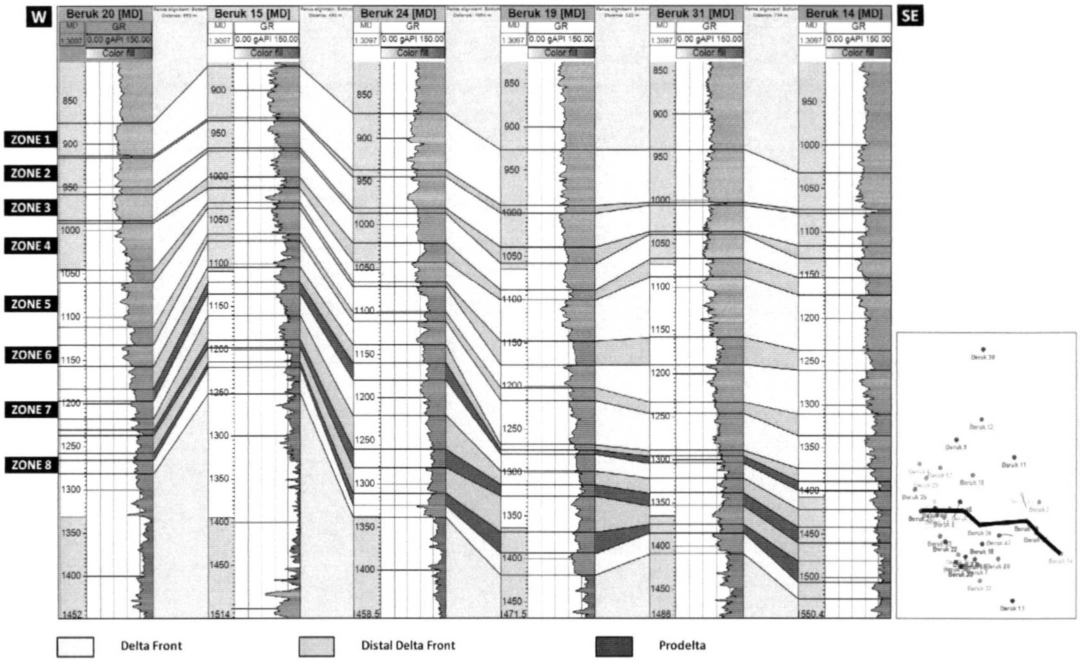

(a)

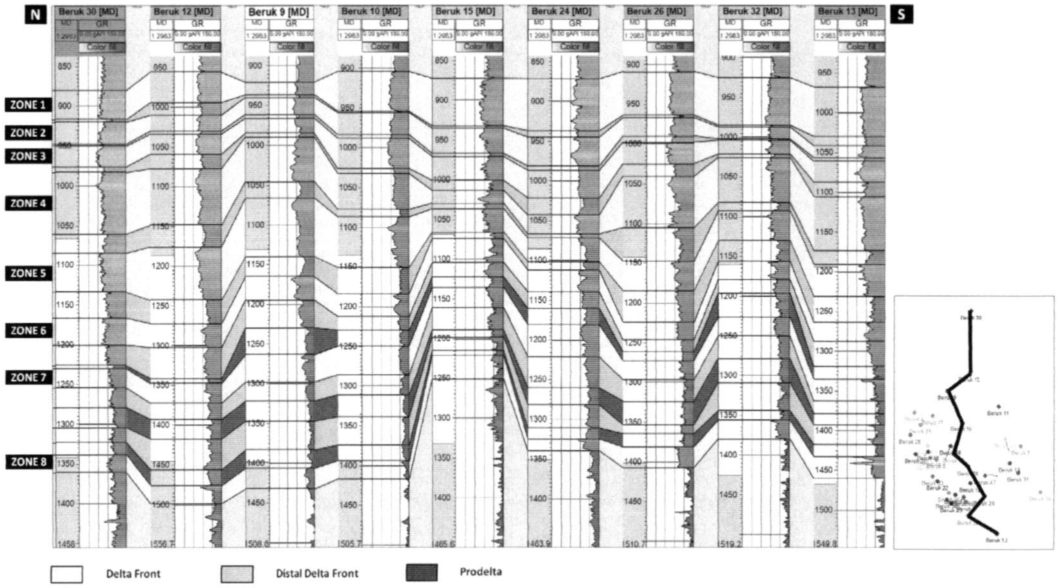

(b)

Figure 2. Well log correlation identifies 8 zones of delta front facies model and on the top of zone 1 bounded by Telisa Formation from west to south-east (a) and north to south (b). From zone 1 to 6 or 7, delta front facies series interbedded by distal delta front facies and shale break. When getting into zone 7 or 8, there are interval of prodelta facies

Commonly, the facies comprises mouth bar sandstone, inter-distributary mouth bar sandstone. The thickness of delta front facies varies laterally from north (proximal) to south-east (distal) likewise west

IOP Conf. Series: Earth and Environmental Science 132 (2018) 012033 doi:10.1088/1755-1315/132/1/012033

to east. Zone 1 has good facies distribution from proximal to distal compared to zone 2 and zone 3 (Figure 3a). Zone 2 and 3 become less from proximal to distal and the facies is relative thin or little different in the western and eastern part (Figure 3b and 3c). For zone 4 and 5, delta front facies is thin in the middle but rather thick in proximal and distal (Figure 3d and 3e). Each of thickness distribution of delta front facies of zone 7 and 8 (Figure 3g and 3h) is thin from north to south and west to east but rather thicker in the eastern part of zone 6 (Figure 3f).

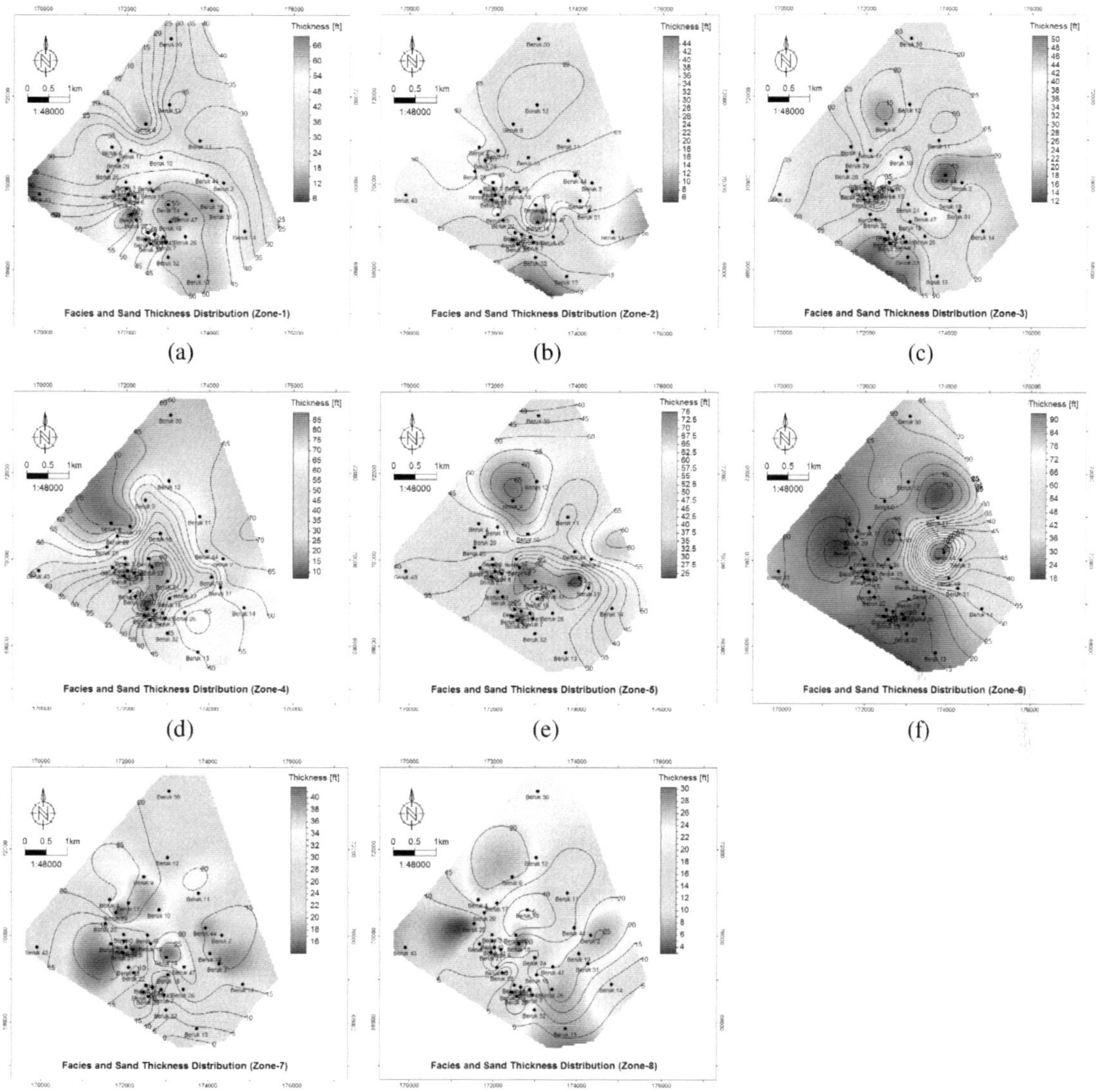

Figure 3. Facies map distribution of zone 1, 2, 3, 4, 5, 6, 7, and 8, labeled by (a), (b), (c), (d), (e), (f), (g), and (h), respectively

3. Theory

CEEMD is an extension method of Ensemble Empirical Mode Decomposition (EEMD) which can separate the high-frequency noise from the raw data by adding the white noise, but the low-frequency noise cannot be reduced. This method is an adaptive signal analysis approach based on the signal characteristic of local extrema. It decomposes a time signal into series of Intrinsic Mode Function (IMF) components, and each IMF satisfies the following conditions:

- Over the entire time range, the number of zero-crossings must be equal or differ by one at most;
- At any point, the mean value of the envelope defined by the local maxima and the envelope defined by the local minima is zero [7].

CEEMD process should be initiated by Empirical Mode Decomposition (EMD) process which its steps can be shown as follows:

- Find all the maxima and minima of the signal and use cubic splines to match the upper envelope U(t) and the lower envelope L(t), respectively. Then, calculate the mean value of the upper and lower envelope M(t):

$$M(t)=\frac{U(t)+L(t)}{2} \quad (1)$$

- Subtract the average envelope M(t) from the original signal S(t) to obtain a new series N(t):

$$N(t) = S(t) - M(t) \quad (2)$$

- Use the above IMF decision criteria to determine the sequence N(t). Repeat 1 and 2 step until sequence $N_k(t)$ meets the conditions of IMF. $N_1(t)$ is the first IMF component $K_1(t)$.
- Subtract $K_1(t)$ from the signal S(t) to get a difference signal $R_1(t)$. The sequence $R_1(t)$ is treated as a new preprocessing signal to decompose, and repeat step 1,2, and 3 until got a second-order IMF component. Stop the process if the difference sequence is constant or a monotonic function.

Each IMF correspondingly contains different frequency components of the signal from high to low frequency, representing inherent mode characteristics of the nonlinear signal, and they will vary with the change of the signal. The residual function stands for the average trend of the signal, and the original signal can expressed as a sum of the IMFs and the residual function:

$$S(t) = \sum_{i=1}^{N} K_i(t) + R_n(t) \quad (3)$$

The procedure of CEEMD was described as follows [8]:

- Add noise onto raw signal and obtain the first EMD component of the data with noise.
- Repeat the decomposition and add different noise realization
- Compute ensemble average to define it as the IMF-1:

$$K_1(t) = \frac{1}{N}\sum_{i=1}^{N} G_1\left[S(t) + \alpha T_i\right] \quad (4)$$

Here, S(t) is the original signal, $T_i(t)$ is the different white noise, α is a ratio coefficient, $G_i[\ ]$ denotes generating the *i* th IMF component, and N is the number of realizations. Next, calculate the first different signal $d_1(t)$:

$$d_1(t) = S(t) - K_1(t) \quad (5)$$

Then, set $d_1(t) + \alpha G_1[T_i(t)]$ (i = 1,2, …, N) as the new signal for decomposition. To obtain the first IMF component, the ensemble average needs to be calculated and the result is the second IMF component:

$$K_2(t) = \frac{1}{N}\sum_{i=1}^{N} G_1\,[d_1(t)\; + \alpha\, G_1\,[T_i(t)]] \tag{6}$$

Finally, repeat the above steps, thus, it can be obtained:

$$S(t) = \sum_{i=1}^{N} K_i(t) + \text{Res}(t) \tag{7}$$

Where, S(t) is original data, $K_i(t)$ is *i* th IMF component, and Res(t) is the last residual function.

4. CEEMD-Well Log Curve Type

The result of well log data decomposition in IMF series reflects different sedimentary cycle grades. If the inter-correlation of sequence interfaces can be identified, it can be considered as the standard to split the sequence of well log data. Base level occurs when the river reaches a balance under the dynamics (potential energy surface). Sedimentary boundary will extend towards the base level through sedimentary or erosion action to get a new balance which is called sequence formation cycle [9, 10]. Various cycle of sequence stratum can be seen in well log data as regular and or irregular pattern of mineral grains granularity and lithology. This pattern can be considered as different amplitude and frequency characteristic in well log curve likewise signal pattern. Then, the break points of curve are fixed as a sequence interfaces [11]. Besides, it is useful to make the mirror image on response characteristic of CEEMD curve of well log parameters reflecting all kinds of sedimentary environment which is summarized in five basic types (See on Table 1).

Table 1. Basic type of sedimentary environment corresponding to CEEMD curve of well log

Curve type		Indicating type	Analysis of curve type
Gradually varied type	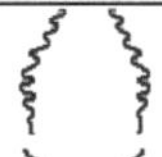	Top	Poorly sorted. To the top, amplitude becomes less gradually and the curve turns graded bed sequence
		Bottom	Poorly sorted. To the bottom, amplitude becomes less gradually and the curve turns reverse grain size order.
Abrupt change type		Top	Well sorted. An abrupt change at the top. Stacked upward-fining cycle.
		Bottom	Well-sorted. An abrupt change at the bottom. Stacked downward-fining cycle
Fluctuating type		Top	Granularity turns from coarser below to finer and the amplitude becomes less dras- tically. It belongs to typical transgression and shows positive grain size order
		Bottom	Granularity turns from finer below to coarser and the amplitude becomes bigger drastically. It belongs to typical regression and shows reverse grain size order
Thin interbedding type	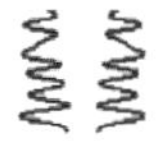	Inter- bedding	Multi-lithology alternately appears and changes quickly. It often can be seen in thin inter-bedding
Block combination type		Simplex lithology	The lithology is simplex, so the grain size of sediment changes are small and stable

IOP Conf. Series: Earth and Environmental Science **132** (2018) 012033 doi:10.1088/1755-1315/132/1/012033

5. Methodology

CEEMD decomposes the original data into IMF with different frequency and residual. Each IMF describes characteristic dimension and information. First and second IMF tend to high frequency reflecting the short-term cycle characteristic while the next IMFs (fourth and so on) have middle - low frequency reflecting the middle and long term stable cycle. Residual refers to the data trend or zero drift of instrument [12]. Based on this rule, correlation of IMF obtained after CEEMD can be determined as follows:

- Calculate the correlation coefficient for every IMF(i) and its original signal. Mostly, those IMFs which have a good correlation with the original signal are the best signal components reflecting stable sedimentary environment.
- Analyze the correlation between different IMFs and combine those IMF(i) with good correlation. This indicates that good correlation reveals will be a consistent in variation style. Combining them can reduce interference factors and strengthen signals that reflect geological characteristic.

Gamma ray (GR) log is a common logging method to measure intensity of natural gamma ray in rock along the well bore. GR curve reflects the granularity, sorting and clay minerals content, which can be used to judge lithology, compare stratum, and estimate the shale content. Spontaneous potential (SP) log is an effective logging method for analyzing geological profile in borehole, which can be used to measure shaft spontaneous potential changes in an open hole. SP curve is often used to divide the sandy-shaly profile into sandstone and shale lithology, compare the stratum, define the interface position of filtering layer, calculate the content of shale of the stratum, and spread the sedimentary face to study, etc. The presented analysis indicates that these two well logging data (GR and SP) are the sensitive parameters that are suitable for sequence division of the study area. At the same time, its curve indication has consistent rule: at the profile of sandstone and shale section well, the value of these two curves is low at sandstone while the value is high at shale. So, these parameters can be combined. Normalization treatment on natural gamma ray and spontaneous potentials curves was done to eliminate the influence of units and select linear normalization formula [13]:

$$\mathrm{GR_{norm}} = \frac{\mathrm{GR} - \mathrm{GR_{min}}}{\mathrm{GR_{max}} - \mathrm{GR_{min}}} \tag{8}$$

$$\mathrm{SP_{norm}} = \frac{\mathrm{SP} - \mathrm{SP_{min}}}{\mathrm{SP_{max}} - \mathrm{SP_{min}}} \tag{9}$$

Average values of the normalized parameter (GR and SP) were combined to gain a new parameter marked $\mathrm{GR} - \mathrm{SP}$:

$$\mathrm{GR} - \mathrm{SP} = \frac{\mathrm{GR_{norm}} + \mathrm{SP_{norm}}}{2} \tag{10}$$

6. Result and Discussion

The decomposition of GR – SP data, for instance well BRK-30, are manifested to each of IMF presented in Figure 4. IMF-1 to IMF-5 reflects strong vibration and high frequency while IMF-6 up to residual (R) signal more low frequency. A different IMF component can be obtained through the decomposition of CEEMD. Superposing of decomposed components, the original signal may be reproduced. Using such characteristic, the universe accumulation of the IMF-1 up to IMF-10 and trend item R of each stages of GR – SP is repeated, namely, the combination from the component with low frequency to high frequency (see on Figure 5).

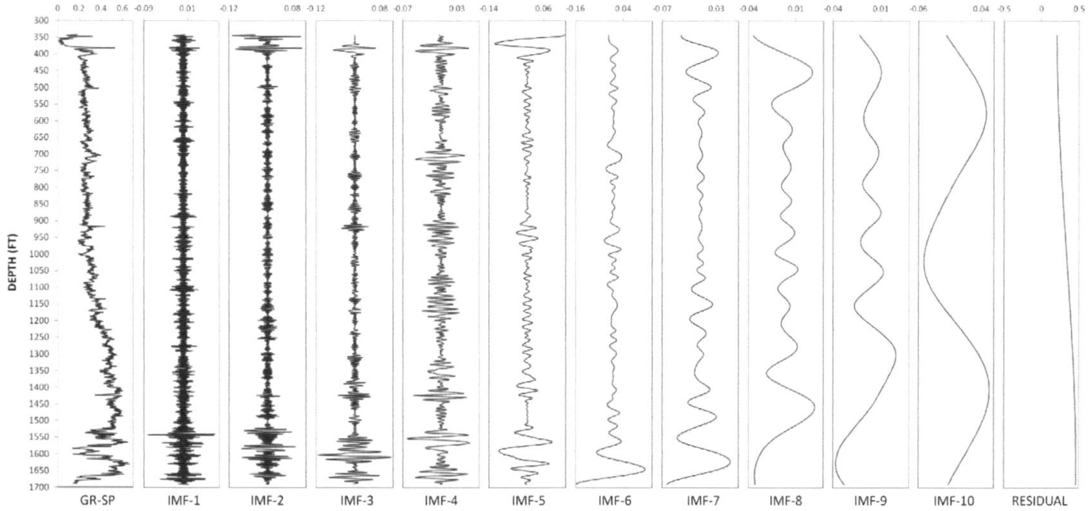

Figure 4. The composition of GR – SP (well BRK-30) into IMFs 1-10 which is the new combined parameter between GR and SP. RES panel represents the final residual which showed the data trend

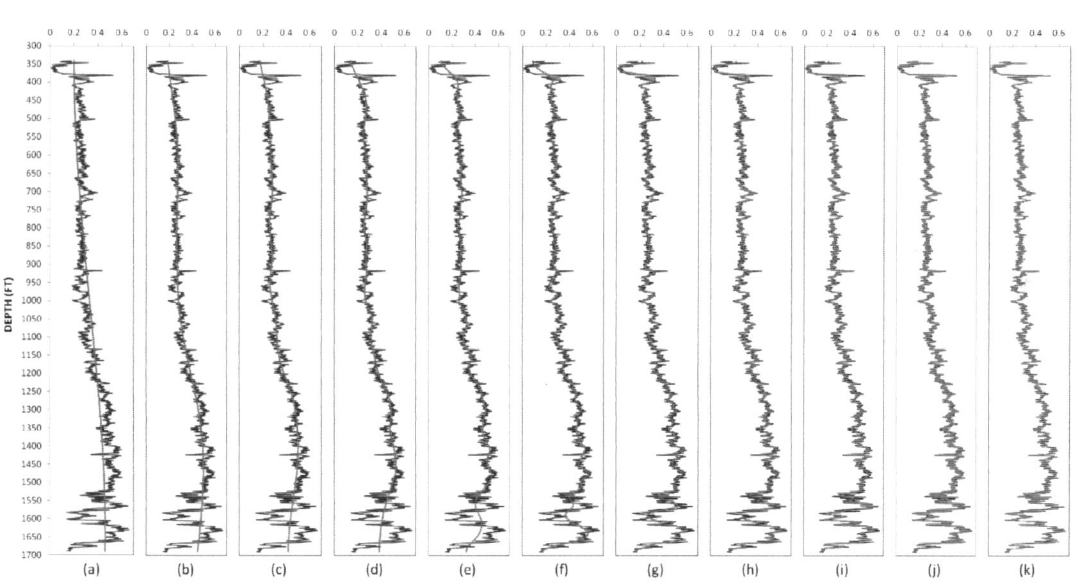

Figure 5. Reconstruction of GR – SP (well BRK-30) components. Blue solid line is GR – SP and red broken line is correlation coefficient of reproduced sequence with GR – SP. (a) Residual (red broken line with $R^2 = 0.755$) (b) Residual + IMF10 (red broken line with $R^2 = 0.819$). (c) Residual + IMF-10 + IMF-9 (red broken line with $R^2 = 0.837$). (d) Residual + IMF-10 + IMF-9 + IMF-8 (red broken line with $R^2 = 0.849$). (e) Residual + IMF-10 + IMF-9 + IMF-8 + IMF-7 (red broken line with $R^2 = 0.871$). (f) Residual + IMF-10 + IMF-9 + IMF-8 + IMF-7 + IMF-6 (red broken line with $R^2 = 0.926$). (g) Residual + IMF-10 + IMF-9 + IMF-8 + IMF-7 + IMF-6 + IMF-5 (red broken line with $R^2 = 0.955$). (h) Residual + IMF-10 + IMF-9 + IMF-8 + IMF-7 + IMF-6 + IMF-5 + IMF-4 (red broken line with $R^2 = 0.969$). (i) Residual + IMF-10 + IMF-9 + IMF-8 + IMF-7 + IMF-6 + IMF-5 + IMF-4 + IMF-3 (red broken line with $R^2 = 0.984$). (j) Residual + IMF-10 + IMF-9 + IMF-8 + IMF-7 + IMF-6 + IMF-5 + IMF-4 + IMF-3 + IMF-2 (red broken line with $R^2 = 0.996$). (k) Residual + IMF-10 + IMF-9 + IMF-8 + IMF-7 + IMF-6 + IMF-5 + IMF-4 + IMF-3 + IMF-2 + IMF-1 (red broken line with $R^2 = 1$).

IMF-5, IMF-6, IMF-7 and IMF-10 acted importantly in original sequence (Figure 5). The changes of original sequence were mainly caused by the oscillation of these four functions. They all have consistent change pattern. This pattern reflects the characteristics of stable sedimentary environment of the stratum. Specifically, the wave scope of IMF-5, IMF-6, and IMF-7 covered several meters to tens of meters and the wave width of IMF-10 is over one hundred meters. The cycles with different base levels commonly are divided according to time distribution. In practical layers, the thickness of short-term cycle is of several meters to dozens of meters. The middle term cycle is dozens of meters to nearly one hundred meters and the long term cycles are nearly one hundred meters to several hundred meters. Therefore, IMF-5, IMF-6, and IMF-7 reflect the short-term circle characteristics, while IMF-10 the middle and long term characteristics.

According to sedimentology, the stratum in different periods may have different sedimentation rules while the stratum in the same period may have similar characteristic. At well BRK-30, combination of IMF-5 plus IMF-6, IMF-7, IMF-10 describes a short-term sedimentation rules of the stratum (Figure 6a to 6c) while IMF-6 plus IMF-7, IMF-10 or IMF-7 plus IMF-10 most likely middle-term sedimentation (Figure 6d to 6f). For long-term sedimentation is illustrated on Figure 6g which IMF-10 describes the stratum boundary of inter-distributary mouth bar. At well BRK-13, short-term sedimentation and middle-term can be depicted on Figure 6h and Figure 6i or 6j, respectively. The long-term or period lies on IMF-10 (Figure 6k) and consistently showing the boundary of inter-distributary mouth bar.

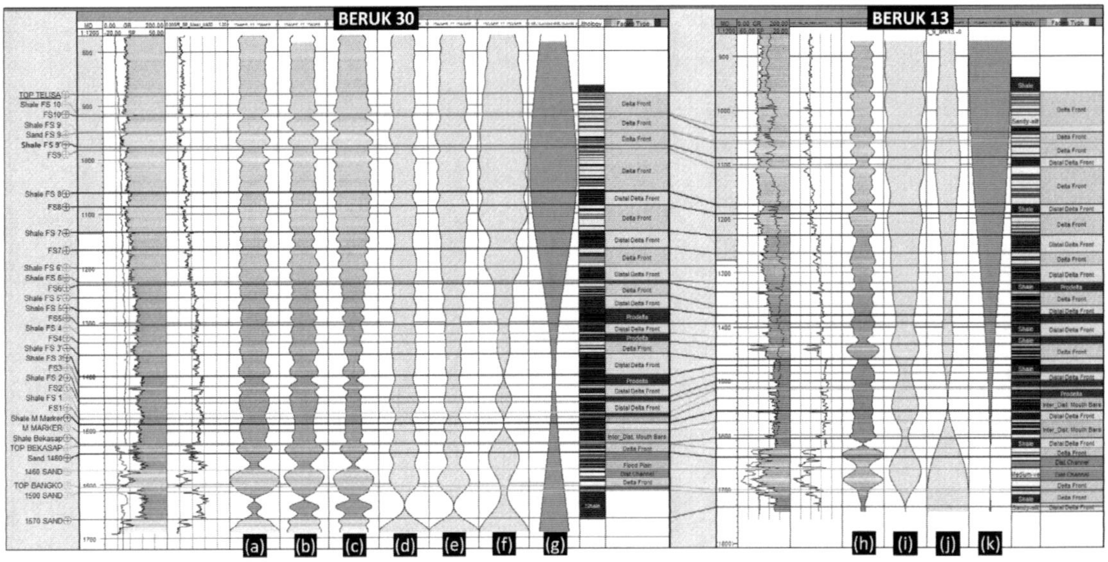

Figure 6. Illustration and identification of short, middle, and long-term cycle sequence stratum of well BRK-30 correlated to well BRK-13 using CEEMD method (a). Well BRK-30 :IMF-5 + IMF-6 (b). IMF-5 + IMF-7 (c). IMF-5 + IMF-10 (d). IMF-6 + IMF-7 (e) IMF-6 + IMF-10 (f) IMF-7 + IMF-10 (g) IMF-10 (h) Well BRK-13: IMF-5 + IMF-6 (i) IMF-7 + IMF-10 (j) IMF-8 + IMF=9 (k) IMF-10

7. Conclusion

CEEMD is robust signal analysis method to decompose the inherent characteristics in non-linear and non-stationary signal which this trend also occurs in well log recording. This method is helpful for identifying sequence strata with different frequency intrinsic (IMF). Combination of these IMFs can represent short, middle, and long-term sedimentation through characteristic curve (Table 1). In this study, short-term and middle-term describes the facies of distal front and delta front. These cycles were identified at the combination of IMF-5, IMF-6, and IMF-7 and calibrated to facies analysis at well. Commonly, the low frequency (IMF-9 and IMF-10) trend represents the long-term sedimentation and the cycle was detected at the boundary of inter-distributary mouth bar facies. Thus, the combination of IMFs determines the different sedimentary layer interface and better identification of the cycle of stratigraphic base level. Furthermore, downscaling sequence characteristic can be applied to the seismic data for lateral distribution.

Acknowledgments

The authors thank to BOB PT. Bumi Siak Pusako - Pertamina Hulu Energi for the permission to publish this study results

References

[1] Neal J and Abreu V 2009 Sequence stratigraphy hierarchy and the accommodation succession method *Geology* **37** p 779-782
[2] Catuneanu O, Abreu V, Bhattacharya J P, Blum M D, Dalrymple R W, Eriksson P G, Fielding C R, Fisher W L, Galloway W E, Gibling M R, Giles K A, Holbrook J M, Jordan R, Kendall C G, Macurda B, Martinsen O J, Miall A D, Neal J E, Nummedal D, Pomar L, Posamentier H W, Pratt B R, Sarg J F, Shanley K W, Steel R J, Strasser A, Tucker M E and Winker C 2009 Towards the standardization of sequence stratigraphy *Earth-Science Reviews* vol. 92 p 1-33.
[3] Van Wagoner J C, Mitchum R M, Campion K M and Rahmanian V D 1990 Siliciclastic Sequence Stratigraphy in Well Logs, Cores, and Outcrops: Concepts for High-Resolution Correlation of Time and Facies *AAPG Methods in Exploration Series*, No. 7 (The American Association of Petroleum Geologists, Tulsa, USA)
[4] Huang N E, Shen Z, Long S R, Wu M C, Shih H H, Zheng Q, Yen N C, Tung C C and Liu H H 1998 The empirical mode decomposition and the Hilbert spectrum for nonlinear and non-stationary time series analysis, *Proc. R. Soc. Lond. A* **454** p 903-995,
[5] Huang N E, Shen Z, and Long S R 1999 A new view of nonlinear water waves: the Hilbert spectrum *Ann. Rev. Fluid Mech.* **31** p 417-457
[6] Huang N E and Wu Z 2004 A review on Hilbert–Huang transform: Method and its applications to geophysical studies *Rev. Geophys.* **46** 2 RG2006
[7] Li Y J, Hu X Y and He Z X 2010 Application situation and prospects of Hilbert-Huang Transform in geophysics, *Chin.J.Eng.Geophys.***7** p 537-543
[8] Torres M E, Colominas M A, Schlotthauer G, and Flandrin P 2011 A complete ensemble empirical mode decomposition with adaptive noise: *ICASSP IEEE* p 4144–4147.
[9] Zheng R C, Yin S M and Peng J 2000 Sedimentary dynamic analysis of sequence structure and stacking pattern of base-level cycle, *Acta Sediment. Sin.* **18** 3 p 369-375 (in Chinese).
[10] Zheng R C, J. Peng and Wu C R 2001 Grade division of base-level cycles of terrigenous basin and its implications, *Acta Sediment. Sin.* **19** 2 p 249-255 (in Chinese).
[11] Serra O and Abbott H T 1982 The contribution of logging data to sedimentology and stratigraphy, *Soc. Petrol.Eng.J.* **22** 1 p 117-131
[12] Zhang T and Nie C Y 2011 Information extraction of acoustic logging for oil layers based on EMD time-frequency analysis, *J. Changchun Univ.* **21** 8 p 15-18 (in Chinese).
[13] Zhao N and Li R 2015 EMD Method Applied to Identification of Logging Sequence Strata. *Acta Geophysica* Vol.63 **5** pp 1256-75

HydrocarbonPotential in Sandstone Reservoir Isolated inside Low Permeability Shale Rock (Case Study: Beruk Field, Central Sumatra Basin)

Shidqi ADiria[1,a],Junita T Musu[1],Meutia FHasan[1], Widyo Permono[1], Jakson Anwari[1], Humbang Purba[1], Shafa Rahmi[1],Ory Sadjati[2], Iyep Sopandi[2], Fadli Ruzi[2]

[1] PPPTMGB "Lemigas" Jakarta, Indonesia
[2] BOB PT. Bumi Siak Pusako - Pertamina Hulu Energi, Indonesia
[a]shidqi.anugrah@gmail.com

Abstract. Upper Red Bed, Menggala Formation, Bangko Formation, BekasapFormation and DuriFormationareconsidered as the major reservoirsin Central Sumatra Basin (CSB). However, Telisa Formation which is well-known as sealwithin CSB also has potential as reservoir rock. Field study discovered that lenses and layers which has lowto high permeability sandstoneenclosed inside low permeability shaleof Telisa Formation. This matter is very distinctive and giving a new perspective and information related to the invention of hydrocarbon potential in reservoir sandstone that isolated insidelow permeability shale.This study has been conducted by integrating seismic data, well logs, and petrophysical datathroughly. Faciesand static model are constructed to estimate hydrocarbon potential resource. Faciesmodel shows that Telisa Formationwas deposited in deltaic systemwhile the potential reservoirwas deposited in distributary mouth bar sandstone but would be discontinued beddingamong shale mud-flat.Besides, well log data shows crossover between RHOB and NPHI, indicated that distributary mouth bar sandstone is potentially saturated by hydrocarbon. Target area has permeability ranging from 0.01-1000 mD, whereas porosity varies from 1-30% and water saturation varies from 30-70%. Thehydrocarbon resource calculationapproximates 36.723 MSTB.

Keywords : Telisa Formation, low permeability, Central Sumatra Basin, deltaic system.

1. Introduction

Hydrocarbon is one of the primaryenergyfuels that has been used all over the world. Since the world crude oil price shows declining trend recently,selling price become lower compare to production and operational cost. Therefore, it became economically not-efficient. One of the options to minimize this efficiency problem is by re-evaluate active oil/gas field, to find another reservoir productive layer.

Central Sumatra Basin (CSB) is one of the biggest productive hydrocarbon field in Indonesia.InCSB,main reservoir formationsareUpperRedBed, MenggalaFormation, BangkoFormation, BekasapFormation and Duri Formation.However in Rokan Block (South Balam Sub-Basin), Telisa Formation as reservoir Formation has producing hydrocarbon.Even TelisaFormation dominated by shale regionally(Yarmanto, et al., 2010) it still has potential as reservoir rock.

IOP Conf. Series: Earth and Environmental Science **132** (2018) 012034 doi:10.1088/1755-1315/132/1/012034

2. Regional Geology

Central Sumatra Basin is sedimentary basin which deposited by syn-rift graben that located under postrift sequence in Central Sumatra (William and Eubank, 1995 in Doust and Noble, 2008). Generally, hydrocarbons accumulated above or close to the synrift graben, caused the accumulation is shallow and post rift sequence remainimmature (Figure 1).

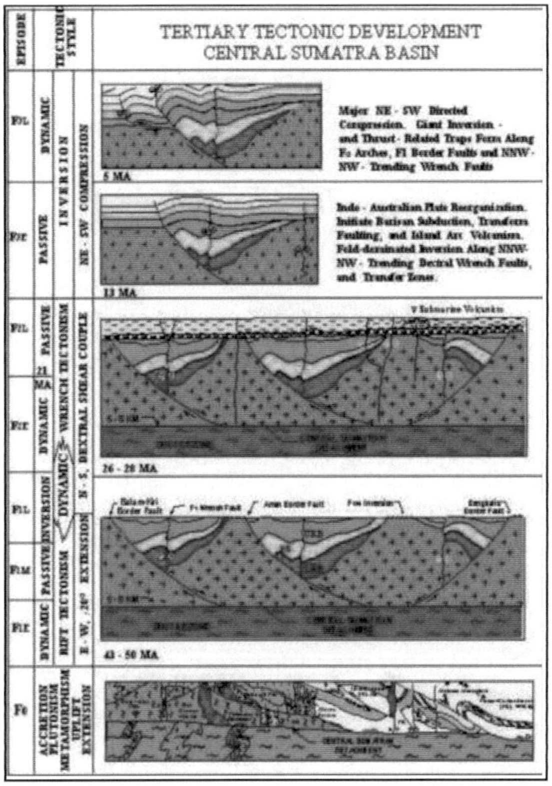

Figure 1. Tectonostratigraphy of Central Sumatra Basin(Heidrick and Aulia, 1993)

There are five productive grabensin CSB (Figure 1), include Bengkalis Through, Aman Through, Balam Through, Tanjung Medan Through and Kiri/Rangau Through. The grabens have relatively proximalstratigraphic succession, formed along the trend of the Pra-Tertiary structure (N-S and WNW- ESE).Thosegrabenoriginally are half graben that formed at oblique extension stress regime (William and Eubank, 1995 in Dovist and Noble, 2008). The tectonostratigraphy events that form the basin are Early Syn-Rift, Late Syn-Rift – Early Post-Rift, Early Post-Rift and Late Post-Rift.

Early Synrift (EarlyEocene - Oligocene) depositioned byPematang and KelesaFormation thatconsists of composed by association of alluvial, lacustrine and fluvial deltaic environment, with lithology oflaminated shale, silt and sand with coal and conglomeratic rock. Late Synrift - EarlyPostrift (Late Oligocene–Early Miocene),this sequence equivalent with Sihapas Group, including some of paralicfacies which recordgradually transgressive event. Menggala Formation deposited at fluvialenvironment and the upper part atsandy shallow marine environment(Bekasap Formation) and argillaceous (Bangko Formations).

Early Postrift (Early Miocene– Middle Miocene) which deposited by distal marine faciesfrom Sihapas Group,then at the end of the transgressive event, sand and clayin upper delta front facies from Duri Formation are deposited, followed by Tertiary maximum flooding period (shale and silt of Telisa

Formation).Late Postrift (Middle Miocene - Quarternary), this stage represents the basinfillingprocess atLate Tertiary, including regressivedeltaicand alluvial sediments with some unconformity.

3. Methodology

Data thatused in this study are seismicdata, well log data, petrophysicalevaluation and well report. Well log data of36 wells (Figure 2.a) consist of gamma ray, caliper, neutron porosity, density, resistivity and sonic. While seismic data is Post Stack Time Migration with 235 seismic line(Figure 2.b).

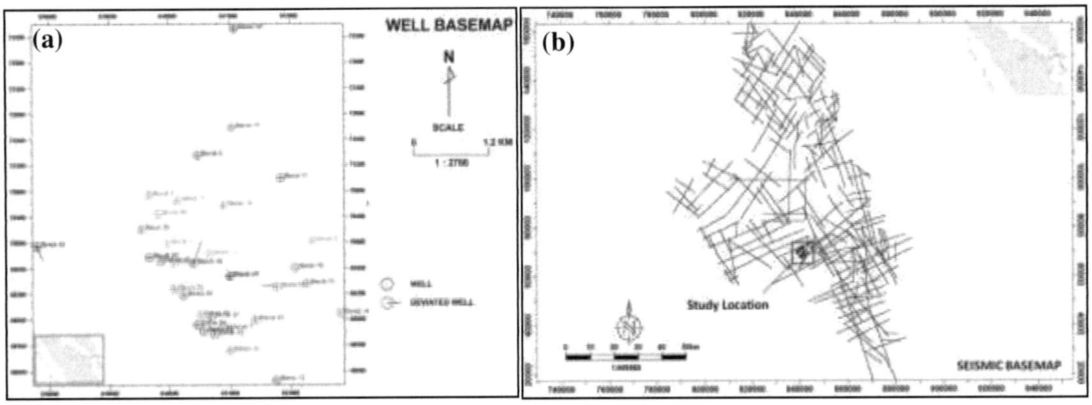

Figure2. Well (a) and seismic (b) basemap

On seismic data especially 2D, amplitude difference often found in process of composites one seismic lineto another. Therefore the pattern of reflector must be consideretely noticed.Validation ofanalysis and interpretation of well log data using well report data such as description of core, cutting, mud log and side wall core (SWC) to improve the accuracy of interpretation.

Well to seismic tieusingcheckshots data and synthetic seismogram was made to determine characters of Top Formation on well log in reflector of sesimic. After the seismic well tie done,horizon could be picked to know lateraldevelopment and also picking ofgeological structure.

Petrophysical evaluation was done after analysis and the interpretation of well log data,usingdeterministic approach. Workflow of petrophysical evaluationisdata preparation, determination of shale volume (Vsh), determination of total porosity, determination of water saturation (Sw), determination of permeability (k), determination of cut-off and determination of fluid contact.

4. Results

4.1. Depth Structure Map

The results that obtained from seismic data is depth structure map in Top Telisa Formation (Figure 3.a) and Top Bekasap Formation (Figure 3.b) to observing lateral continuity, depth and thickness of the Formation. The map built by horizon which picked in each of Top Formation that exist in each seismic line, while Top Formation on seismic line obtained from Top Marker in well log with seismic reflector (well-seismic tie). The results of correlation between well-seismic tie is 0.61

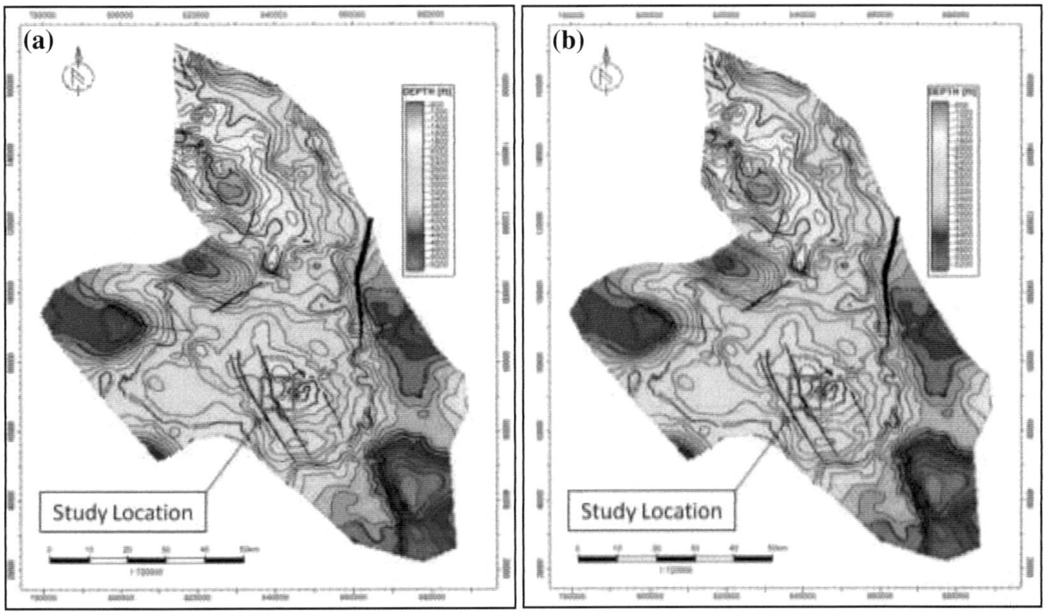

Figure3. Telisa Formation (a) and Bekasap Formation (b) depth structure map

Horizon in the seismic still in time domain,to change it into the depth domain required velocity model in each Formation. In this study, velocity model obtained on top of Telisa Formation is 2300 m/s and on top of Bekasap Formation is 2850 m/s. From time to depth conversion, the depth of Top Telisa Formation is 600 - 1200 feet and Top Bekasap Formation located at depth of 1200 - 2000 feet. On the west, presentfault zone withN-S trend.

4.2. Well Log Analysis
Lithologyinterpretation on well validated by cutting and mud log data and some well top obtained from master log fromwell report, including of Top Formation and "M" Marker. While FS1 - FS10 is the Top of Parasequence markerwhich interpretedfrom log pattern (electrofacies), cutting and mud log and the line Shale/Sand FS is the boundaryof shale lithologyand sandstone in parasequence. The well log divided into 8 parasequences (zones) with coarsening upward pattern (Figure 4.b).

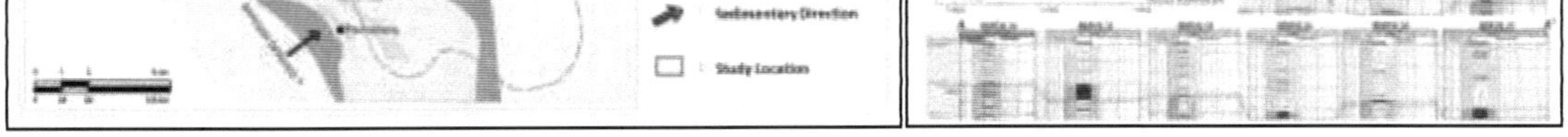

Figure4. Gross Depositional Environment map (modification of Barber, et al., 2005; Cameron, 1983; Yarmanto, et al., 2010; Lemigas Unpublished Report) (a) and Well log interpretation (b)

Based on Figure 4.aand facies map,depositional environment of this studylocated in deltaic system consisting of distributary mouth bar, inter distributary mouth bar, distal delta front and prodelta. Facies

IOP Conf. Series: Earth and Environmental Science **132** (2018) 012034 doi:10.1088/1755-1315/132/1/012034

that being the target of this study is distributary mouth bar facies, because the area dominated by sandstone with good enough porosity.Based on lateral distribution of lithology, facies modelcan be build (Figure 5.a), sand in delta front area may be the distributary mouth bar which have good total porosity and good permeability that isolated between shale which have badporosity and bad permeability (Figure 5.b).

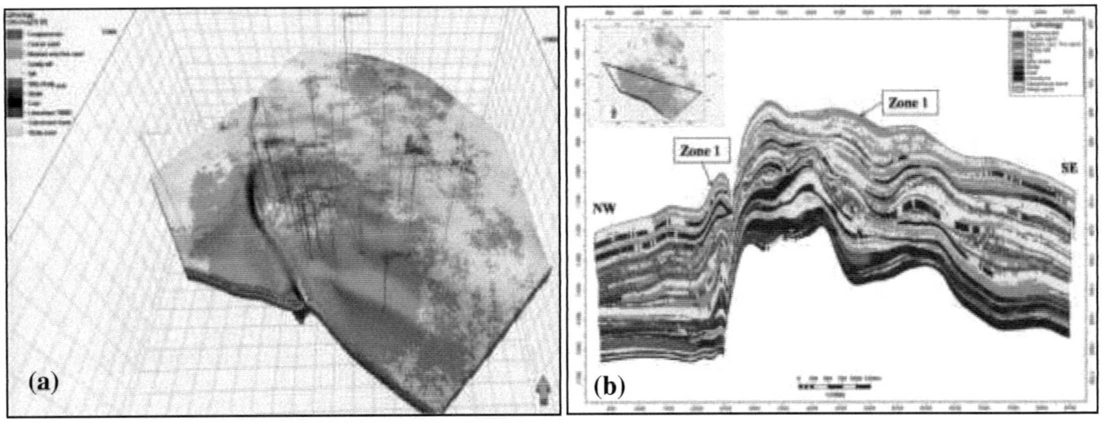

Figure5. Distribution of lithology (a) and Intersection lithology in Zone 1 (b)

Based on well log correlation, Zone 1 potential to be hydrocarbon reservoir.Delta frontsandin this zone also become thicker and widely spread,it shows supply of sediment in this zone has increased. Generallydistributary mouth bar sand in Zone 1 has thickness between 30 - 69 feet and delta front facies which considered as a distributary mouth bar between 35 - 55 feet. In Zone 1,facies gradually thickening from north to south therefore it shows the direction of the sedimentationrelatively dominant from north to south.

Zone 1 has crossover between NPHI log (49-59%) and RHOB (1.75 - 1.95 gr/cc). The log response indicate that Zone 1 possible contains hydrocarbon, because crossover RHOB and NPHI log built by high hydrogen atoms medium thereforeNPHI values relatively high and RHOB relatively low because bulk density (ρ) of sandstone which contains hydrocarbon tend to have low ρ value. That suspect supported by resistivity log patternwhichrelatively high between 1.34 - by 2.09 Ωm. In Zone 1 the value of gamma ray is relatively low (59.75 - 72.8 API).

4.3. Petrophysical Analysis

About 50 % value of Vshale is used as cutoff parameter. Petrographic analysis shows that for 50% clay content, porosity value is about 22% (Lemigas unpublished report, 2015). Therefore, 22% value of porosity is used as cutoff parameter. Pickett plot curve were used for determine cementation factor (m), saturation exponent (n) and tortuosity factor (a).

Cementation factor (m) represent the relation between pore connection and water resistivity, usually ranging from 1.7-4.1, where m value for shally sand is 1.33. Exponential saturation (n) potrays the presence of non-conductive fluid at pore space and related with rock wettability. Value of n that usually used is 2 and tortuosity factor (a) used for correcting variety of compaction, structure, and grain size. Water saturation calculated using Indonesia Equation because clay type in the study area is dispersed clay.

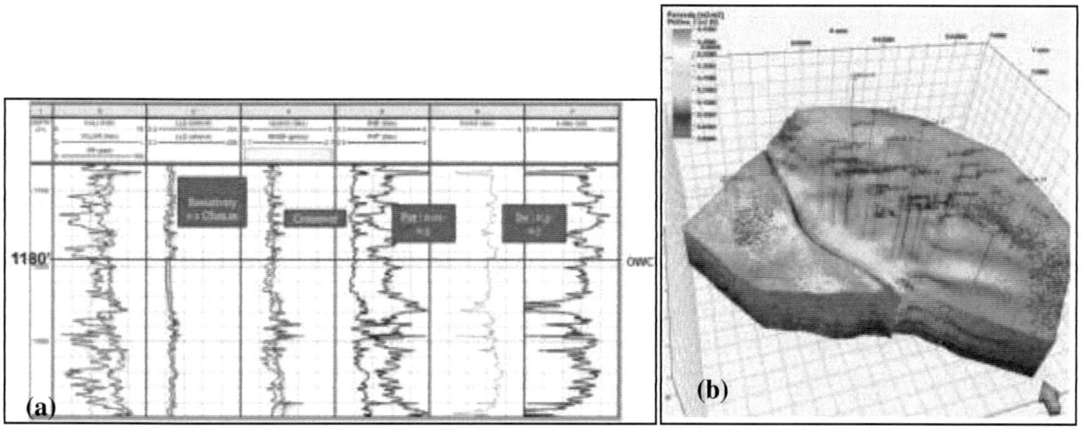

Figure6. Petrophysical evaluation (a) and distribution of porosity (b).

Fluid contact obtained from resistivity log trend of Beruk-14 well at depth of 1840 ft, since that depth shows low resistivity trend (1 Ohm.m). Therefore, at depth of 1180 ft interpreted as oil-water contact (OWC) (Figure 6.a). Petrophysical analysis result (Figure 6.a) shows that the study area has porosities ranging from 1-30% (Figure 6.b).

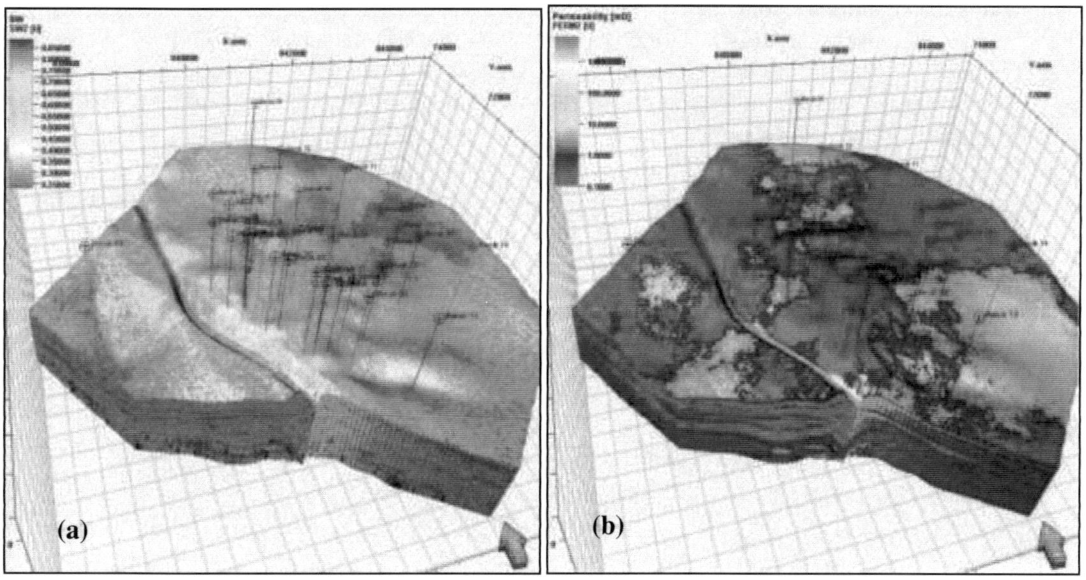

Figure7. Distribution of Saturation Water (a) and permeability (b).

Permeabilities varies from 0.01-1000 mD (Figure7.a), saturation water varies from 0.3-0.7 (Figure 7.b). Shale permeability value ranging from 0.01-10 mD and classified as low permeability rock (Widarsono et al, 2007), meanwhile permeability of distributary mouth bar sandstone is about 10-1000 mD. Zone -1 as potential zone of study area has hydrocarbon resource approximate36.723 MSTB based on volumetric calculation.

5. Conclusion

Zone 1 that located at distributary mouth bar facies with gamma ray value varies from 59.75-72.8 API, resistivity varies from 1.34 – 2.09 Ωm, *crossover* betweenNPHI logwith value varies from49 - 59 %

and density log with value varies from 1.75 – 1.95 gr/cc. Petrophysical evaluation at zone 1 shows permeability value ranging 1-10 mD, water saturation varies from 30-40% and porosity value varies from 10-20%, with volumetric estimation of 36.723 MSTB. It conclude that Zone 1 of study area has potential as low permeability hydrocarbon reservoir.

Acknowledgement
The authors wish to thank to PPPTMGB "LEMIGAS" and BOB PT. Bumi Siak Pusako for the permission to publish the data and study results.

6. References

[1] Asquith, G., and Krygowski, D. Basic Well Log Analysis. AAPG, Oklahoma.(2004).
[2] Doust, H., and Noble, R. A. Petroleum systems of Indonesia. ScienceDirect, 103-129. (2008).
[3] Heidrick, T.L., and Aulia, K. A Structural and Tectonic Model of The Coastal Plains Block, Central Sumatra Basin, Indonesia. Proceedings Indonesian Petroleum Association, 22nd Annual Convention, Jakarta. (1993).
[4] Morris, R. B. Using Log Derived Values of Water Saturation and Porosity. 1-26. (1967).
[5] Serra, O. Sedimentary Environment FromWireline Logs. (1985).
[6] Timur, A. An Investigation of Permeability, Porosity and Residual Water Saturation Relationships for Sandstone Reservoirs. SPWLA, 8-17.(1968).
[7] Widarsono, B., Muladi, A., and Jaya, I. Buletin IATMI. Retrieved from Website IATMI http://www.iatmi.or.id/assets/bulletin/pdf/2007/2007-09.pdf. (2007).
[8] William, H.H., and Eubank, R.T. Hydrocarbon Habitat in the Rift Graben of the Central Sumatra Basin. (1995).
[9] Walker, R.G. Facies Models 1st and 2nd edition. Geoscience Canada.(1979, 1984).
[10] Yarmanto, Noeradi, D., and Hendar. Telisa Depositional Model in Central Sumatra Basin, Proceedings of the Indonesian Petroleum Association, 34th Annual Convention & Exhibition, IPA 10-G-204. (May 2010).

IOP Conf. Series: Earth and Environmental Science **132** (2018) 012035 doi:10.1088/1755-1315/132/1/012035

Passive Seismic for Hydrocarbon Indicator : Between Expectation and Reality

Riky H B Pandito[1]
[1]Pertamina Hulu Energi / JOB PPEJ, Indonesia

*Corresponding author's email : rikpandita@gmail.com

Abstract. *In between 5 – 10 years, in our country, passive seismic method became more popular to finding hydrocarbon. Low price, nondestructive acquisition and easy to mobilization is the best reason for choose the method. But in the other part, some people are pessimistically to deal with the result. Instrument specification, data condition and processing methods is several points which influence characteristic and interpretation passive seismic result. In 2010 one prospect in East Java Basin has been measurement constist of 112 objective points and several calibration points. Data measurement results indicate a positive response. Furthermore, in 2013 exploration drliing conducted on the prospect. Drill steam test showes 22 MMCFD in objective zone, upper – late oligocene. In 2015, remeasurement taken in objective area and show consistent responses with previous measurement. Passive seismic is unique method, sometimes will have difference results on dry, gas and oil area, in field production and also temporary suspend area with hidrocarbon content.*

1. Introduction

East Java basin is one of the oil and gas production field in Indonesia. Almost the entire formations are proven in containing of hydrocarbon. There are many play types developed in this basin, from the Eocene formation up to Pleistocene formation are proven to be oil and gas reservoir. At the end of the year 2013, the drilling activity in the exploration well was established in the one of the prospect location. The prospect is located relatively close to the coast and borders the working area. Furthermore, it also be challenging since the prospect surrounded by the many dry hole wells. The total depth that can be aimed by the wells is approximately 7800 ft with the vertical drilling direction. Based on the well test, it was recorded that 22 MMCFD of gas out from an objective zone. Even though the zone was shifted from the previous prediction, which was previously expected that the prospect would be generating oil, however the prospect still interesting to be studied and assessed especially related with the petroleum system which is the background of the area. From this paper, the writer want to review the new play potential from the well evidences based on passive seismic acquisition.

Geologically, East Java Basin is located in the active tectonic zone which has at least three tectonic phases [1]. This basin elongated in the direction of W-E with a length about 250 km and bordered by Karimun Java arc, Sunda Shelf, Meratus high, and Masalembo high. Generally, the basement rocks of this basin is pre-tertiary. The rifting occurred in the lower Eocene, after that in the middle Miocene the tectonic is reactivated and made a few of the folding block. The tectonic phase begins in the Pleistocene until recent which is based on the many references, there are uplift phase and in some zone were occurred wrench fault. Basement rock in the East Java basin is related with pre-tertiary metasediment which its characteristic consist of folding and layering, that can be found in the seismic pattern. The cutting result from some wells that penetrated the basement shows that the basement can be divided based on the minerals namely, slate, volcaniclastic, metasilstone, metagraywacks, diorite, marble, metaquartzille, vitric tuff, schist, andesite, phyllite and vitric tuffbasalt. Most of the petroleum system in the basin formed at the East Java graben system, as a result of collisions that occur in the back arc basin. According to some studies, the source rocks are located in the Ngimbang Formation with the age of Eocene to Oligocene. The formation has two main facies, namely fluvio deltaic and lacustrine. The fluvio deltaic facies is dominantly by calcareous shale and coal, which is type three of kerogen. The lab analysis shows that TOC from the formation is around 2-17% on the shale and 20-80% on the coal, also the HI reach to 100. Otherwise, on the lacustrine facies the TOC result is around 2-4% with HI in 400. The hydrocarbon filling process occurred in the Miocene until Pleistocene. In general, the source rocks from East Java Basin are mature, although in some places are immature, which happens because the temperature and pressure are not working ideally. The dominant trap mechanism in the basin is stratigraphy and structural.

2. Methodology

Passive seismic is a method to find hydrocarbon. Initially, the measurements are performed in the geotechnical field. Along with the development processes, passive seismic measurements are performed in areas containing hydrocarbon that shows a certain frequency (2-6 Hz) there is a pattern and a distinctive character on the spectral amplitude vs. frequency. Whereas in the certain frequency above the reservoir zone that contain hydrocarbon will be shows the changing of amplitude [4]. On the lowest frequency, the natural movement from the oceanic crust in the whole of the earth will be made dominant noise between 0.05 and 0.2 Hz [3]. Furthermore, the human activities and wind movement will be recorded on the higher frequency (>1 Hz). There are several hypotheses about the causes of the emergence of passive seismic measurement anomalies in the areas which containing hydrocarbons.

As an example of the signal anomaly is the result of amplification or resonance scattering of hydrocarbons in the subsurface. Whereas the assumption is interaction between the hydrocarbon fluid and porous rocks which can distort natural signals from in the earth [2]. The ideal method to find hydrocarbon is not only depends on the one method, but there are many methods that can be tried to support the existence hypothesis of the hydrocarbon. The benefit of the passive seismic method is the data that can be used are from the measurement which is relatively easy, quick, low risk, cheap, and nothing the destructive vibrating source. The measurement can be carried out on the frontier zone with the relatively big population. By calibrating of the precise locations or wells, misinterpretation of passive seismic data can be minimized. The data result is also affected by technical specifications of seismometers. The bandwidth wide and sensitivity of the tools have a little role in the method. Ideally, using three components seismometer, response frequency 0.03-50 Hz, the sensitivity about 2000-3200 V/ms-1, and the dynamic range higher than 145 dB. Dealing with the issue about the noise, it can be minimized in several ways such as measurement time plus, time windowing, or stacking process time series data. It happens because the general noise that appears periodically, while the data tend to be quasi stationary. Process time windowing carried out to improve the S / N ratio.

This research has a goal to test the consistency of the data by doing some variations in the time of the measurement between 90 minutes, 3 hours, and 24 hours. Furthermore, the data processing following the steps namely, demeaning and detrending, windowing, FFT, stacking, smoothing, and attribute analysis. The author uses some passive seismic attributes to test for the presence and consistency of the method. PSD (Power Spectral Density) attribute 3Hz, PSD-IZ, and Attribute V/H ratio are some of the attribute that are used by the author in this research. Attribute PSD 3 Hz is done by getting the value of the Z component (up and down) at a frequency of 3 Hz. In general, the presence of hydrocarbons will be directly related with the signal on this attribute. PSD IZ is the integral of the signal spectrum Z component which has a frequency range 1.7 - 3.7 Hz. This attribute can be used for the highest noise condition. The last attribute is V/H ratio, which is used to reduce the effects of radiation on daily measurements in every measurement. Generally, the hydrocarbon existence anomaly is related with the high of the V/H attribute ratio. Furthermore, all of the attribute will be showed the same pattern and consistency.

As a point for calibration, will be conducted the measurement on more than a few key points, namely wells containing oil, gas, oil and gas wells as well. There are approximately 12 keys as reference point. Measurement is not only done in the area of oil and gas production, but also to plug the well with temporary conditions.

IOP Conf. Series: Earth and Environmental Science **132** (2018) 012035 doi:10.1088/1755-1315/132/1/012035

In this study, the data used are time series recorded on three components seismometer. Each data sets consists of three time series data toward East-West, North-South, Up-Down. In general, data recorded with duration 90 minutes and sampling frequency 100 Hz. Hydrocarbon characterization is done by extracting some passive seismic attribute data. Ideally some results of processing some attribute show consistency trend.
In detail, the processing is done with several stages:
a) Windowing is one way to identify the signal and noise. This step is performed to obtain free noise data. Is based on the ratio between the short-time average (STA) and the long time average (LTA), which signals with minimum noise will have ratio STA / LTA ~ 1. Windowing is done by selecting the signal with a length of approximately 1 minute.
b) Fast Fourier Transform, the process of transforming the signal in another form, in this case the recorded signal is the domain of time series. For further analysis of the signal is converted to a frequency series.
c) Stacking to improve the signal to noise ratio. Several methods to reduce noise is by stacking on the entire received signal. This is assuming that the incoming noise discrete, not stationser. So it will be reduced if stacking.
d) Smoothing performed for doing moving average
e) Horizon spectra is to combine spectral components with East West and South North.
f) Calculate vertical and horizontal ratio.

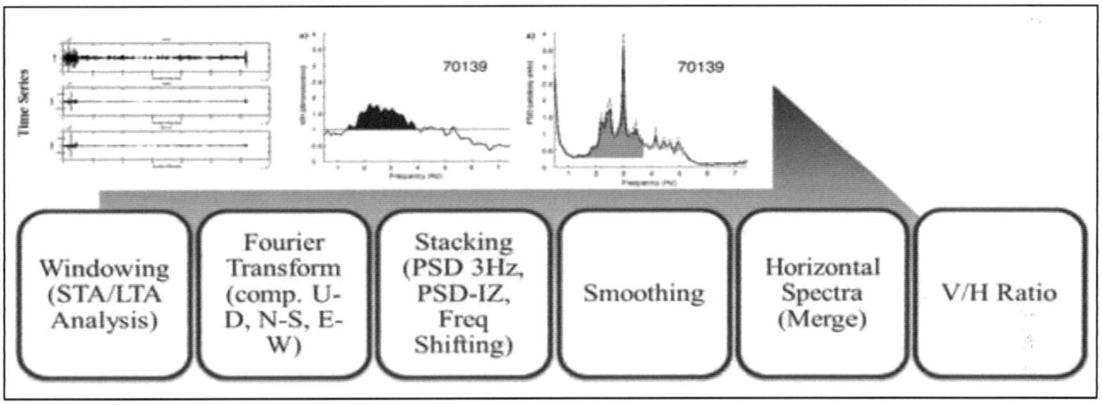

Figure 1. Methodology and Data Processing

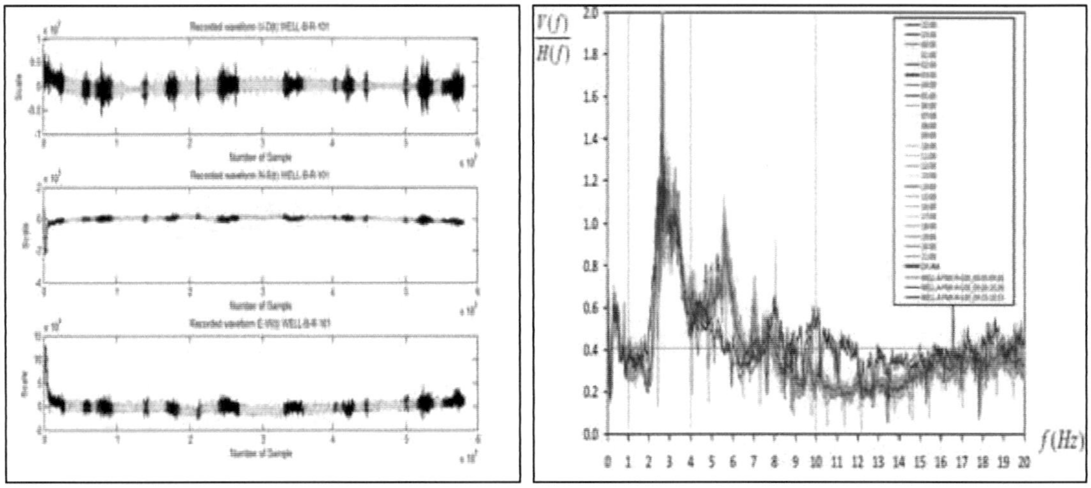

Figure 2. Raw Data of Passive Seismic (left) and ratio V/H with several time

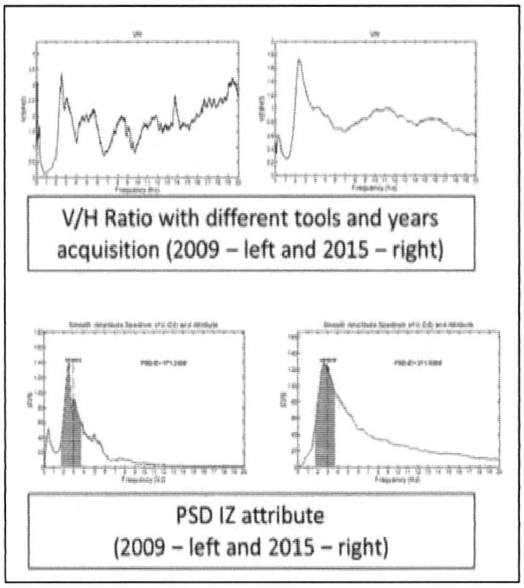

Figure 3. Time lapse recording in discovery well, showing the consistency of hidrocarbon signal in low frequencey 2-4 Hz V/H ratio and PSD IZ

3. Conclusion

Passive seismic measurements at several points wells provide quite interesting results. In wells with hydrocarbon content, obtained upside amplitude at PSD attribute IZ, 3 Hz and the ratio of V/H. Increase amplitude, if it can be said to be an anomaly, that are in the range of 2- 7 Hz, both in well with oil alone and well layer of oil and gas. For dry wells, on a plot of amplitude with a frequency not showing an increase in both : the amplitude attribute IZ and the ratio of V/H. In this study, also conducted time lapse measurement, i.e at one carbonate prospect. Data obtained at this point is quite interesting. The first data measured in 2009. In the first measurement have amplitude anomaly at all of passive seismic attribute. Subsequently, in 2013 carried out drilling in the those prospect area. The well penetrated the basement, which is dominated by andesite. At the age of Late Oligocene Carbonate results from DST 22 MMCFD gas. This proves that the measurement of passive seismic is valid. Furthermore, in 2015 the passive seismic measurements were taken back to the same well. Conditions at that time in a state of temporary plug wells. The measurement results show the response of passive seismic anomaly that is relatively similar to previous measurements. Although when we look in more detail, there is little difference in curves. But probably caused by a different instrument specifications, especially sensitivity of the instrument. From this research, it can be concluded that the method of measurement and analysis of passive seismic is one method that can not be ignored. With proper calibration wells is helpful to interpret the data acquisition. Instrument and measurement determine the quality of data to be acquired.

References :

[1] Burgon, G. E., The Petroleum Potential of the Tuban Block, East Java, Indonesia., Internal Report JOB PTT, Jakarta, 1993.
[2] Holzner R, Eschle P, Dangel S, Frehner M, Narayanan C, Lakehal D 2006 Hydrocarbon Microtremors Interpreted as Oscillations Driven by Oceanic Background Waves Communications in Nonlinear Science and Numerical Simulation 14 (1) 160-173
[3] Oliver, J., Ewing, M., Microseisms in the 11- to 18-second period range. Bull. Seismol. Am., 47 (2), 111-127, 1957.
[4] Saenger, E. H., Torres, A., Rentsch, S., Lambert, M., Schmalholz, S. M., A hydrocarbon microtremor survey over a gas field: Identification of seismic attributes. 76th SEG Annual Meeting Expanden Abstract, 1277-1281, 2007.

Acknowledgments

Thanks are also due to Purwo Hadi Suseno, Moh Imron, Isnaini, JOB PPEJ Subsurface Team, Pusat Studi Energi UGM, Andri Dian - LAPI ITB and M N Krisnayadi - SKKMIGAS for supporting this research.